The ARRL
Ham Radio License Manual

All You Need to Become an
Amateur Radio Operator

Third Edition
For use with ARRL's *Online Exam Review for Ham Radio*

By Ward Silver, NØAX

Contributing Editor
Mark Wilson, K1RO

ARRL Education Services Manager
Debra Johnson, K1DMJ

Production Staff

Maty Weinberg, KB1EIB, Editorial Assistant

David Pingree, N1NAS, Senior Technical Illustrator

Jodi Morin, KA1JPA, Assistant Production Supervisor: Layout

Sue Fagan, KB1OKW, Graphic Design Supervisor: Cover Design

Michelle Bloom, WB1ENT, Production Supervisor: Layout

Front Cover Photos: Robert Wood, W5AJ, operates using Morse code during the Midland (Texas) Amateur Radio Club Field Day event. Every June Field Day brings Amateur Radio operators – old and new – together for a weekend of enjoyment on the air. By earning your license, you'll be able to take part in exciting activities like these, and many more! (Photo courtesy of Alan Sewell, N5NA)

Shown at the upper left is Yvette Cendes, KB3HTS, at the W8EDU Field Day site in Huntsburg, Ohio. (Photo courtesy of Jim Galm, W8WTS)

Back Cover Photo: Rebecca Rubsamen, KJ6TWM, uses a portable dual-band Yagi antenna to make a digital communications contact with the International Space Station. (Photo by Reid Rubsamen, N6APC)

ARRL *The national association for* **AMATEUR RADIO®**

225 Main Street, Newington, CT 06111-1494

Feedback: We're interested in hearing your comments on this book and what you'd like to see in future editions. Please email comments to us at **pubsfdbk@arrl.org**, including your name, call sign, email address and the title, edition and printing of this book.

This book may be used for Technician license exams given beginning July 1, 2014.

QST and the ARRL website (**www.arrl.org**) will have news about any rules changes affecting the Technician class license or any of the material in this book.

We strive to produce books without errors. Sometimes mistakes do occur, however. When we become aware of problems in our books (other than obvious typographical errors), we post corrections on the ARRL website. If you think you have found an error, please check **www.arrl.org/ham-radio-license-manual** for corrections. If you don't find a correction there, please let us know by sending e-mail to **pubsfdbk@arrl.org**.

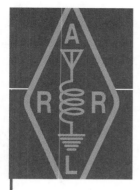

Contents

Foreword

Welcome to the diverse group of individuals who make up Amateur Radio! There are more than 700,000 amateurs, or "hams," in the United States alone and 3,000,000 around the world. Hams come from all walks of life, all ages and every continent. Hams are busily communicating without regard to the geographic and political barriers that often separate humanity. This is the power of Amateur Radio — to communicate with each other directly, without requiring any other commercial or government communications systems.

Amateur Radio was born along with radio itself. Marconi, the father of radio, considered himself "an amateur" and many of the wireless technologies and systems we take for granted today had their origins in the workshops and imaginations of amateurs. Governments make room for Amateur Radio when valuable radio spectrum is allocated because they know and respect the flexibility and inventiveness of hams. Amateur Radio is an excellent training and educational opportunity for a country's students and technicians. In the face of emergencies and disasters, the amateur's ability to innovate and adapt is legendary.

Hams came to Amateur Radio from many walks of life and many interests. Perhaps you intend to provide emergency communications for yourself and your community. Technical experimentation might be your interest or you might be one of the burgeoning "do-it-yourself" community, discovering the pleasures of building, testing, using and learning. Making new friends via the radio, keeping in touch as you travel, or exploring where a wireless signal can take you — these are all valuable and valued parts of the Amateur service.

A time-honored ham tradition is that of helping newcomers learn about the ways and skills of Amateur Radio. These helpers — known as "Elmers" by other hams — are everywhere. You are almost certainly near another ham and probably an entire ham radio club! They'll gladly help you get started. There's more information in Chapter 1 about connecting with them. If you need assistance, the staff here at ARRL Headquarters will be more than happy to help, too.

As you read this book, getting ready to pass your first ham radio licensing exam, you will find that there is a lot more material here than just the answers to exam questions. That's the ARRL way of going the extra mile to help you learn about Amateur Radio. "Of, By and For the Amateur" is the ARRL's motto. By providing this extra information, we help you learn the "why" behind each question so that you are prepared when ready to get on the air. Keep this book handy as a reference to help you understand how ham radio "works" and you'll have more fun and be a more effective operator.

Most active radio amateurs in the United States are ARRL members. They realize that since 1914, the ARRL's training, sponsorship of activities and representation both nationally and internationally are second to none. The book you're reading now, *The ARRL Ham Radio License Manual* is just one of many publications for all levels and interests in Amateur Radio. The ARRL will be there to extend a helping hand throughout your ham radio activities. You don't need a license to join the ARRL — just be interested in Amateur Radio and we are interested in you. It's as simple as that!

David Sumner, K1ZZ
Chief Executive Officer
Newington, Connecticut
March 2014

New Ham Desk
ARRL Headquarters
225 Main Street
Newington, CT 06111-1494
(860) 594-0200

Prospective new amateurs call:
800-32-NEW-HAM (800-326-3942)
You can also contact us via e-mail: **newham@arrl.org**
or check out **ARRLWeb: www.arrl.org**

ARRL Membership Benefits

Public Service

Advocacy

Education

Technology

Membership

QST Monthly Magazine

QST covers new trends and the latest technology, fiction, humor, news, club activities, rules and regulations, special events, and much more. Here is some of what you will find every month:

- Informative product reviews of the newest radios and accessories
- A monthly conventions and hamfest calendar
- A public service column that keeps you up to date on the public service efforts hams are providing around the country and shows you how you can join in this satisfying aspect of our hobby
- Eclectic Technology, a monthly column that covers emerging Amateur Radio and commercial technology
- A broad spectrum of articles in every issue ranging from challenging topics to straightforward, easy-to-understand projects

ARRL members also get preferred subscription rates for *QEX*, the ARRL Forum for Communications Experimenters.

Members-Only Web Services

- ***QST* Digital Edition**
  All ARRL members can access the online digital edition of *QST*. Enjoy enhanced content, convenient access and a more interactive experience. Apps for *iOs* and *Android* devices are also available.
- ***QST* Archive and Periodicals Search**
  Browse ARRL's extensive online *QST* archive. A searchable index for *QEX* and *NCJ* is also available.
- **Free E-Newsletters**
  Subscribe to a variety of ARRL e-newsletters and e-mail announcements: ham radio news, radio clubs, public service, contesting and more!
- **Product Review Archive**
  Search for, and download, *QST* Product Reviews published from 1980 to present.
- **E-Mail Forwarding Service**
  E-mail sent to your **arrl.net** address will be forwarded to any e-mail account you specify.
- **Customized ARRL.org home page**
  Customize your home page to see local ham radio events, clubs and news.
- **ARRL Member Directory**
  Connect with other ARRL members via a searchable online Member Directory. Share profiles, photos and more with members who have similar interests.

Technical Information Service (TIS)

Get answers on a variety of technical and operating topics through ARRL's Technical Information Service. Our experts can help you overcome hurdles and answer all your questions.

Member Benefit Programs and Discounts

- **ARRL "Special Risk" Ham Radio Equipment Insurance Plan**
  Insurance is available to protect you from loss or damage to your station, antennas and mobile equipment by lightning, theft, accident, fire, flood, tornado, and other national disasters.
- **MetLife® Auto, Home, Renters, Boaters, Fire Insurance and Banking Products**
  As an ARRL member you could enjoy up to a 10% discount on various insurance programs.
- **The ARRL Visa Signature® Card**
  Show your ham radio pride with the ARRL Visa credit card. You earn great rewards and every purchase supports ARRL programs and services.

Outgoing QSL Service

Let us be your mail carrier and handle your overseas QSLing chores. The savings you accumulate through this service alone can pay your membership dues many times over.

Continuing Education

Find classes to help you prepare to pass your license exam or upgrade your license, learn more about Amateur Radio activities, or train for emergency communications or public service. ARRL also offers hundreds of books, CDs and videos on the technical, operating, and licensing facets of Amateur Radio.

Regulatory Information Branch

Reach out to our Regulatory Information Branch for information on FCC and regulatory questions; problems with antenna, tower and zoning restrictions; and reciprocal licensing procedures.

ARRL as an Advocate

ARRL supports legislation in Washington, D.C. that preserves and protects access to existing Amateur Radio frequencies as a natural resource for the enjoyment of all hams. As a member, you contribute to the efforts to preserve our privileges.

ARRL *The national association for* **AMATEUR RADIO®**

www.arrl.org/join

AMERICAN RADIO RELAY LEAGUE • 225 MAIN ST • NEWINGTON, CT, USA 06111-1494 • TELEPHONE 860-594-0200 • FAX 860-594-0259

Join ARRL and experience the BEST of Ham Radio!

ARRL Membership Benefits and Services:

- *QST* magazine — your monthly source of news, easy-to-read product reviews, and features for new hams! Members also enjoy a new digital edition and *QST* Apps for Apple and Android devices.
- Technical Information Service — access to problem-solving experts!
- Members-only Web services — find information fast, anytime!
- Best amateur radio books and software

FREE Book Offer!

I want to join ARRL.
Send me the FREE book I have selected
(choose one)
☐ The ARRL Repeater Directory
☐ Get on the Air with HF Digital
☐ Small Antennas for Small Spaces

Name		Call Sign	
Street			
City		State	ZIP
Phone		Email	

Circle your membership option:

☐ **1 year $49**   ☐ **2 years $95**   ☐ **3 years $140**

☐ Total amount enclosed, payable to ARRL $ _____
☐ Charge to: ☐ VISA   ☐ MasterCard   ☐ Amex   ☐ Discover

Card Number	Expiration Date
Cardholder's Signature	

Blind, youth and life membership rates are available. Contact ARRL for more details. US Memberships include $21 per year for subscription to *QST*. Dues are subject to change without notice and are non-refundable.

❏ I do not want my name and address made available for non-ARRL related mailings.

Call Toll-Free (US) **1-888-277-5289**
Join Online **www.arrl.org/join/HRLM** or
Clip and send to:

ARRL The national association for **AMATEUR RADIO®**
225 Main Street
Newington, CT 06111-1494 USA

Web Code HRLM

When to Expect New Books

A Question Pool Committee (QPC) consisting of representatives from the various Volunteer Examiner Coordinators (VECs) prepares the license question pools. The QPC establishes a schedule for revising and implementing new Question Pools. The current Question Pool revision schedule is as follows:

Question Pool	Current Study Guides	Valid Through
Technician (Element 2)	*The ARRL Ham Radio License Manual,* 3rd Edition *ARRL's Tech Q&A, 6th Edition*	June 30, 2018
General (Element 3)	*The ARRL General Class License Manual,* 7th edition *ARRL's General Q&A, 4th Edition*	June 30, 2015
Amateur Extra (Element 4)	*The ARRL Extra Class License Manual,* 10th Edition *ARRL's Extra Q&A, 3rd Edition*	June 30, 2016

As new Question Pools are released, ARRL will produce new study materials before the effective date of the new pools. Until then, the current Question Pools will remain in use, and current ARRL study materials, including this book, will help you prepare for your exam.

As the new Question Pool schedules are confirmed, the information will be published in *QST* and on the ARRL **website at www.arrl.org**.

Online Review and Practice Exams

Use this book with ARRL Exam Review for Ham Radio to review chapter-by-chapter. Take randomly-generated practice exams using questions from the actual examination question pool. You won't have any surprises on exam day! Go to **www.arrl.org/examreview**

How to Use this Book

The ARRL Ham Radio License Manual is designed to help you learn about every topic in the Technician exam question pool. Every page presents information you'll need to pass the exam and become an effective operator. This book goes well beyond the answers to exam questions — it also contains explanations, guidelines, and helpful information to help you remember and use what you learn on the air.

The book is organized to help you learn about radio and operating in easy-to-understand, bite-sized steps. You'll begin by learning about the basics of radio signals and simple ham radio equipment. The next steps cover the principles of electricity and an introduction to electrical components. You'll then learn how a simple station is assembled and some basic operating procedures. At that point, you'll be ready to understand the rules and regulations of ham radio. The final section is all about ham radio safety.

At the back of this book you'll also find a large glossary of ham radio words, a detailed index, a supplement to help you choose a radio, and a selection of advertisements from some vendors of ham radio equipment and supplies.

Conventions

Throughout your studies keep a sharp eye out for words in *italics*. These words are important so be sure you understand them. Many of them are included in the glossary. Another thing to look for is the web mouse symbol, indicating that there is supplemental information on the *Ham Radio License Manual* website (**www.arrl.org/ham-radio-license-manual**) to accelerate and broaden your understanding. If a web or e-mail address is included, it will be printed in **boldface type**.

 As you read the book, you will see question designators in square brackets, such as [T1A01]. These are references to the question ID in the exam's question pool. This will help you find the material that addresses a specific question. The question pool also includes a page reference where each topic is discussed.

The Exam Question Pool

The complete Technician exam question pool is included at the back of this book. The 35 questions you'll answer on the exam will be drawn from this question pool. Yes, these are the actual questions on the exam but resist the temptation to just memorize the answers! Memorizing without learning the subject is likely to leave you "high and dry" when you begin using your new operating privileges. Do yourself a favor and take the time to understand the material.

A *Study Guide* version of the question pool has been prepared by the ARRL in which each question is presented in the order it is covered in this book. The *Study Guide* version can be downloaded from the *Ham Radio License Manual* website.

 When using the question pool for exam practice, each question also includes a cross-reference back to the page of the book covering that topic. If you don't completely understand the question or answer, please go back and review that material.

Self-Study and Classroom Tips

For self-study students, the material in the book is designed to be studied in order from beginning to end. Read the material and then test your understanding by answering the questions at the end of each section. Use the supplemental material on the Ham Radio License Manual website if you need extra help.

The ARRL's New Ham Desk can answer questions emailed to newham@arrl.org. Your question may be answered directly or you might be directed to more instruction material. The New Ham Desk can also help you find a local ham to answer questions. Studying with a friend makes learning the material more fun as you help each other over the rough spots and you'll have someone to celebrate with after passing the exam!

If you are taking a licensing class, the instructors will guide you through the material. Help your instructors by letting them know where you need more assistance. They want you to learn as thoroughly and quickly as possible, so don't hold back your questions. Similarly, if you find their explanations particularly clear or helpful, tell them that, so it can be used in the next class!

At the end of each section is a short list of exam questions covered in that section. This is a good time to pause for a short review session. Be sure you understand the material by answering the questions before moving to the next section. It is a lot easier to learn the material section-by-section than by rushing ahead and you'll remember it more clearly. For a focused discussion on each exam question, pick up a copy of the *ARRL's Tech Q&A*. Every question is included with the correct answer and a short explanation.

To make the best use of the on-line reference material:
- Bookmark the *Ham Radio License Manual* website to use as an online reference while you study.
- Download the *Study Guide* version of the question pool from the website.

 The *Ham Radio License Manual* web page lists other resources organized by section and chapter to follow the book. Browse these links for extra information about the topics in this book.

Online Review and Practice Exams

As you complete each chapter of this book, use ARRL's online Exam Review for Ham Radio to help prepare you for exam day. This web-based service uses the question pool to construct chapter-by-chapter reviews. Once you've finished this book, use the online service to take practice exams with the same number and variety of questions that you'll encounter on exam day. You can practice taking tests over and over again in complete privacy (even print practice exams!). These exams are quite realistic and you get quick feedback about the questions you missed. When you find yourself passing the practice exams by a comfortable margin, you'll be ready for the real thing. To find out more about ARRL Exam Review for Ham Radio, visit the *Ham Radio License Manual* web page (**www.arrl.org/hrlm**) or go to **www.arrl.org/examreview**.

Chapter 1

Welcome to Amateur Radio

In this chapter, you'll learn about:

- **What makes Amateur Radio unique**
- **Why the FCC makes the rules**
- **What activities you'll find in Amateur Radio**
- **Where you can find other hams**
- **The Technician license — what it is and how to get it**
- **Ready? Set? Go!**

When you see the mouse, you'll find more information at www.arrl.org/ham-radio-license-manual

Welcome to the *Ham Radio License Manual*, the most popular introduction to Amateur Radio of all! In this study guide, not only will you learn enough to pass your Technician license exam, you'll also learn what ham radio is all about and how to jump right in once you're ready to get on the air.

If you want to know more about amateur or "ham" radio before you start preparing to get a license, you'll find your answers in sections beginning with "What Is Amateur Radio?" If you already know about ham radio and are anxious to get started, you're in good company — there are thousands of other folks getting ready to become a "ham" radio operator. Jump ahead to section 1.4 — "Getting Your Ham Radio License" and get started!

1.1 What is Amateur Radio?

Amateur Radio will surprise you with all its different activities. If you've encountered Amateur Radio in a public service role or if someone you know has a ham radio in their home or car, then you already have some ideas. Maybe you have seen ham radio in a movie or read about it in a book. Are you a part of the growing "do-it-yourself" or maker communities? If so, you'll really enjoy getting involved with one of the most "hands-on" hobbies of all. Amateur Radio is the most powerful communications service available

Jerry Clement, VE6AB, demonstrates that you don't need much equipment to make contacts through a ham radio satellite.

to the private citizen anywhere on Earth — or even above it!

Amateur Radio is a recognized national asset, providing trained operators, technical specialists and emergency communications in time of need. It was created for people just like you who have an interest in radio communications. Some hams prefer to focus on the technology and science of radio. Competitive events and award programs hold the interest of others. Some train to use radio in support of emergency relief efforts or to keep in touch with family. There are many hams who simply like to talk with other hams, too. This introductory section of the *Ham Radio License Manual* will give you a broad overview of Amateur Radio so you can understand how radio works and why hams do what they do. Let's start at the beginning, shall we?

BEGINNINGS OF HAM RADIO

Amateur Radio has been around since the beginning of radio communications. It wasn't long after Marconi spanned the Atlantic in 1901 before curious folks began experimenting with "wireless." Amateur Radio more or less invented itself, right along with broadcasting and wireless telegraphy. The very first amateur licenses were granted back in 1912 and the number of "hams" grew rapidly. Early stations used "spark," literally a vigorous and noisy electrical arc, to generate radio waves. Inefficient and hazardous, spark was soon replaced by far more effective vacuum tube transmitters. By the end of the 1920s both voice and Morse code could be heard on the airwaves. Radio became very popular, instantly connecting communities and individuals as they had never been before.

Ham operators (from left) Rochelle, AE7ZQ, Kevin, N2LGN, Greg, NF7H, and ICS Commander Tina Birch staffed the Multnomah County (Oregon) EOC during a recent Simulated Emergency Test. [Nathan Hersey, N9VCU, photo]

As radio communication became widespread, the Federal Communications Commission (FCC) was created to regulate the competing radio uses, including broadcasting, commercial message and news services,

Hams have been building "OSCAR" satellites for decades. OSCAR stands for Orbiting Satellite Carrying Amateur Radio. These photos show Lance Ginner, K6GSJ, holding the first OSCAR launched in 1961 and a flight model of today's FUNcube-1 satellite.

military communications, and public safety. The Amateur service (the legal name for Amateur Radio) was created in 1934 and has expanded in size and capability ever since.

Amateurs, skilled in the ways of radio, played crucial roles during World War II as operators and radio engineers. After the war, thousands of hams turned to radio and electronics as a profession, fueling the rapid advances in communications during the 1950s and 60s. Amateur Radio evolved right along with industry — spanning the globe was commonplace! With Morse code as popular as ever, the amateur airwaves were also filling with voice and radioteletype signals. Hams even invented a new form of picture transmission called slow-scan television that used regular voice transmitters and receivers. The first satellite built by amateurs, called OSCAR-1, was launched in 1961, transmitting a simple Morse message back to Earth for several weeks.

Through the 1970s amateurs built an extensive network of relay "repeater" stations to provide regional communications with low-power mobile and handheld radios. In the 1980s and 1990s, microprocessors were quickly applied to radio, greatly increasing the capabilities of amateur equipment and ushering in a new era of digital communications. Packet radio, an adaptation of computer network technology, was developed by hams and is now widely used for commercial and public safety communications.

The personal computer, as in many other fields, gave amateurs a powerful new tool for design, modeling, station automation, and recordkeeping, as well as making Amateur Radio computer networks a reality. Finally, the Internet arrived and hams quickly adapted the new technology to their own uses just as they had many times before. At each step in the development of today's communication-intensive world, hams have contributed either as part of their profession or as individuals pursuing a personal passion.

That's Why It's Called "Amateur" Radio

In order to keep businesses or municipalities from unfairly exploiting the amateur bands, amateurs are strictly forbidden from receiving compensation for their activities. That means you can't talk with a co-worker about an assignment, for example. If you provide communications for a parade or charity activity, you can't accept a fee. This keeps radio amateurs free to explore and improve and train — it's worked well for many years.

HAM RADIO TODAY

Here we are a century later and wireless is still very much at the forefront of communications technology. Far from being eclipsed by the Internet, ham radio continues its tradition of innovation by combining the Internet with radio technologies in new ways. Hams have created their own wireless data networks, position reporting systems, and even a radio-based e-mail network that enables the most solitary ham to "log in" from anywhere in the world. Voice communications hop between Internet and radio links to connect hams on the opposite sides of the globe using only handheld transmitters less powerful than a flashlight!

Don't let anyone tell you that Morse code is finished! It's still very much alive in

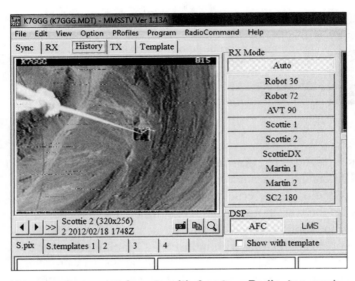

Hams perform experiments with Amateur Radio, too, such as this high-altitude balloon that carried a slow-scan TV transmitter to 88,000 feet! [Courtesy Gary Miller, K7GGG]

Edison High School (Queens, NY) students Kolsuma Begum, KD2DME, and Karl Anthony Singh, KD2DMF, are holding CSCEs that document they just passed their Technician exams! Their mentor (or Elmer) at left is Fred, N2EGQ, and the Volunteer Examiner team (left to right) are Rich, W2RB, Pete, K2IQK, and Mike, W2RT.

The Goldfarb Scholarship winner for 2013 was Calvin Darula, KØDXC. Calvin was selected as the WØ Young Ham of the Year, served as the ARRL Minnesota Youth Assistant Section Manager and has been a member of the USA High Speed Telegraphy Team.

Operating with a portable station as part of a hike or camping trip is a lot of fun. Tommy, W4TZM, operated from the summit of Wesser Bald (North Carolina) during the North Carolina QSO Party. [Tom Mitchell, photo]

Amateur Radio where its simplicity and efficiency continue to make it popular. Amateurs also speak to each other directly using sophisticated radios that are grown-up versions of the Citizen's Band (CB) and Family Radio Service (FRS) radios available at local electronics stores. Computers are a big part of ham radio today as hams chat "keyboard-to-keyboard" or send pictures via radio. You'll even find some hams assembling their own TV stations and transmitting professional-quality video!

In step with the telecommunications industry, hams also look to the skies for their communications. There are many Amateur Radio satellites whirling through orbit, connecting hams on the ground by voice, Morse code and data signals. There is even a ham station on the International Space Station used by astronauts (most astronauts have ham licenses) and ground-based hams. Ham-written software allows signals to be bounced off the Moon and even meteor trails in the Earth's atmosphere.

When disaster strikes, you will find hams responding quickly and capably in support of public safety agencies and relief organizations such as the Red Cross and the National Hurricane Center. Amateur Radio is an important part of many disaster relief efforts. Between emergencies, hams turn out in great numbers to provide communications for parades, sporting events, festivals and other public occasions.

While Amateur Radio got its start long ago as a collection of tinkerers in basements and backyard "shacks," it has grown to become a worldwide communications service with millions of licensees. The tinkerers are very much still with us, of course, creating new and useful ways of putting radio to work. You will find that ham radio has more aspects than you could imagine and they're growing in number every day.

WHO CAN BE A HAM?

Are you ready to join us? Anyone can become a ham! It doesn't matter how old you are or how much you know about radio when you begin. One of ham radio's most enjoyable aspects is that you're on a first-name basis with every other ham, whether you're an elementary school student, a CEO, an astronaut or a long-distance truck driver. There are also thousands of people with disabilities for whom ham radio is a new window to the world.

Hams range in age from six to more than one hundred years old. While some are technically skilled, holding positions as scientists, engineers, or technicians, all walks of life are represented on the airwaves.

Musicians? Try Patti Loveless, KD4WUJ, country music superstar, or Joe Walsh, WB6ACU, guitarist for the Eagles. Nobel Prize winner? Meet Joe Taylor, K1JT. Athlete? Now active in radiosport competitions, Joe Rudi, NK7U, is a retired major league outfielder and MVP! Why not imagine your name and call sign up there in lights?

Taking a sun-drenched break from winter in the Midwest, Glenn Johnson, WØGJ, was one of 30 skilled operators who recently put Desecheo Island on the air as K5D. Glenn is a pediatric surgeon and sometimes combines ham radio with his travels to donate medical

HANDI-HAMS — No Barriers to Ham Radio

Amateur Radio presents an opportunity to communicate and participate for those with disabilities. Disabled hams often use their unique talents to make valuable contributions over the airwaves, even though they might not be physically present or active. The Courage Center, of Golden Valley, Minnesota, sponsors the HANDI-HAM system (**www.handiham.org**) to help people with physical disabilities obtain amateur licenses. The system provides materials and instruction to persons with disabilities interested in obtaining ham licenses. The Center also provides information to other hams, dubbed "verticals," who wish to help people with disabilities earn a license.

Rachel, KI6PJY (left), and Christina, KI6QLR, helped build a radio from kit with the rest of the Granite Bay Montessori School students.

Ocean sailors Jan, KF4TUG and Mike, KM9D, keep in touch with home through ham radio on the high seas.

Microwave contesters such as Tony, WA8RJF, often travel to special locations to make long-distance contacts.

services. He has been the ARRL's Humanitarian of the Year and has operated from as far away as Bhutan and the Andaman Islands.

Rachel Finerman, KI6PJY, and Christina Soltero, KI6QLR, worked together as a two-student team to help build an Elecraft K2 HF transceiver kit at Granite Bay Montessori School in California. The 4th through 8th-grade students also assembled and installed a GAP Titan vertical antenna and operate the K6GBM repeater on the UHF band of 70 centimeters. As part of learning about science, more than 20 students have earned their Technician license and several have upgraded to General.

One of radio contesting's most enjoyable challenges is operating mobile or portable, putting rare counties or other locations on the air. At VHF and UHF, "hilltopping" is a popular way of greatly extending a station's range. Microwave enthusiasts like Tony

Emanuele, WA8RJF, also look for DX (long-distance) contacts from special locations that offer regular tropospheric paths over dozens or hundreds of miles, such as a beachfront park on Lake Erie.

Toni Linden, OH2UA, usually operates from his home in Finland, but as part of the Radio Arcala Group he occasionally travels to the Azores, placing at the top of international radiosport competitions using the call sign CU2X. Toni races cars and is active in other sports, too. He and teammate OH4JFN placed in the top third of nearly 50 teams competing in the 2010 World Radiosport Team Championships, held in Moscow.

Combining a love of sailing and ham radio, husband-and-wife team Mike Young, KM9D, and Jan Heaton, KF4TUG, operate from aboard their sailboat. On a recent trip they traveled around the South Pacific, operating from the Solomon Islands as H44MY and H44TO. They've visited many other "DX" locations, such as the island of Pohnpei in Micronesia, Java, Wallis Island, and more, staying in contact with ham friends worldwide all along the way.

Taking ham radio to new heights, astronaut Richard Garriott, W5WKQ, operated from aboard the International Space Station (ISS), using the call sign NA1SS. Richard contacted more than 500 different hams from the ISS in a 12-day visit to space. Richard's father Owen, W5LFL, was the first ham in space from aboard the Space Shuttle in 1984. Nearly all of the US and Russian astronauts have ham licenses, contacting both individual hams and students in school classrooms from orbit.

Richard, W5WKQ, made hundreds of contacts from orbit aboard the International Space Station (ISS).

Toni, OH2UA, travels to the Azores to participate in (and win!) international radio competitions as part of the Radio Arcala team.

Glenn, WØGJ, took a break from his surgery practice to operate from Desecheo Island as K5D.

The most popular ham radio event of all is called ARRL Field Day, held on the fourth full weekend of June every year. More than 30,000 hams participate across the United States, Canada and the Caribbean. (European clubs hold their own Field Day on different weekends.) The goal is to exercise your abilities to set up and operate radio equipment "in the field" as if an emergency or disaster occurs. Field Day takes on all the aspects of a competition, emergency exercise and club picnic — it's all three! Visitors, particularly prospective hams like you, are welcome at these events and there's no better way to learn about ham radio. See the ARRL website for more information.

Members of the Ski Country Amateur Radio Club of Carbondale, Colorado, had a great time on the Western Slope during the ARRL Field Day. [Peter Buckley, NØECT]

WHAT DO HAMS DO?

There are so many things that hams do that it's almost easier to tell you what they don't do! The image most people have of a ham is someone with headphones hunched over a stack of glowing radios, listening to crackling voices from around the globe. We certainly do that — thousands of contacts span the oceans every day — but there is so much more!

Talking

Hams talk, literally, more than any other way of communicating. After all, it's the way humans are built. For that reason, most radios that hams use are made for voice communication. Relay stations called repeaters allow hams using low-power radios to talk with each over a wide region, 50 miles or more. Hams also bounce signals off the upper layers of the atmosphere, making contacts around the globe. Even computer-to-computer speech is available on ham radio, across town or to a distant repeater system on another continent.

Byte-ing

Hams also use digital "codes" to communicate. By connecting a computer to a radio, digital information can be sent over the air waves. This is one of the most exciting facets of 21st-century ham radio. One popular style is simple "keyboard-to-keyboard" communications similar to mobile-phone texting. Hams also use higher-speed digital systems to transfer files and web pages. Not only are hams sending digital data, they're inventing new ways of doing it, too!

Sending

The oldest method hams use is the venerable Morse code, now well into its second century of use. Far from being a useless antique, Morse transmissions are simple to generate and very efficient. Hams also love the musicality of "the code" and have devised many ways to generate Morse, from the original "straight key" to fancy electronic versions and even via computer keyboards. Morse has a language all its own, both in its clear, pure tones and in the mannerisms with which hams connect and converse.

Building

Unlike many other types of radio communications, you are allowed — no, encouraged — to build and repair your own equipment, from the radio itself to the antennas and any accessory you can think of. Hams call this build-it-yourself ethic "homebrewing" and are proud to use equipment and accessories they built themselves. You'll find hams constructing anything from high-power, signal-boosting amplifiers to a miniature hand-built radio that fits in a metal candy tin! Hams have been responsible for numerous advances in the state of the antenna art as they tinker and test, too. If you like to know what's "under the hood" you'll find many like-minded friends in ham radio.

Watching

Hams have devised many ways of exchanging pictures and video. Many years ago, hams used teletype to send pictures made from text characters. Today a ham can set up a video camera and transmit pictures every bit as good as professional video. This is called ATV for amateur television, and there are even ATV transmitters flying in model aircraft and balloons that ascend to the edge of space! Hams pioneered the use of regular voice transmitters to send still photographs as "slow-scan television" over long-distance paths to hams thousands of miles away.

Hams Respond in Times of Need

Hurricane Sandy was the largest Atlantic hurricane ever recorded and the second most costly of all time. Before, during, and after the storm, ARRL Amateur Radio Emergency Service (ARES) volunteers were active in ARRL sections from Delaware through Eastern Massachusetts. Amateurs worked closely with the National Weather Service and the National Hurricane Center to report on local conditions as the storm moved inland. Since mobile and landline telephones, electrical power, cable TV, and Internet service were knocked out, amateur voice and digital links were put to work at EOCs (Emergency Operations Centers), Red Cross shelters, and supporting numerous other agencies and facilities across the region. The amateur response lasted well beyond the duration of the storm.

In the Louisiana, Mississippi, Alabama, and South Texas ARRL sections, ARES volunteers were heavily engaged in the Hurricane Katrina recovery effort. Winds and flooding from the huge storm wreaked havoc after Katrina came ashore. In Louisiana alone, ARES Section Emergency Coordinator Gary Stratton, K5GLS, told the ARRL that some 250 ARES members had been working with relief organizations and emergency management agencies from the beginning. Overall, more than 1000 amateurs assisted during that year's disastrous hurricane season.

While these two major storms were incredibly damaging and costly, amateurs respond to many other events that can be just as devastating on a local scale. Across the Midwest, hams are called when tornados strike, such as in Joplin, Missouri and Moore, Oklahoma. In the West, hams support wildfire responders in the incredibly rugged mountain terrain that stretches from San Diego to Colorado and Montana. Around the world, hams were there when earthquakes leveled cities across Lushan Province in China and they coordinated disaster relief from flooding in Bangkok.

These stories offer just a tiny glimpse of how Amateur Radio operators respond when their communities are damaged or destroyed. You'll find that becoming a radio amateur gives you an opportunity to learn the skills and techniques necessary to be of service when called upon. It's one of the most valued features of the Amateur service.

Emergency Communications and Public Service

One of the reasons Amateur Radio continues to enjoy its privileged position on the airwaves is the legendary ability of hams to organize and respond to disasters and emergencies. Because ham radio doesn't depend on extensive support systems, ham stations are likely to be able to operate while the public communications networks are recovering from a hurricane or earthquake. Hams are also self-organized in teams that train to respond quickly and provide communications wherever it's needed. It's not necessary to have a big emergency for hams to pitch in. We also provide public service by assisting with communications at parades and sporting events, or by serving as weather watchers.

Hams recognize the value of Amateur Radio to their communities and have created training programs such as the ARRL Amateur Radio Emergency Service (ARES®) that promote readiness. The ARRL offers emergency communications training classes over the Internet. Hams often work closely with other citizen volunteer teams. Many are also certified as emergency response workers with a wide variety of skills such as first aid, search-and-rescue and so forth. Hams hone their message-handling skills in "traffic" nets that pass routine messages around the country so that they'll be ready when called upon for real.

AMATEUR RADIO CLUBS AND ORGANIZATIONS

Helping newcomers is one of ham radio's oldest traditions. After all, we are all "amateurs," learning and training together. Nearly every ham has mentored or *Elmered* another ham at one time or another. You'll be amazed at the amount of sharing within the ham community. Your ham radio support group comes in many forms — a fellow student or classmate, a nearby ham, a club or even a nationwide organization. All of them are resources for you, not only during your studies for the licensing exam, but also after you have your call sign and are learning how to be a ham.

If you haven't already found a local radio club, you can find one by using the ARRL's *Affiliated Club Search* on the ARRL website. There are several types of clubs; some specialize in one type of operating or public service. Most are "general interest" clubs for the members to socialize, learn and help each other out — a good first choice.

Eagle Scout Carey, N5RM, demonstrates ham radio to another Scout at the John Nickels Scout Ranch (Oklahoma) during the Jamboree On the Air (JOTA).

Ross, KR4USA, and his daughter Hope look for some contacts on the "magic band" — 6 meters — during Field Day at K4PJ, the Oak Ridge (Tennessee) Amateur Radio Club.

What's an Elmer?

A ham radio "Elmer" is a person who personally guides and tutors a new ham through the learning process, both before and after getting a license. It doesn't refer to anyone in particular, just the more experienced hams who lend a helping hand to newcomers. Just about everyone has an Elmer at one point and sometimes several! It's one of ham radio's highest compliments to be someone's Elmer.

ARRL — The National Association for Amateur Radio

The American Radio Relay League (ARRL) has been an integral part of Amateur Radio from the very beginning. The ARRL offers more assistance to potential and licensed hams than any other organization, including operating the largest of the Volunteer Examiner Coordinators and working on behalf of all hams with the FCC and Congress. The core missions of the ARRL are:

• Public Service

The ARRL actively promotes the public-service aspects of Amateur Radio, a tradition that has earned respect through decades of service. The ARRL's legacy of public service began in 1935 with the creation of the Amateur Radio Emergency Service, better known as ARES, to provide communication support during natural and man-made disasters.

• Advocacy

The ARRL represents Amateur Radio at the local, state, federal and international levels. Thanks to the efforts of the ARRL, Amateur Radio has been able to thrive despite repeated attempts to restrict its growth. The ARRL serves as a voice for Amateur Radio before regulatory agencies such as the Federal Communications Commission (FCC) and the International Telecommunication Union (ITU).

• Education

The educational mission of the ARRL is twofold. (1) To recruit new amateurs, the ARRL publishes books and study guides such as this one for Amateur Radio license exams, maintains a mentor program for new hams and much more. (2) The ARRL also promotes ham radio in school classrooms, advocating its use as a tool to teach science and technology. To that end, the ARRL assists teachers with appropriate instructional materials and training.

• Technology

The ARRL promotes technical skills and training for all amateurs. It publishes more Amateur Radio technical material — in print and online — than any other amateur organization in the world. The *ARRL Handbook* and *ARRL Antenna Book* have been amateur and professional references for decades. The ARRL's online *Technical Information Service* provides numerous resources on technical topics. *QEX* is the ARRL's magazine for advanced technology and presents state-of-the-art information in each issue.

• Membership

The majority of active amateurs belong to the ARRL, and for good reason! In addition to all the member services listed above, ARRL members receive *QST* magazine each month and can subscribe to an entire roster of newsletters and bulletins. Members also have full access to the extensive ARRL website. The ARRL also sponsors the largest *radiosport* program in the world with an event for every mode of operating. Membership does not require a license and costs about the same as a couple of large pizzas, with considerably longer-lasting benefits!

• *QST* — The Amateur's Magazine

QST is the authoritative source for news and information on any topic that's part of, or relates to, Amateur Radio It is available in print and digital formats, whichever you prefer. In each colorful issue you'll find technical articles and informative Product Reviews of the newest radios and accessories from handheld and mobile FM radios, to home station transceivers, antennas and even shortwave radios. Each month's Coming Conventions and Hamfest Calendar columns show you who's getting together at hamfests, conventions and swap meets in your area. Whether you're interested in radiosport contests, DXing, or radios, accessories and antennas you can build at home, *QST* covers them all: New trends and the latest technology, news, club activities, rules and regulations, special events and much more.

Don't hesitate to make use of the contact info and attend a meeting! Many clubs make an extra effort to offer special assistance to aspiring and new hams.

Once you decide on a club, you'll get a lot more than just study help by participating in the club activities. Log on to the club website. Take advantage of open houses, work parties or operating events, and maybe attend those informal lunches or breakfasts. Be sure to introduce yourself to the club officers and let them know you're a visitor or new member. Is there another new member? Buddy up! Soon you'll be one of the regulars.

WHAT MAKES AMATEUR RADIO DIFFERENT?

There are lots of other types of two-way radios you can buy in a store — Citizen's Band, handheld FRS/GMRS "walkabouts," marine radio for boaters — what is it about Amateur Radio that sets it apart? In a word — variety. You'll find that each of the radio types listed in **Table 1.1** are designed for just a few purposes and they might do that well. Amateur Radio, on the other hand, is tremendously flexible with many different types of signals and radio bands. As a ham, you're not restricted to any one combination; you can experiment and try different things as much as you want to get the job done.

Unlicensed Personal Radios

The most popular personal radios are the FRS/GMRS handheld radios that are seemingly sold everywhere. FRS stands for Family Radio Service and GMRS stands for General Mobile Radio Service. These radios use a set of 22 channels in a narrow frequency band best suited for short-range, line-of-sight communications. (You may be unaware that using the GMRS channels and features of the radio requires a license! It's in the manual's fine print.) Without the GMRS license, your maximum ½ watt of transmitter output power limits you to communications over a few hundred yards to a couple of miles. You can't extend your range with repeaters, nor can you use more powerful mobile radios.

Although the "Good Buddy" CB fad of the 1970s is history, Citizens Band remains popular in the applications for which it was originally intended. Mobile radios in vehicles, boats and farm equipment provide useful, medium-range radio communications to other vehicles or with radios at home or at work. Handheld radios are also popular. CB radios have 40 channels (more with selectable sidebands) and communication is fairly reliable over a range of several miles.

Boaters will be familiar with marine VHF radios used for boats to communicate with each other and with stations on shore. These radios can use up to 50 channels for communicating around harbors and for short-range needs during both fresh and salt-water travel.

All three of these radios are designed to use a set of channels selected for a single type of communications as shown in Table 1.1. They do their designated job well. Amateurs have access to a much broader range of communications options and create new ways of communicating that are more powerful and flexible than those of the unlicensed radio

Table 1.1

Types of Personal Radio

Service	Channels	Intended Use	Range
Citizens Band (CB)	40	Private/Business	10 miles +
Marine VHF	50	Maritime	20 miles +
Family Radio Service (FRS)	22	Personal	2 miles
Multi-Use Radio Service (MURS)	5	Personal	5 miles+

services. If you find your personal radio interesting, but limited, then Amateur Radio is definitely the place for you.

Business and Public Safety Radio

Every day you can see police and firemen using handheld and mobile radios as part of their jobs. Many businesses also have their employees use similar radios. How do these relate to Amateur Radio? The FCC has created radio services for public agencies and private businesses. These organizations, public and private, are all licensed, just like hams must be, although the individual users need not have a license. The electronics in these radios are very similar to those of radios that hams use, sometimes identical. In fact, many a ham has converted a surplus commercial or public safety radio to ham radio use. While business and public safety radio users are restricted to just a few "channels," though, hams can use their radios on hundreds of channels and for far more varied uses.

1.2 The FCC and Licensing

The Federal Communications Commission (FCC) is charged with administering all of the radio signals transmitted by US radio stations. The FCC also coordinates these transmissions with other countries as part of the International Telecommunication Union (ITU). While you may not need a license to use an FRS or CB radio, the vast majority of radio users must have a license or be employed by a company that has a license. This section explains how licensing works for Amateur Radio.

WHY GET A LICENSE?

Amateurs are free to choose from many types of radios and activities — that's what you get in return for passing the license exam. If you can learn the basics of radio and the rules of Amateur Radio, then the opportunities of ham radio are all yours! Just remember that the license is there to ensure that you understand the basics before transmitting. This helps keep Amateur Radio useful and enjoyable to everyone.

Why don't people just buy radios and transmit anyway? (This is called "bootlegging" or "pirating.") First of all, it's quite apparent to hams who has and who hasn't passed a license exam. You'll find yourself attracting the attention of the FCC but more importantly, you won't fit in and you won't have fun.

One of the most important benefits to being licensed is that you have the right to be protected from interference by signals from unlicensed devices, such as consumer electronics. Your right to use the amateur bands is similarly protected. The protection doesn't work perfectly all the time, but nevertheless, as a licensed amateur operator, your license is recognized by law. This is a big improvement over unlicensed radio users. It's definitely worth the effort to get that license!

LICENSING OVERVIEW

The FCC has a different set of rules for each type of radio use. These uses are called *services*. Each service was created for a specific purpose — Land Mobile, Aviation and Broadcasting, for example. Nearly all services require that a license be obtained before transmissions are made. These are called licensed services and the Amateur service is one such service.

Most services do not require an examination to be licensed. This is because the FCC sets strict technical standards for the radio equipment used in these services and restricts how those radios may be used. This tradeoff reduces the training required for those radio users. Licensing in these services is primarily a method to control access to the airwaves.

Amateurs, on the other hand, have great latitude in how we use radios. We can build and repair our own radios. The procedures we use to communicate are completely up to us. We can operate however we want, with few restrictions. This flexibility, in order to not cause interference to other radio services, requires that amateurs be more knowledgeable than the typical user in other services. That is why amateurs have to pass a licensing examination.

Amateur Licenses

Once upon a time, the FCC gave the exams for Amateur Radio licenses. In those days, hams often had to travel long distances to get to a regional Federal Building, stand in lines for hours, sit on uncomfortable chairs, and sweat their way through exams graded by grim-faced proctors. It's a wonder any of them survived the experience!

Today, amateurs give and grade the exams ourselves under the guidance of a Volunteer Examiner Coordinator (VEC). There are currently 14 different clubs or organizations recognized as VECs by the FCC. These make up the National Conference of Volunteer Examiner Coordinators (NCVEC). The NCVEC elects representatives to write the questions used for the license exam *question pool*. The representatives make up the Question Pool Committee. There is one question pool for each class of amateur license.

Each VEC also certifies Volunteer Examiners (VE) who actually administer the exam sessions. The VEC then handles the paperwork for each license exam and application. That doesn't mean you won't sweat a little bit, but the examination process is not as imposing as it seems.

The result of passing the exam is an *operator license* (or "ticket") granted by the FCC after it receives the necessary paperwork from the VEC that administered your exam session. The license also specifies a call sign that becomes your radio identity.

There are three classes of license being granted today: the Technician, General and Amateur Extra. The exam for each of the three license classes is called an *element*. Passing each of the elements grants the licensee more and more *privileges* allowed by the FCC's Amateur Service rules. **Table 1.2** shows the elements and privileges for each of the Amateur license classes as of early 2014.

You'll learn the privileges of the Technician class license as you study this book. For now, all you need to remember is that Technician licensees are granted privileges on the radio airwaves referred to as the "VHF and UHF bands" and a few privileges on the "HF bands." You'll learn what those terms mean as you study.

Table 1.2
Amateur License Class Examinations

License Class	Exam Element	Number of Questions	Privileges
Technician	2	35 (passing grade is 26 correct)	All VHF and UHF privileges, with some HF privileges
General	3	35 (passing grade is 26 correct)	All VHF, UHF and most HF privileges
Amateur Extra	4	50 (passing grade is 37 correct)	All amateur privileges

1.3 Amateur Radio Activities

Ham radio has a lot to offer, but the many activities can be confusing. To help you understand why we operate in certain ways or why rules are written the way they are, this section presents some basics of ham radio. Later, you'll learn more details, but this introduction present some of the fundamentals that are present in almost all ham radio communications.

IDENTIFICATION AND CONTACTS

On the airwaves, your everyday identity gets something new — a *call sign*. Instead of "Steve" or "Mary," your primary radio identity becomes "Steve WB8IMY" or "Mary K1MMH." Hams become known by their call signs and often keep them for life. Your call sign or call is completely unique among all the radio users anywhere in the world! There is only one N6BV (Dean in USA) just like there is only one G4BUO (Dave in England) and one PP5JR (Sergio in Brazil) and one JE1CKA (Tack in Japan). By transmitting your call, other hams know who you are and your nationality. Identifying yourself with your call sign is known as signing. Because you can't see other hams except when using video transmissions, your call sign is very important. In fact, you're required to transmit your call regularly during every contact so that everyone knows whose transmissions are whose.

Speaking of contacts, any conversation between hams over the air is called a contact and starting a conversation is making contact. Attempting to make contact by transmitting your call sign is making a call or calling. If you're making a "come in anybody" call to which any station can respond, that's calling CQ. ("CQ" means "a general call.")

Once you establish contact, the next step is to exchange more information such as a *signal report* that lets the other station know how well you are receiving or copying them. Name and location are exchanged after that — then you're off to whatever business is at hand. A long conversation is known as a *ragchew*. At the end of a contact, you sign off.

Hams use repeaters to relay signals from low-power radios over a wide area. Repeaters are a popular on-the-air meeting place for hams.

Ham Shorthand

Like any activity that has been around for a while, such as sailing or flying, radio has a special jargon of its own. Many of these conventions originate from the days of the telegraph. In those days, every word took up precious time so the operators developed an extensive series of abbreviations and special characters (called prosigns) that kept the information (or traffic) flowing quickly and smoothly. For example, you may have heard the word "break" or "breaker" used over the radio. The word originally referred to a telegraph operator disconnecting or breaking the telegraph line so that no characters could be sent. This got the attention of every other operator along the line and to do so was breaking in. The word is still in use 150 years later!

Later, as radio became a worldwide tool, operators that didn't speak the same language were aided by the creation of Q-signals. For example, "QTH?" means "What is your location?" and in response "QTH Seattle" means "I am located in Seattle." Many of these procedures and abbreviations are still in place today, because they work!

USING YOUR VOICE

By far, the most popular method or *mode* of making contacts is by voice. There are a number of ways to transmit voices via radio signals and you'll learn about those in the next chapter. It's easy and natural to converse this way, using the proper procedures. Voice is widely used by amateurs for short-distance and long-distance contacts. It is the most popular mode for hams on the go and during public service or emergency response.

As a Technician licensee, you'll be able to make voice contacts directly with other hams and also by using stations called *repeaters* that relay signals from low-power mobile or handheld transmitters across a wide area. Hams have also built Internet-linked repeater systems that send digitized voices around the world so that local repeater users can communicate world-wide with just a low-power, handheld radio!

Hams usually use English as a common language when making international contacts, although when communicating within a country hams use their native languages. Voice contacts can be a good way to learn or polish your foreign language skills.

EXCHANGING DIGITAL DATA

The recent availability of inexpensive sound card hardware and signal processing software for personal computers has brought about a surge of interest in the *digital modes* where the conversation is carried out as streams of characters sent over the airwaves. A data interface is used to connect the radio to the computer. Most digital contacts are keyboard-to-keyboard, meaning that the operators take turns typing, just as in an Internet chat or messaging system.

Hams have been blazing trails in developing new methods of converting the computer characters into radio signals and back again. The methods are called *protocols* and are referred to by their initials, such as RTTY, PSK31, PACTOR or MFSK to name just a few. Different protocols are applied to different types of radio communication because of the effects that transmission and reception have on the radio signals.

One of the digital radio systems developed by hams exchanges e-mail over Amateur Radio. It's called Winlink and it looks a lot like regular e-mail on the computer screen. It's used daily by thousands of hams who are unable to access the Internet while traveling, at sea, or camping in remote locations.

USING MORSE CODE

Morse is still quite popular in Amateur Radio. Because all of a signal's energy is concentrated in a single on-and-off signal, Morse works very well in the presence of interference or when signals are weak. Morse signals can be generated by extremely simple transmitters — all you need is something to generate a radio signal and something else to turn it on and off! Receiving and decoding Morse (called copying the code) requires only a basic receiver and a human ear.

Many operators enjoy the rhythm and musicality of "the code," as well. Aside from its utility as a communications protocol, it's a skill like playing a musical instrument that you can enjoy for its own sake. Listening to a skilled Morse operator is quite a treat!

EMERGENCIES AND PUBLIC SERVICE

One of the reasons ham radio is so valuable (and maybe a reason you're reading this book) is that hams are good at helping. Communications is key to making any kind of organized effort work, whether it's a small parade or major emergency response to a natural disaster. While the day-to-day telecommunication systems are recovering, hams can quickly set up networks that support public safety and government operations. Why?

Table 1.3
Emergency Response Organizations

ARES	Amateur Radio Emergency Service — organized by the ARRL
RACES	Radio Amateur Civil Emergency Service (works with civil defense agencies)
SATERN	Salvation Army Team Emergency Radio Network
HWN	Hurricane Watch Net — works with the National Hurricane Center
SKYWARN	Severe weather watch and reporting system — works with the National Weather Service

Because there are lots of hams and we are skilled in basic communication techniques that don't depend on a pre-existing telecommunications system.

Can Technician class licensees help out? You bet they can! By learning how to use your radio and taking some simple training classes, such as the ARRL's Emergency Communications training courses, you'll be ready to join and practice with other hams. The largest ham emergency organization is the Amateur Radio Emergency Service (ARES) which is organized by the ARRL. You can join a local ARES team to receive training and practice providing emergency communications support. **Table 1.3** lists several ham radio emergency response groups.

Hams can pitch in and help in many ways. Not everyone has to be on-site to make a contribution. Whatever your personal capabilities and license class, there is a need you can fill.

- From home — Use base station radios and antennas to provide long distance communications, relay messages and act as a net control to coordinate communications.
- From a vehicle — From a personal car or a communications-equipped van, portable stations provide valuable relay and net control functions in the field.
- On foot — Go where the action is to provide status reports and relay supply and operations messages between the control centers and workers in the field.

One of the most important functions, repeated three times in the list above, is to relay communications. It is no coincidence that the third letter in ARRL stands for "relay," one of the oldest and most highly valued radio functions. Relaying information requires accuracy and efficiency. Hams that provide emergency communications pride themselves on both.

AWARDS AND CONTESTS

Hams keep their skills sharp not only by training, but by getting on the air and having fun! Just as sports and recreational activities keep your body fit and in good health, there are competitive radio activities, as well! There are many suitable for Technician license holders.

There are operating achievement awards for almost anything you can imagine, such as working (contacting) every state or different countries, contacting satellites, and making low-power contacts. **Table 1.4** shows several examples of awards and operating events, but

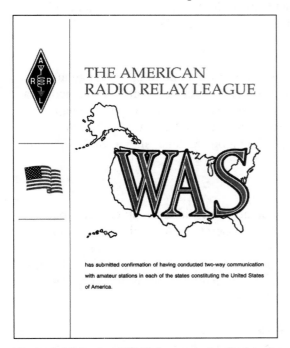

Worked All States (WAS) is a popular first major award pursued by hams around the world.

Table 1.4

Awards and Operating Events for Technician Licensees

OSCAR Satellite Communications Achievement Award: Contact 20 different states, Canadian provinces, or countries using amateur satellites. Sponsored by AMSAT.

VHF/UHF Century Club (VUCC): Contact grid squares using VHF and UHF bands (100 grids on 50 and 144 MHz; 50 grids on 222 and 432 MHz; fewer on the higher bands)

ARRL and CQ VHF Contests:
Several contests throughout the year that make use of the VHF and UHF bands

ARRL Field Day:
The largest on-the-air event in ham radio

thousands of awards are available. Collecting these colorful certificates and other prizes can be addictive!

Contests, also called *radiosport*, fill the ham bands with rapid-fire contacts as amateurs strive to make as many short contacts as possible within a limited time. There are a tremendous variety of contests, from sprints lasting only a few hours to international contests that last for 48 hours! Working hard to compete and improve provides a strong incentive to become skilled in both technical and operating capabilities.

NOVEL ACTIVITIES

As if all this wasn't interesting enough, hams are famous for pushing the envelope and either inventing an entirely new technology or adopting a commercial technology in an unexpected way. It was mentioned before that hams have their own satellites and radio/ Internet networks, but that's just the start. Here are some examples:

SSTV and ATV

Slow-scan television (SSTV) was invented by hams to send pictures over regular voice radios. Black-and-white or color pictures and images can be sent and received by a computer with a sound card. Broadcast TV-style video is the domain of the amateur TV (ATV) enthusiasts. They hook up a regular video camera to an ATV transmitter and *voila!* they're on the air, beaming video that looks just like a professional signal. Both ATV and SSTV are increasingly used in emergencies.

Packet Radio

As computer networks became common, amateurs immediately adapted the popular X.25 computer-to-computer protocol to over-the-air operation. A special kind of data interface called a *terminal node controller* (TNC) takes characters from a computer and re-packages them into data packets which are transmitted by a regular, unmodified radio, usually on the VHF or UHF bands. Packet stations and networks are used to support emergency communications and a variety of other uses.

APRS — Automatic Packet Reporting System

Invented by Bob Bruninga, WB4APR, APRS integrates GPS position data and other information with packet radio. Amateurs with a GPS and a mobile radio can send their

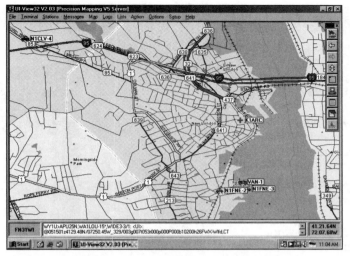

The Automatic Packet Reporting System (APRS) was invented by hams to show location and travel path by using ham radio.

position to a local APRS relay point or gateway and on to Internet-based servers. Other users can log on to the APRS servers and find the location of anyone sending position data by tracking their movements on maps of various detail levels. It's fascinating!

Meteor Scatter and EME

Perhaps the most exotic of all ham activities is making contacts via meteor scatter or Earth-Moon-Earth (EME) reflections. The trail of a meteor can reflect radio signals during the few seconds it lasts. A skilled amateur can use that trail to make short contacts! The biggest reflector in the sky is the Moon and hams can bounce their signals off the Moon and hear them when they complete the round trip back to Earth. Nobel Prize-winner Joe Taylor, K1JT, wrote software that uses a computer's sound card to both send and receive data in a highly specialized code that enables even modest stations to operate both meteor scatter and EME. Does that sound exciting? It is!

1.4 Getting Your Ham Radio License

THE TECHNICIAN LICENSE

The Technician license is the first license for newcomers to ham radio. There are more Technician licensees than of any other class, about 50% of all US hams. You'll be able to communicate with thousands of other hams in many of the ways amateurs use the airwaves.

Once you gain some experience, you'll be ready to *upgrade* your license to General class and beyond to the top-of-the-line Amateur Extra class. These licensees gain more privileges on the traditional HF or "shortwave" bands of Amateur Radio. They all started just like you, taking the basic exams and getting on the air.

OBTAINING A LICENSE

The first step is in your hands right now! To get your license, you'll need to pass a 35-question, multiple-choice exam on the rules of ham radio, simple operating procedures and basic electronics. You can study on your own or you can enroll in a licensing class. Log on to the ARRL website where you can search for classes being held near you. There are more than 2000 instructors throughout the United States who are part of the ARRL-sponsored training program. By joining a class, you can take advantage of the experience of these ham radio experts and learn in the company of other students.

Want More Information?

Looking for more information about ham radio in your local area? Interested in taking a ham radio class? Ready for your license exam? Call 1-800-32 NEW HAM (1-800-326-3942). Do you need a list of ham radio clubs, instructors or examiners in your local area? Just let us know what you need!

You can also contact us via e-mail:
newham@arrl.org

Or check out our website: **www.arrl.org**
Find some special content just for *Ham Radio License Manual* readers at

www.arrl.org/ham-radio-license-manual

You can even write to us at:
New Ham Desk
ARRL Headquarters
225 Main St
Newington, CT 06111-1494

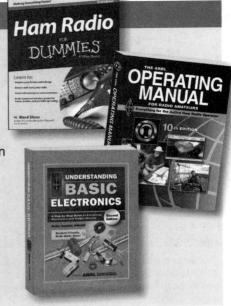

When you are ready to take your exam, it will be time to locate an exam session. If you're part of a study class, the instructor will make the necessary arrangements. For solo students, you can find an exam session by visiting the ARRL website. Use the exam search page to find exam sessions near you, including complete contact information. All Amateur Radio exams are given by ham radio operators acting as volunteer examiners.

After you pass your exam, the examiners will give you a *Certificate of Successful Completion of Examination* (CSCE) that documents your achievement. They will also file all of the necessary paperwork so that your license will be granted by the Federal Communications Commission (FCC). In a few days, you will be able see your new call sign in the FCC's database via the ARRL's website. Congratulations — you're authorized to get on the air! Later, you'll receive a paper copy of your license by mail.

WHAT WE ASSUME ABOUT YOU

The only thing you'll really need to succeed is a strong interest in Amateur Radio and a willingness to learn. You don't have to be a technical guru or an expert operator to get your license! As you progress through the material, you'll encounter some basic science about radio and electricity. There will be a simple bit of math here and there. When we get to the rules and regulations you'll have to learn some new words and maybe memorize a few numbers. That's it! It will help if you have regular access to the Internet. You should have a simple calculator — you'll be allowed to use it during the license exam.

ADVANCED STUDENTS

Perhaps you've used other types of radios, such as Citizen's Band or a business band radio at work. You might have a technical background or have experience as a radio operator. If so, we suggest that you jump to the question lists at the end of each section. If you find it easy to answer the questions correctly, you can skim or skip the corresponding section of the book. Regardless of your background, be sure to review the chapters on Licensing Regulations and Operating Regulations since ham radio rules and procedures are probably different than what you're used to.

ONLINE PRACTICE EXAMS

When you feel like you're nearly ready for the actual exam you can get some good practice by taking Amateur Radio practice exams using the ARRL's online tests or one of the other Amateur Radio practice exam websites listed on this book's web page. These websites use the question pool to construct an exam with the same number and variety of questions that you'll encounter on exam day. You can take them over and over again in complete privacy.

These exams are quite realistic and you get quick feedback about the questions you missed. When you find yourself passing the online exams by a comfortable margin, you'll be ready for the real thing!

A note of caution: Be sure that the questions used are current — the Technician question pool is completely rewritten every four years. The set of questions put in place in July of 2014 will be replaced in 2018. (The General and Amateur Extra class question pools will expire in 2015 and 2016.)

TESTING PROCESS

The final step is to find a *test session*. If you're in a licensing class, the instructor will help you find and register for a session. Otherwise, you can find a test session by using the ARRL's web page for finding exams. If you can register for the test session in advance, do so. Other sessions, such as those at hamfests or conventions, are available to "walk-ins," that is anyone who shows up. You may have to wait for an available space though, so go early!

Bring two forms of identification including at least one photo ID, such as a driver's license, passport or employer's identity card. Also know your Social Security Number (SSN). You can bring pencils or pens and a calculator, but any kind of computer or online device is prohibited. (If you have a disability and need these devices to take the exam, contact the session sponsor ahead of time.) Once you're signed in, you pay the test fee (check with the test session administrator) and get ready.

Amateur Radio licensing test sessions are administered by volunteer examiners — hams just like you will be. They grade the exams, help you fill out the necessary forms and take care of all the paperwork for your ham radio license.

The Technician test usually takes less than an hour. You will be given a question sheet and an answer sheet. As you answer each question, mark a box on the answer sheet. Once you've answered all 35 questions, the volunteer examiners (VE) will grade and verify your test results. Assuming you've passed (congratulations!) you'll fill out a Certificate of Successful Completion of Examination (CSCE) and a NCVEC Form 605. The exam organizers will submit your results to the FCC while you keep the CSCE as evidence that you've passed your Technician test. As soon as your name and call sign appear in the FCC's database of licensees, typically a week to ten days later, you can start transmitting!

If you don't pass, don't be discouraged! You might be able to take another version of the test. Ask the session organizers about a second try. Even if you decide to take the test again at a later date, you now know just how the test session feels — you'll be more relaxed and ready next time. The ham bands are full of hams that took their tests more than once before passing. You'll be in good company!

US Amateur Radio Technician Privileges

This chart shows privileges and band plan recommendations for each of the frequencies, as granted by the FCC to the Technician licensee. It is good amateur practice to follow the band plan established by the Amateur Radio community. The band plan is developed so that spectrum allocated for our use is used most effectively. You'll find a complete description of the band plan online at www.arrl.org/band-plan.

Published by:
ARRL The national association for **AMATEUR RADIO**
www.arrl.org

Effective Date March 5, 2012

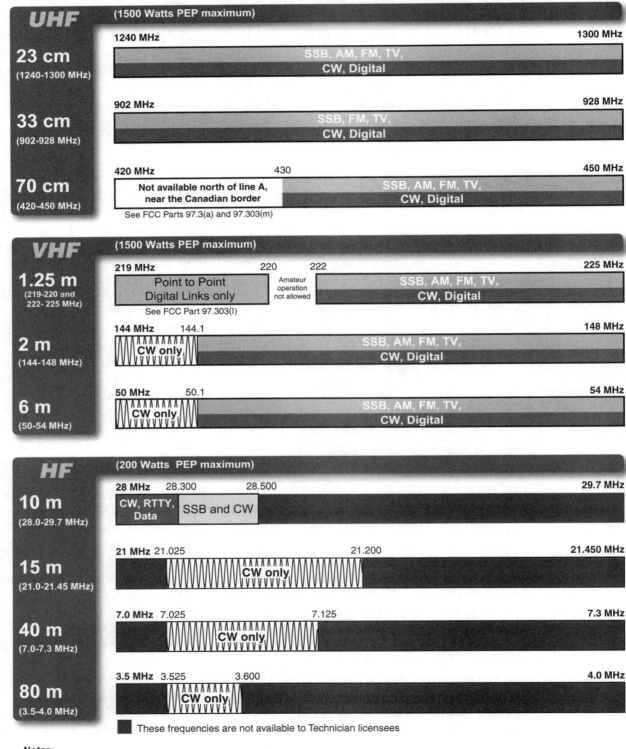

UHF (1500 Watts PEP maximum)

23 cm (1240-1300 MHz)
1240 MHz — 1300 MHz
SSB, AM, FM, TV, CW, Digital

33 cm (902-928 MHz)
902 MHz — 928 MHz
SSB, FM, TV, CW, Digital

70 cm (420-450 MHz)
420 MHz — 430 — 450 MHz
Not available north of line A, near the Canadian border
See FCC Parts 97.3(a) and 97.303(m)
SSB, AM, FM, TV, CW, Digital

VHF (1500 Watts PEP maximum)

1.25 m (219-220 and 222-225 MHz)
219 MHz — 220 — 222 — 225 MHz
Point to Point Digital Links only
See FCC Part 97.303(l)
Amateur operation not allowed
SSB, AM, FM, TV, CW, Digital

2 m (144-148 MHz)
144 MHz — 144.1 — 148 MHz
CW only
SSB, AM, FM, TV, CW, Digital

6 m (50-54 MHz)
50 MHz — 50.1 — 54 MHz
CW only
SSB, AM, FM, TV, CW, Digital

HF (200 Watts PEP maximum)

10 m (28.0-29.7 MHz)
28 MHz — 28.300 — 28.500 — 29.7 MHz
CW, RTTY, Data
SSB and CW

15 m (21.0-21.45 MHz)
21 MHz — 21.025 — 21.200 — 21.450 MHz
CW only

40 m (7.0-7.3 MHz)
7.0 MHz — 7.025 — 7.125 — 7.3 MHz
CW only

80 m (3.5-4.0 MHz)
3.5 MHz — 3.525 — 3.600 — 4.0 MHz
CW only

▓ These frequencies are not available to Technician licensees

Notes:
Technician Licenses may use up to 1500 Watts PEP on the VHF and higher bands, but are limited to 200 Watts on the HF bands.
You also have privileges to explore these microwave bands with CW, Digital, SSB, AM, FM and TV:

2300-2310 MHz	2390-2450 MHz	3300-3500 MHz	5650-5925 MHz	10.0-10.5 GHz	24.0-24.25 GHz
47.0-47.2 GHz	76.0-81.0 GHz	122.25-123.0 GHz	134-141 GHz	241-250 GHz	All above 275 GHz

rev. 3-7-14

Chapter 2

Radio and Signals Fundamentals

In this chapter, you'll learn about:

- **What is a radio signal**
- **The characteristics of radio signals**
- **How modulation adds information to radio signals**
- **Types of modulation**
- **Names and types of basic radio equipment**

When you see the mouse, you'll find more information at www.arrl.org/ham-radio-license-manual

This is the real beginning of your Amateur Radio adventure! In this section, we dive into what makes radio work — starting with the signals themselves — and then basic radio equipment. We'll then go on to basic electronics, operating, rules and safety in the following chapters. The material in each chapter is presented in a "here's what you need to know" style. References will be provided so that you can learn more about topics that interest you. Start by bookmarking the ARRL's *Ham Radio License Manual* web page, **www.arrl. org/ham-radio-license-manual**. Don't forget that there's a comprehensive glossary in the back of the book for unfamiliar terms.

Covering technical topics first makes it easier for you to understand the chapters about operating and the rules and regulations. You'll also be a better and safer operator. Relax — we'll start at the beginning and learn one step at a time!

2.1 Radio Signals and Waves

We'll start our discussion with the signals that travel back and forth between radios, carrying voices, data and Morse code. These signals are made up of *radio waves* that travel at the speed of light. A radio wave begins its journey in an *antenna* as an electrical signal that constantly reverses direction. The rate at which the signal changes direction — backward and forward, over and over — determines the signal's *frequency*. [T5A12] The radio wave then travels away from the antenna into space, vibrating or *oscillating* at the same frequency as the electrical signal.

As the wave passes other antennas, it creates replicas of the original electrical signal. A radio then turns the received signal back into a voice, digital data or even Morse code.

The process of turning the transmitter's output signal into radio waves that leave the antenna is called *radiating* or *radiation*. The word "radiation" should not concern you. Radiation from an antenna is not the same as *ionizing radiation* from radioactivity, nor does it interact with living organisms in the same way. You'll learn more about this subject in the **Safety** chapter.

The units of measurement employed in radio use the *metric system* of prefixes. The metric system is used because the numbers involved cover such a wide range of values. **Table 2.1** shows metric prefixes, symbols, and their meaning. The prefixes expand or shrink the units, multiplying them by the factor shown in the table. For example, a *kilo*meter (km) is one thousand meters and a *milli*meter (mm) is one-thousandth of a meter.

The most common prefixes you'll encounter in radio are pico (p), nano (n), micro (μ), milli (m), centi (c), kilo (k), mega (M) and giga (G). It is important to use the proper case for the prefix letter. For example, M means one million and m means one-thousandth. Using the wrong case would make a big difference!

If you're already familiar with the metric system, review the questions at the end of the chapter to be sure you have it mastered. If the metric system is unfamiliar to you, the *Ham Radio License Manual* web page has a detailed discussion of how the prefixes work and examples for you to learn from. Review that material until you are comfortable with the examples and definitions. [T5B01 to T5B08, T5B12 and T5B13]

Table 2.1
International System of Units (SI) — Metric Units

Prefix	Symbol	Multiplication Factor	
Tera	T	10^{12} =	1,000,000,000,000
Giga	G	10^{9} =	1,000,000,000
Mega	M	10^{6} =	1,000,000
Kilo	k	10^{3} =	1000
Hecto	h	10^{2} =	100
Deca	da	10^{1} =	10
Deci	d	10^{-1} =	0.1
Centi	c	10^{-2} =	0.01
Milli	m	10^{-3} =	0.001
Micro	μ	10^{-6} =	0.000001
Nano	n	10^{-9} =	0.000000001
Pico	p	10^{-12} =	0.000000000001

Before you go on, study these Technician exam questions from the question pool included at the back of this book or as a downloadable Study Guide version on the web:
T5B01 through T5B08 T5B12 T5B13
If you have difficulty with any question, review the preceding section.

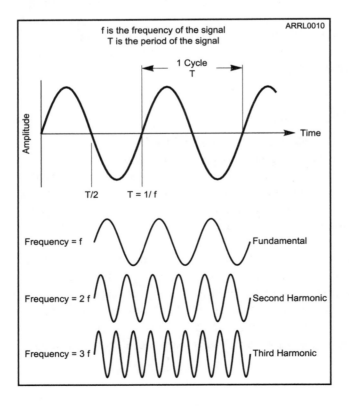

f is the frequency of the signal
T is the period of the signal

ARRL0010

1 Cycle
T

Amplitude

Time

T/2   T = 1/ f

Frequency = f — Fundamental

Frequency = 2 f — Second Harmonic

Frequency = 3 f — Third Harmonic

All radio equipment is designed to generate or manipulate *radio signals*. A radio signal can be a transmitted wave traveling around the world or electrical energy inside radio equipment. No matter how you communicate by ham radio — with voice, Morse code or computer — those are radio signals, usually referred to as just *signals*.

There is also a lot of radio energy bouncing around out there that isn't used for communication: static from the atmosphere, electrical noise from computer equipment and motors, buzzing and humming from the power lines to name just a few sources. When hams talk about signals, though, they're referring to the electrical energy they use to exchange information, inside or out of a radio.

Figure 2.1 — The frequency of a signal and its period are reciprocals. Higher frequency means shorter period and vice-versa. Harmonics are signals with frequencies that are integer multiples of a fundamental frequency.

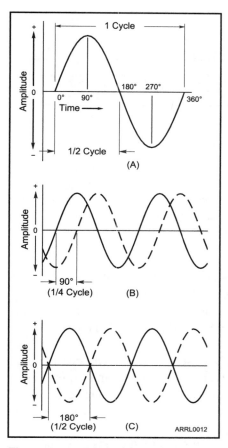

Figure 2.2 — Phase is used as a measure of time within the signal. Each cycle of a sine wave is divided into 360 degrees of phase (A). Parts (B) and (C) show two special cases. In (B) the two signals are 90 degrees out of phase, and in (C) they are 180 degrees out of phase.

FREQUENCY AND PHASE

As the signal oscillates, each complete back-and-forth sequence is called a *cycle*. The number of *cycles per second* is the signal's frequency, represented by a lower-case *f*. The unit of measurement for frequency is *hertz*, abbreviated Hz. [T5C05] One cycle per second is one hertz or 1 Hz. As frequency increases, it becomes easier to use units of kilohertz (1 kHz = 1000 Hz), megahertz (1 MHz = 1000 kHz = 1,000,000 Hz), and gigahertz (1 GHz = 1000 MHz = 1,000,000,000 Hz).

The strength or *amplitude* of a radio signal oscillates like the *sine wave* shown in **Figure 2.1**. The *period* of the cycle (represented by capital *T*) is its duration. The reciprocal of the period, $1/T$, is the signal's frequency, *f*.

A *harmonic* is a signal with a frequency that is some integer multiple (2, 3, 4 and so on) of a *fundamental* frequency. The harmonic at twice the fundamental's frequency is called the *second harmonic*, at three times the fundamental frequency the *third harmonic*, and so forth. There is no "first harmonic."

Harmonics are used by radio designers to create signals at new frequencies. Harmonics can also cause problems. Transmitter output signals include small harmonic signals as well as the desired signal. These unwanted or *spurious emissions* can cause interference to other signals.

Let's return to the sine wave signal of Figure 2.1. Every cycle of the signal has the same basic shape: rising and falling and returning to where it started. Position within a cycle is called *phase*. Phase is used to compare how sine wave signals are aligned in time.

Phase is measured in *degrees* and there are 360 degrees in one cycle of a sine wave. If two sine waves have a phase difference of 180 degrees so that one wave is increasing while the other is decreasing, they are *out of phase*. Waves that have no phase difference so that they are increasing and decreasing at the same time are *in phase*. This is illustrated in **Figure 2.2**.

THE RADIO SPECTRUM

If connected to a speaker, signals below 20 kHz produce sound waves that humans can hear, so we call them *audio frequency* or *AF* signals. Signals that have a frequency greater than 20,000 Hz (or 20 kHz) are *radio frequency* or *RF* signals. [T5C06] The range of frequencies of RF signals is called the *radio spectrum*. It starts at 20 kHz and continues through several hundred GHz, a thousand million times higher in frequency!

For convenience, the radio spectrum of **Figure 2.3** is divided into ranges of frequencies that have similar characteristics as shown in **Table 2.2**. Frequencies above 1 GHz are generally considered to be *microwaves*. Microwave ovens operate at 2.4 GHz, for example. Hams primarily use

Table 2.2
RF Spectrum Ranges

[T3B08-T3B10]

Range Name	Abbreviation	Frequency Range
Very Low Frequency	VLF	3 kHz – 30 kHz
Low Frequency	LF	30 kHz – 300 kHz
Medium Frequency	MF	300 kHz – 3 MHz
High Frequency	HF	3 MHz – 30 MHz
Very High Frequency	VHF	30 MHz – 300 MHz
Ultra High Frequency	UHF	300 MHz – 3 GHz
Super High Frequency	SHF	3 GHz – 30 GHz
Extremely High Frequency	EHF	30 GHz – 300 GHz

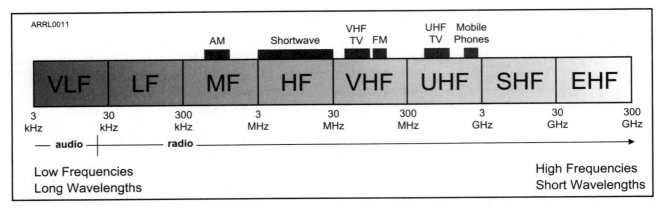

Figure 2.3 — The radio spectrum extends over a very wide range of frequencies. The drawing shows the frequency ranges used by broadcast stations and mobile phones. Amateurs can use small frequency bands in the MF and higher frequency regions of the spectrum.

frequencies in the MF through UHF and microwave ranges. [T3B08 to T3B10]

A specific range of frequencies in which signals are used for a common purpose or have similar characteristics is called a *band*. The AM broadcast band extends from 550 to 1700 kHz and the FM broadcast band covers 88 to 108 MHz. Frequency bands used by amateurs are called *amateur bands* or *ham bands*.

Figure 2.4 shows how a typical AM broadcast receiver "sees" the AM broadcast band. Starting at the lowest frequency on the left, if the receiver is tuned higher in frequency it encounters first a signal at 570 kHz, tunes "past" it to find the next signal, and so forth. The receiver is designed to recover information (the station's programming) from only one signal at a time — the one with the right frequency.

Figure 2.4 also shows a new way of looking at signals. Instead of showing how the signal's amplitude varies with time from left to right, as in Figure 2.1, the graph organizes the signals according to their frequencies. The horizontal axis represents frequency and the vertical axis shows signal strength. This type of graph is a common way to describe the radio spectrum or a specific signal.

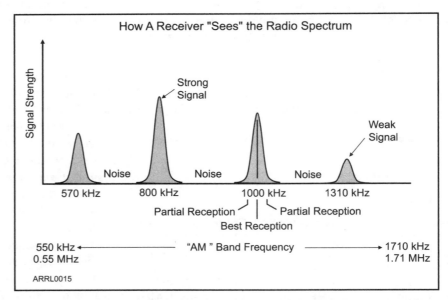

Figure 2.4 — As a radio receiver is tuned across the AM broadcast band, starting at the left, it encounters each signal in turn. Between signals, only noise is received. Although signals can be received slightly lower and higher in frequency, the signal is received best when the receiver is tuned exactly to the signal's frequency.

WAVELENGTH

The *wavelength* of a radio wave is the distance that it travels during one complete cycle. [T3B01] Wavelength is represented by the Greek letter lambda, λ. **Figure 2.5** shows the relationship between the wave's frequency, wavelength and speed.

All radio waves travel at the speed of light (represented by a lower-case c) in whatever medium they are traveling, such as air. [T3B04] The speed of light in space and air is very close to 300 million meters per second (300,000,000 or 3×10^8 meters per second) [T3B11]. In water or glass, along wires, and inside cables, the light and radio waves travel slower, so c is lower.

A radio wave can be referred to by wavelength or frequency since the two are related by the speed of light. Because radio waves travel at a constant speed (in one media), one wavelength, $\lambda = c / f$. This can also be stated as $f = c / \lambda$.

The formula $\lambda = c / f$ also illustrates two important relationships between frequency and wavelength. First, as frequency increases, wavelength decreases and vice-versa. [T3B05] This is because the wave is moving at a constant velocity. A higher frequency wave takes less time to complete one cycle and so doesn't move as far during that time. Waves at very high frequencies have very short wavelengths — such as microwaves with frequencies above 1 GHz.

Second, if you know the frequency of a radio wave, you automatically know its wavelength! This means that you can use the most convenient way of referring to a wave and still be accurate. It's very common for the amateur bands to be referred to by wavelength. You'll often hear something like this, "I'll call you on 2 meters. Let's try 146.52 MHz." The frequency band is referred to as "2 meters" because the radio waves are all approximately that long. [T3B07] The exact frequency then tells you precisely where to tune in the band.

For waves in air or space, the formula for wavelength in meters is:

$$\lambda \text{ in meters} = \frac{300,000,000 \text{ meters per second}}{f \text{ in hertz or cycles per second}}$$

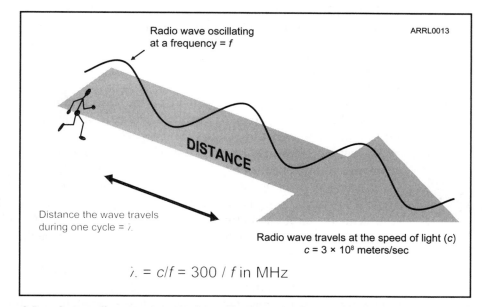

Figure 2.5 — As a radio wave travels, it oscillates at the frequency of the signal. The distance covered by the wave during the time it takes for one complete cycle is its wavelength.

We often refer to frequencies in terms of megahertz (1,000,000 or 10^6 hertz), so an easier version of this relationship to remember is:

$$\lambda \text{ in meters} = \frac{300}{f \text{ in MHz}} \qquad\qquad \text{[T3B06]}$$

For example, the wavelength of a 1 MHz radio wave from an AM broadcast station is

$$\lambda = \frac{300,000,000}{1,000,000} = 300 \text{ meters}$$

Using scientific notation, this becomes:

$$\lambda = \frac{300 \times 10^6}{1 \times 10^6} = 300 \text{ meters}$$

This is how we arrive at the simpler version of the formula with the frequency expressed in MHz:

$$\lambda = \frac{300}{f} = \frac{300 \text{ in meters per second}}{f \text{ in megahertz } (10^6 \text{ cycles/second})}$$

Clearly, the shorter form is more convenient when working with radio frequency signals. In the US, where we still use the English system to measure distances, we often need to convert from meters to feet by multiplying the wavelength in meters by 3.28. To get the wavelength in meters, divide feet by 3.28. To convert from meters to inches, multiply by 39.37. For example, the wavelength of an 80 meter signal in feet is:

80 meters × 3.28 feet per meter = 262.4 *feet*

and the wavelength of a 70 cm signal in inches is:

0.7 meters × 39.37 inches per meter = 27.6 *inches*

Before you go on, study these Technician exam questions from the question pool included at the back of this book or as a downloadable Study Guide version on the web:

T3B01 T3B04 T3B05 T3B06 T3B07 T3B08 T3B09 T3B10 T3B11
T5A12 T5C05 T5C06

If you have difficulty with any question, review the preceding section.

2.2 Modulation

A simple radio signal just sitting there on the radio spectrum not doing anything isn't very useful and doesn't do much communicating. To communicate, information must be added to the radio signal. How does that happen?

The simplest radio signal at one frequency whose strength never changes is called a *continuous wave*, abbreviated CW. Adding information to a signal by modifying it in some way is called *modulation*. Modulation is what enables us to communicate information using radio signals. A signal that doesn't carry any information is called *unmodulated*. Recovering the information from a modulated signal is called *demodulation*. Understanding modulation is very important to understanding the various techniques that radio amateurs use to communicate.

The simplest type of modulation is a continuous wave turned on and off in a coded pattern, such as the Morse code. In fact, Morse code radio signals are called CW for that reason.

If speech is the information used to modulate a signal, the result is a *phone* or *voice mode* signal. If data is the information used to modulate a signal, the result is a *data mode* or *digital mode* signal. *Analog* modes carry information that can be understood directly by a human, such as speech or Morse code. Digital or data modes carry information as individual characters between two computers.

The three characteristics of a signal that can be modulated are the signal's amplitude or strength, its frequency and its phase. All three types of modulation are used in ham radio. You already know about two forms of modulation — AM and FM — the different types of signals you can receive from broadcast stations on a home audio system or car radio. These abbreviations refer to the different types of modulation used to carry the station programming.

AMPLITUDE MODULATION

Varying the power or amplitude of a signal to add speech or data information is called *amplitude modulation* or *AM*. (Morse code is the simplest form of AM.) If you have watched a meter jump in response to your voice or music, an AM signal's amplitude changes in the same way to carry the information in your voice. An AM transmitter adds your voice to the unmodulated signal by varying its amplitude just as the jumping meter suggests. The information is contained in the outline or *envelope* of the resulting signal. **Figure 2.6** shows the result of using a tone to create an AM signal.

All a receiver has to do to recover your voice from an AM signal is to follow the signal's amplitude variations and your voice reappears! The process of recovering speech or music by following the envelope of an AM signal is called *detection* and it can be performed by very simple circuits. In fact, the first AM receivers were nothing more than a crystal of galena (lead ore), a thin steel sliver (called a "cat's whisker") and a sensitive pair of headphones! AM is widely used because it is simple to transmit and receive.

An actual AM signal is made up of three separate components working together — a *carrier* and two *sidebands*. The total energy in an AM signal is divided among those three

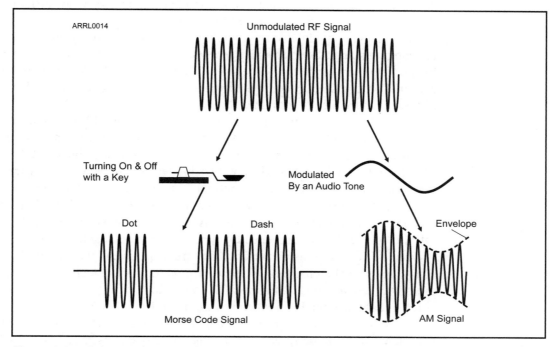

Figure 2.6 — Information can be added to an RF signal by modulating the signal's amplitude. Turning the signal on and off in a pattern such as the Morse code is a very simple form of amplitude modulation. A tone or speech can also be used to modulate the signal, resulting in a signal whose shape or envelope contains the information from the tone or speech.

Composite signals, both radio and audio, are made up of groups of individual signals that combine to create a single signal. The individual signals that make up a composite signal are called *components* and cover a range of frequencies. The difference between the frequency of the lowest and highest component is the signal's *bandwidth*. Of typical amateur signals, CW signals occupy the least bandwidth, followed by SSB, AM and FM.

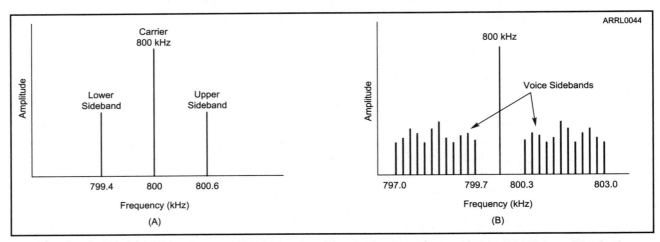

Figure 2.7 — (A) shows an 800 kHz carrier amplitude modulated by a single tone of 600 Hz. This results in the two sidebands 600 Hz away from the carrier. At (B) the many frequencies present in a voice signal from 300 to 3000 Hz are represented as many smaller sidebands, each corresponding to a component of the voice signal.

components. The AM signal's carrier is a continuous wave whose amplitude does not change. It contains no information but simple receivers need the carrier to be able to detect the information which is carried by the sidebands.

An AM signal modulated by a single tone has two sidebands that are present as steady, unchanging signals. The *upper sideband* or *USB* is higher in frequency than the carrier by the frequency of the tone. The *lower sideband* or *LSB* is lower in frequency than the carrier. The tone's information is contained in the amplitude of the sidebands and their difference in frequency from the carrier.

Figure 2.7A shows an AM signal with a carrier of 800 kHz modulated by a single, steady tone of 600 Hz (0.6 kHz). As your receiver tunes across the AM signal from the left, it would first encounter the lower sideband as a steady signal at 800 − 0.6 = 799.4 kHz, then the carrier at 800 kHz, 600 Hz higher in frequency. The carrier would be twice as strong as the sideband. Continuing up in frequency, the upper sideband would be encountered 600 Hz above the carrier at 800 + 0.6 = 800.6 kHz. This set of three components combine to make a single AM signal. Both sidebands each contain the information needed to reproduce the tone used to modulate the signal.

Because voice or music contains many frequencies and the amplitude is constantly changing, an AM signal carrying that information looks much more complex, although the carrier remains steady. Figure 2.7B shows an AM signal carrying the information of a voice signal that contains frequencies from 300 Hz to 3 kHz.

The addition of sidebands during the process of modulation causes the resulting modulated signal to be spread over a range of frequencies called the signal's *bandwidth*. The AM voice signal in Figure 2.7B has a bandwidth of 803 − 797 = 6 kHz. Every modulated signal has some bandwidth. Even a simple CW signal requires a bandwidth of about 150 Hz.

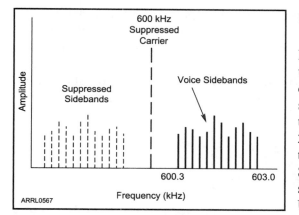

Figure 2.8 — Removing or *suppressing* the carrier and one sideband of an AM signal creates a single-sideband (SSB) signal. All of the signal's power then is concentrated in one sideband. SSB signals work well for long-distance and weak-signal voice contacts.

SINGLE-SIDEBAND (SSB)

From the standpoint of power, an AM signal is inefficient. First, the carrier doesn't carry any information, even though it takes up most of the signal power! In addition, each sideband contains a copy of the modulating signal. It seems like only a fraction of the signal is really needed and that's just what *single-sideband* or *SSB* signals are. **Figure 2.8** shows a single-sideband signal — an AM signal with the carrier and one sideband removed. [T8A01] All of the SSB signal's power can then be devoted to the remaining sideband. The upper sideband (USB) is used on VHF and UHF. Both USB and LSB are used on the MF and HF bands.

Even though generating and receiving SSB signals requires more complex equipment, the performance improvement is worth the trouble. SSB transmissions have a superior range compared to AM because all of the power is concentrated in a single, information-carrying sideband. SSB's smaller bandwidth of less than 3 kHz also makes better use of the radio spectrum because more SSB signals can fit within a fixed range of frequencies without overlapping or interfering with each other.

FREQUENCY AND PHASE MODULATION

The remaining two signal characteristics that can be varied to carry information are frequency and phase. Modes that vary the frequency of a signal to add speech or data information are called *frequency modulation* or *FM*. The frequency of an FM signal varies with the amplitude of the modulating signal as shown in **Figure 2.9**. *Phase modulation* or *PM* is very similar to FM. Phase modulation uses the information signal's amplitude to vary a signal's phase instead of changing its frequency. FM and PM signals have one carrier and many sidebands that all add together. The resulting signal changes frequency (or phase), but its amplitude is constant and its power does not change whether modulated or not.

The amount of frequency variation is called *carrier deviation* or just *deviation*. Speaking louder into the microphone of an FM transmitter increases deviation. [T2B06] As deviation increases, so does the signal's bandwidth, so excessive deviation can cause interference to signals on nearby frequencies. [T2B07]

FM and PM signals are quite similar and receivers can demodulate them using the same circuit. As a result, amateurs refer to either FM or PM signals simply as "FM." If an FM

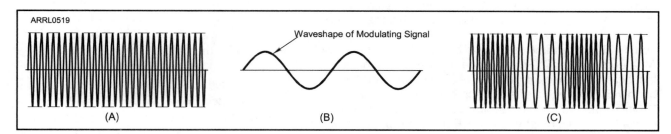

Figure 2.9 — At (A), each cycle of the unmodulated carrier has the same frequency. When the carrier is frequency modulated with the signal at (B), its frequency increases and decreases corresponding to the increases and decreases in amplitude of the modulating signal.

Watch That Band Edge!

Amateurs are allowed to use any frequency within a band, but you have to be careful when operating near the edge of the band. The rule is that *all* of your signal must remain within the band. Since your radio displays the *carrier frequency*, you have to remember to leave room for the signal's sidebands. For example, if your FM voice signal is 15 kHz wide, your carrier frequency (in the center of the signal) should never be less than 7.5 kHz from the band edge. To give yourself a bit of margin, 10 kHz from the edge would be even better. [T1B09]

signal is generated by using the amplitude of the information signal to vary the frequency directly, it is sometimes referred to as "true FM" to distinguish it from various other methods. [T2B05]

Before you go on, study these Technician exam questions from the question pool included at the back of this book or as a downloadable Study Guide version on the web:

T1B09

T2B05 T2B06 T2B07

T8A01

If you have difficulty with any question, review the preceding section.

COMPARING TYPES OF MODULATION

With all the different types of modulation and signals available, how do you choose one over another? What are the strengths and weaknesses of each? What makes one "better" than another? All of these are excellent questions. Luckily, a lot of experimenting has already been done. As a result, hams know which kind of modulation provides the best results for the desired use.

If FM signals occupy more bandwidth than SSB, why is FM used for VHF and UHF voice? [T8A04] Recall that the information in an FM signal is carried only by variations in the signal's frequency. Atmospheric and electrical noises received as amplitude variations are meaningless to an FM signal. The *limiter* circuit in an FM receiver strips away all of the amplitude variations from FM signals, including noise, so that they are not heard in the receiver's output. That's why programs broadcast on AM stations experience static crashes while an FM station's program is static-free. For short-range and regional communications, the lower noise of the FM signal outweighs bandwidth considerations.

FM can also be used for data signals, such as those for packet radio on VHF and UHF. [T8A02] The data is exchanged as audio tones by using the FM radio's speech input and audio output. This allows FM voice radios to be used.

Even though FM may provide better fidelity, SSB is often used where signals are weaker and where available spectrum space is not sufficient to support a large number of FM users. Signals on the HF bands below 30 MHz are almost exclusively SSB (or CW) for these reasons. SSB signals use less bandwidth than FM as shown in **Table 2.3**. [T8A05, T8A08 to T8A11]

Because the SSB signal's power is concentrated into a narrow bandwidth, it is possible to communicate with SSB over much longer ranges and in poorer conditions than with FM or AM, particularly on the VHF and UHF bands. That is why the VHF

Table 2.3
Signal Bandwidths

[T8A05, T8A08 to T8A11]

Type of Signal	Typical Bandwidth
AM voice	6 kHz
AM broadcast	10 kHz
Commercial video broadcast	6 MHz*
SSB voice	2 to 3 kHz
SSB digital	500 to 3000 Hz (0.5 to 3 kHz)
CW	150 Hz (0.15 kHz)
FM voice	10 to 15 kHz**
FM broadcast	150 kHz

* On June 12, 2009 US broadcasters converted all over-the-air TV signals to digital modulation. Within the 6 MHz channel, there may be from four to five digitally-compressed audio-video programs, each with a 1.2 – 1.5 MHz bandwidth. Amateurs will continue to use the older analog format for fast-scan television for the foreseeable future, and the bandwidth of those signals is approximately 6 MHz.
** On 10 meters, FM voice must be narrowband (6 kHz max) below 29.0 MHz. Most VHF/UHF FM voice repeater signals are approximately 15 kHz wide although there is some narrowband equipment with a 5 kHz bandwidth starting to be used.

and UHF "DXers" and contest operators use SSB signals. [T8A03, T8A07]

For even better range, extremely narrow CW signals are the easiest for a human operator to send and receive, particularly in noisy or fading conditions. Even though the signals can be quite strong, hams refer to CW and SSB as *weak signal* modes because they are more effective than FM at low signal strengths.

If an SSB signal can use either an upper or lower sideband — which one should you use? There is no technical reason for choosing USB over LSB. However, in order to make communications easier, ham radio has standardized on the following conventions:

- Below 10 MHz, LSB is used
- Above 10 MHz, USB is used — including all of the VHF and UHF bands [T8A06]

This convention is even programmed into radio equipment as the normal operating mode! There is one exception: amateurs are required to use USB on the five 60 meter band (5 MHz) channels.

Before you go on, study these Technician exam questions from the question pool included at the back of this book or as a downloadable Study Guide version on the web:

T8A02 through T8A11

If you have difficulty with any question, review the preceding section.

2.3 Radio Equipment Basics

You've used radios before, of course — at home, in the car, or at work. That means you're already familiar with some of the topics in this section. Now that you know about radio signals and how modulation is used to carry information, the controls and settings on a radio will make much more sense.

It's a good idea to review even if you have used two-way radios before, since in ham radio the terms might be used a little differently than what you're used to. We'll cover the operating details and procedures later. The goal is to make sure we're using the same words to mean the same thing!

BASIC STATION ORGANIZATION

The three basic elements of a radio station, big or small, are the transmitter, receiver, and antenna as shown in **Figure 2.10**. A *transmitter* (abbreviated XMTR) generates a signal that carries speech, Morse code, or data information. A *receiver* (abbreviated RCVR) recovers the speech, Morse code or data information from a signal. (Figure 2.10 is a *block diagram* that shows how a system of equipment, such as a radio station, is organized without getting into the complex details of every connection and control.) An *antenna* turns the radio signals from a transmitter into energy that travels through space as a radio wave. An antenna also captures radio waves and turns them into

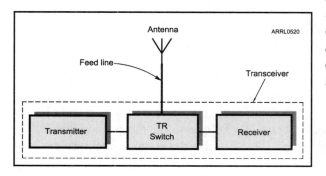

Figure 2.10 — A basic radio station is made up of a transmitter and receiver connected to an antenna with a feed line. The transmit-receive (TR) switch allows the transmitter and receiver to share the antenna. A transceiver includes the transmitter, receiver and TR switch in a single enclosure.

Jargon Alert

Any piece of equipment that can transmit is often called a *rig*. Antennas are often called *skyhooks* and a station with several antennas has an *aluminum farm*. The terms *set* (as in *radio set*) and *aerial* (for antenna) are considered obsolete and are rarely used by hams.

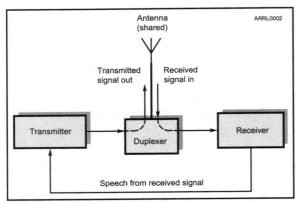

Figure 2.11 — In a repeater, the receiver's output is immediately retransmitted by the transmitter on a different frequency. The duplexer allows the transmitter and receiver to share a common antenna at the same time.

signals for a receiver to work with. A *feed line* connects the antenna to the transmitter or receiver. Feed lines are also called *transmission lines*, just like power lines, because they are used to transfer energy — radio signals in this case.

Most amateur equipment combines the transmitter and receiver into a single piece of equipment called a *transceiver* (abbreviated XCVR). [T7A02] A transceiver shares a single antenna between the transmitter and receiver circuits by using a *transmit-receive (TR) switch*.

REPEATERS

Repeaters are stations that transmit a received signal simultaneously on another frequency or channel. [T1F09]. Repeaters provide local and regional communications between low-power mobile and portable stations. These stations aren't trying to contact the most distant or weakest station — they just want to stay in touch with each other as they move between home, work, and other day-to-day activities. The job of the repeater is to provide a strong, low-noise signal that everyone can hear and understand well, especially in emergencies.

A repeater consists of a receiver and transmitter connected together so that the received signal is retransmitted on a different channel or even multiple channels. Repeaters are located on high buildings, towers or hills for maximum range. **Figure 2.11** shows the basic elements of a repeater station. Most repeaters are designed to relay FM voice signals. (There are repeaters for data and video signals, too. D-STAR repeaters relay voice signals encoded as digital signals, for example.)

Because a repeater receives and transmits at the same time, instead of a transmit-receive switch it uses a *duplexer*. The duplexer allows the strong signal from the transmitter and the tiny signals that the receiver listens for to share a single antenna.

ACCESSORY RADIO EQUIPMENT

With the basic equipment accounted for, now add some useful accessory equipment. **Figure 2.12** shows the most common accessories used with a basic station.

Unless the equipment can use household electrical power directly, a *power supply* is needed to convert it to whatever form is needed by the radio or other accessories. In most cases, that's 13.8 V dc.

A *microphone* (or *mike*, sometimes abbreviated *mic*) turns an operator's voice into an electrical *audio* signal. The transmitter then adds the audio to a radio signal by modulation. The microphone is usually built-in on a handheld transceiver but is a separate piece of equipment for mobile and desktop radios. A Morse *key* is a special switch used by the operator to turn a transmitter's output signal on and off in the patterns of Morse code.

At the receiver's output, a *speaker* turns the electrical audio signal back into audible sound that the operator can hear. A speaker is usually built-in to the radio but external *communications speakers* are specially designed to optimize the understandability of speech and Morse code. *Headphones* make it easier to understand weak or noisy signals and avoid disturbing others.

Amplifiers are circuits or equipment that increase the strength of a signal. *Preamplifiers* (or *preamps* — not shown in the diagram) increase the strength of a signal before it is applied to a receiver. *Power amplifiers* increase the strength of a transmitted signal before it is sent to the antenna.

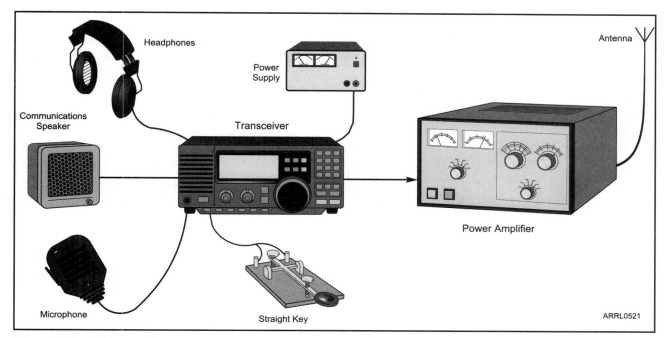

Figure 2.12 — This set of accessory equipment is used to send and receive voice or Morse code signals. Communicating with digital data signals would require other accessories such as a computer and appropriate interface. A power amplifier is used to increase the power of the transmitted signal for better range.

Before you go on, study these Technician exam questions from the question pool included at the back of this book or as a downloadable Study Guide version on the web:
T1F09
T7A02
If you have difficulty with any question, review the preceding section.

Chapter 3

Electricity, Components and Circuits

In this chapter, you'll learn about:

- **Fundamental concepts of electricity and circuits**
- **Voltage and current**
- **Resistance, capacitance and inductance**
- **Reactance, impedance and resonance**
- **Common types of electronic components**
- **How basic types of radios are made**

When you see the mouse, you'll find more information at www.arrl.org/ham-radio-license-manual

3.1 Electricity

Although radios use sophisticated electronics, they are based on fundamental principles of electricity. In this chapter, you'll learn about the basic electrical concepts that apply to everything from the household wall socket to the latest radio or computer.

If you would like some assistance with the math in this chapter, tutorials are available on the book's website, and all exam math problems are worked out for you there, too.

VOLTAGE AND CURRENT

Electric *current* (represented in equations by the symbol I or i) is the flow of *electrons*. [T5A03] Electrons are negatively charged atomic particles. Current is measured in units of *amperes* which is abbreviated as A or amps. [T5A01] Current is always measured as the flow through something, such as a wire or electronic component. An *ammeter* is used to measure current. [T7D04]

Electrons are so small that to light a household 100-watt bulb with a current of 1 ampere, 6.25 billion billion of them must pass through the bulb each second! That quantity, 6,250,000,000,000,000,000 or 6.25×10^{18} electrons, makes up one *coulomb* of electric charge. One ampere is the flow of one coulomb of electrons per second. (If you need to learn about numbers written in scientific notation, such as 6.25×10^{18}, refer to the tutorial on the *Ham Radio License Manual* web page.)

Voltage (represented in equations by the symbol E or e) is the *electromotive force* or *electric potential* that makes electrons move. [T5A05] Voltage is measured in units of *volts* which is abbreviated as V. [T5A11] (Sometimes V or v is used in equations as a symbol for voltage, as well.) Voltage is always measured from one point to another or with respect to some reference voltage. Voltage is measured with a *voltmeter*. [T7D01] The *polarity* of voltage can be either positive or negative. A *negative* voltage repels electrons and a *positive* voltage attracts them. The Earth's surface acts as a universal reference for voltage measurements and is called *earth ground*, *ground potential* or just *ground*.

Electrical Pressure and Flow

Figure 3.1 shows a time-tested analogy that helps newcomers to electricity understand what voltage and current are and how they act. Voltage acts like pressure in a water pipe and current acts like water flow. You can have lots of pressure with no flow — think of a pipe with the faucet closed. You can't have flow with no pressure, though. There must always be something pushing water molecules and electrons before they'll move.

Pressure is always measured between two points in the plumbing or between one point and the open air (atmospheric pressure). Voltage is the same — an electrical force that pushes electrons to flow from one point to another and is measured between these points. Water flow must go through something — a pipe or stream — and current must go through something as well, such as a wire.

You may have noted that while a positive voltage attracts electrons, Figure 3.1 shows the current arrow pointing *away* from the battery's positive terminal. Before the electron was discovered in 1898, scientists assumed that positive charge moved to create current. They based electrical laws and conventions on that assumption, which turned out to be wrong because an electron is negatively charged.

Luckily, the laws of electricity work just as well with negative charges moving backward as they do with positive charges moving forward! Positive charge moving from a positive to a negative voltage as shown in the figure is *conventional current*, the standard in electronics. You'll learn more about conventional current and its mirror-image, *electronic current*, when you upgrade to General.

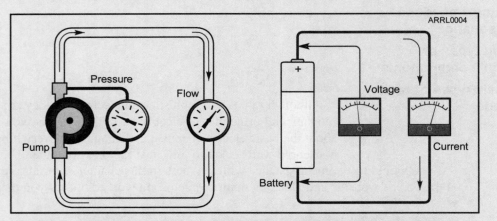

ARRL0004

Figure 3.1 — Voltage acts similarly to pressure and current similarly to flow in a water system. Voltage between two points is what causes the electrons to move between those points. Current is a measure of how many electrons pass through the circuit per second.

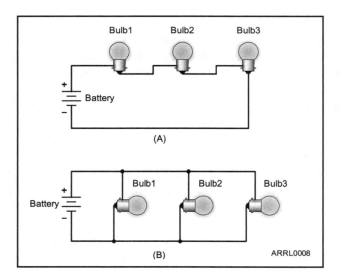

Figure 3.2 — Part A shows three light bulbs and a battery connected in a series circuit. The same current flows from the battery through all three light bulbs. Part B shows the same bulbs and battery connected in a parallel circuit. The same voltage from the battery is applied across each light bulb.

CIRCUITS

A *circuit* is any path through which current can flow. If two or more devices such as light bulbs are connected in a circuit so that the same current must flow through all of them, that is a *series* circuit. If two or more devices are connected so that the same voltage is present across all of them, that is a *parallel* circuit. **Figure 3.2** illustrates the difference between series and parallel circuits.

Voltmeters are connected in parallel with a component (also termed "across a component") or circuit to measure voltage. [T7D02] Ammeters are connected in series with a component or circuit to measure current. [T7D03] Some meters calibrated in amperes are really voltmeters that measure the voltage across a small resistor in series with the current. Be sure to determine which type of meter you have before using it.

A *short circuit* is a direct connection, usually unintentional, between two points in a circuit. An *open circuit* is made by breaking a current path in a circuit.

Before you go on, study these Technician exam questions from the question pool included at the back of this book or as a downloadable Study Guide version on the web:

T5A01 T5A03 T5A05 T5A11

T7D01 through T7D04

If you have difficulty with any question, review the preceding section.

Economies of Scale — The Multimeter

The basic electrical test instruments are simple meters: voltmeters, ammeters and ohmmeters. It is quite inconvenient to use a separate meter for each parameter, so the *multimeter* was invented, short for "multifunction meter." It uses a single meter — analog (with a moving needle) or digital — to measure all three basic electrical values of voltage, current and resistance. [T7D07] Multimeters are often referred to as a VOM (volt-ohm-meter) or DVM (digital volt meter).

The typical digital multimeter shown in **Figure 3.3** uses a switch and different sets of input connections to select which parameter and range of values to measure. The convenience of having all the different functions in a single instrument more than outweighs the extra complexity of learning to use a multimeter. You'll find unexpected ways to interpret a multimeter's readings, too. For example, if you are measuring the resistance of a circuit and the reading starts out low, but gradually increases, that indicates the presence of a large-value capacitor! [T7D10]

The flexibility afforded by the meter also means that it is important to use the meter properly and safely. Trying to measure voltage or connecting the probes to an energized circuit when the meter is set to measure resistance is a common way to damage a multimeter, for example. [T7D06, T7D11] You must also take heed of the meter's voltage rating. Voltages beyond the meter's rating can "flashover" to other pieces of equipment or to you, creating a serious shock hazard. Ensure that the voltmeter and leads are rated for use at the voltages to be measured! [T7D12]

These simple and inexpensive instruments should be part of every ham's tool kit! (More information about the use of test equipment in general is available on the ARRL website's Technical Information Service.)

Figure 3.3 — A multimeter combines three basic instruments into a single package: voltmeter, ammeter and ohmmeter. Multimeters, also referred to as VOMs or DVMs, often include other features, such as continuity checking, diode testers and capacitance or inductance meters. The older term VTVM stands for vacuum-tube volt meter, an older form of multimeter not often seen today.

Before you go on, study these Technician exam questions from the question pool included at the back of this book or as a downloadable Study Guide version on the web:

T7D06 T7D07 T7D10 T7D11 T7D12

If you have difficulty with any question, review the preceding section.

RESISTANCE

All materials impede the flow of electrons to some degree. This property is called resistance, represented by the symbol R. Resistance is measured in ohms which are represented by the Greek letter omega, Ω. Resistance is measured with an *ohmmeter*. [T7D05]

Materials in which electrons flow easily in response to an applied voltage are *conductors*. Metals such as copper are good conductors, and so is salt water. [T5A07] The human body conducts electricity as well! Materials that resist or prevent the flow of electrons are *insulators*, such as glass and ceramics, dry wood and paper, most plastics, and other non-metals. [T5A08]

Georg Ohm, a 19th-century scientist, discovered that voltage, current and resistance are proportional. Ohm's Law states that current is directly proportional to voltage and inversely proportional to resistance. The more a material resists the flow of electrons, the lower the current will be in response to voltage across the material. As an equation, this is stated $I = E / R$. (You will also see this written as $I = V / R$ with V representing voltage.)

If you know any two of I, E or R, you can determine the missing quantity as follows [T5D01, T5D02, T5D03]:

$$R = E / I$$

$$I = E / R$$

$$E = I \times R$$

The drawing in **Figure 3.4A** is a convenient aid to

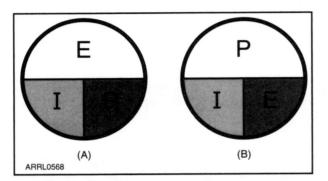

Figure 3.4 — These simple diagrams will help you remember the Ohm's Law (A) and power (B) relationships. If you know any two of the quantities, the equation to find the third is shown by covering up the unknown quantity. The positions of the remaining two symbols show if you have to multiply (side-by-side) or divide (one above the other).

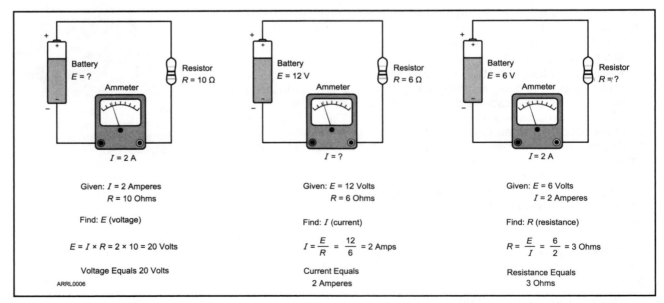

Figure 3.5 — This drawing shows three examples of how to use Ohm's Law to calculate voltage, current, or resistance.

remembering Ohm's Law in any of these three forms. **Figure 3.5** shows several examples of how to use Ohm's Law. [T5D04 through T5D12]

Before you go on, study these Technician exam questions from the question pool included at the back of this book or as a downloadable Study Guide version on the web:
T5A07 T5A08 T5D01 through T5D12
T7D05
If you have difficulty with any question, review the preceding section.

POWER

Power, represented by the symbol *P*, is measured in *watts* which are abbreviated as W. [T5A02] Power is the rate at which electrical energy is used. [T5A10] Power is calculated as the product of voltage and current. [T5C08] An amount of power consumed is often referred to as the *load* that is applied to a circuit. As with Ohm's Law, if you know any two of P, E or I, you can determine the missing quantity as follows:

$P = E \times I$

$E = P / I$

$I = P / E$

Because Ohm's Law links voltage, current and resistance, E and I can be replaced in the power equations with their equivalents:

$P = E \times I$

$P = (I \times R) \times I$, so $P = I^2 \times R$

and

$P = E \times (E / R)$, so $P = E^2 / R$

Here are some examples of how to use these power formulas [T5C09, T5C10, T5C11]:
1) How much power is being used in a circuit when the applied voltage is 12 volts and the current is 10 amperes?

$P = E \times I = 12 \times 10 = 120$ watts

2) How many amperes are flowing in a circuit when the applied voltage is 12 volts and the load is 240 watts?

$I = P / E = 240 / 12 = 20$ amps

3) What is the voltage in a circuit if a 50-watt load draws 5 amps?

$E = P / I = 50 / 5 = 10$ volts

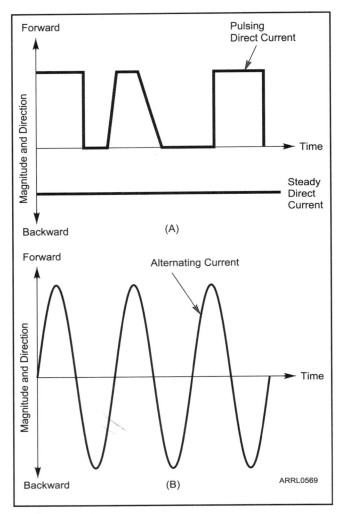

Figure 3.6 — Direct current (A) flows steadily in one direction — forward or backward. Pulsating direct current may stop and start, but always flows in the same direction. Alternating current (B) regularly reverses its direction.

AC AND DC

Electrical current takes different forms, depending on the source that creates the voltage making the electrons move. Current that flows in one direction all the time is *direct current,* abbreviated *dc*. [T5A04] Current that regularly reverses direction is *alternating current,* abbreviated *ac*. [T5A09] **Figure 3.6** shows the difference between ac and dc. Just like current, a voltage that has the same *polarity* (the same direction from positive to negative voltage) all the time is a *dc voltage*. A voltage that regularly reverses polarity is an *ac voltage*. Batteries and solar cells are a source of dc voltage and current. Household power is supplied by an electrical utility in the form of ac voltage and current.

Just as we discussed in the section on radio waves in Chapter 2, a complete sequence of ac current flowing, stopping, reversing and stopping again is a cycle. The number of cycles per second is the ac current's frequency, as well. The same is true for an ac voltage building up to a positive voltage from zero, then reversing to negative polarity, and then returning to zero in a *sine wave*, just like the radio signal. In fact, a radio signal in a cable or wire is ac current! The frequency of household ac voltage is 50 or 60 Hz, while radio signals used by amateurs have frequencies in the MHz and GHz range.

Before you go on, study these Technician exam questions from the question pool included at the back of this book or as a downloadable Study Guide version on the web:

T5A02 T5A04 T5A09 T5A10
T5C08 through T5C11

If you have difficulty with any question, review the preceding section.

3.2 Components and Units

Electronic circuits are made from *components* and the connections between them. Each component performs a discrete function: storing or dissipating energy, routing current or amplifying a signal. In this chapter you'll learn about the different types of common components and their functions. We begin by reviewing the way components and signals are measured.

BASIC COMPONENTS

The three most basic types of electronic components are resistors, capacitors and inductors (coils). These have their own units of measurement and have a different effect on voltage and current. The wires that are used to make connections to the component are

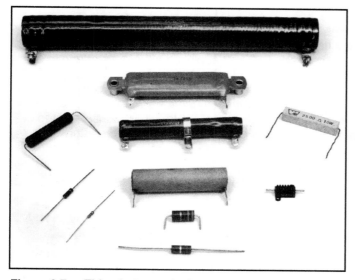

Figure 3.7 — This photograph shows some of the many types of resistors. Large, power resistors are at the top of the photo. The small resistors are used in low-power circuits.

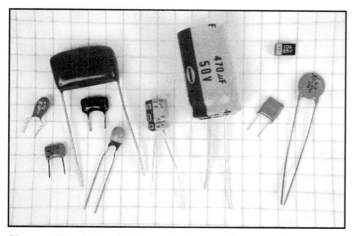

Figure 3.8 — This photograph shows a few styles of capacitors that are used in most common electronic equipment. Capacitors used in transmitters and high-power circuits are larger than those shown here.

called *leads* (pronounced "leeds") and have only small effects on the performance of the component.

Resistors have a certain resistance specified in ohms (Ω), kilohms (kΩ), or megohms (MΩ). The function of a resistor is to oppose the flow of electrical current, just as a valve in a water pipe restricts the flow through the pipe. [T6A01] Resistors, like valves, control flow or current. The electrical current flowing through the resistor loses some of its energy as heat, so resistors also dissipate energy, like an electrical brake. **Figure 3.7** shows different types of resistors.

Capacitors store electrical energy in the *electric field* created by a voltage between two conducting surfaces (such as metal foil) called *electrodes* [T6A04] separated by an insulating *dielectric*. [T6A05] Storing energy this way is called *capacitance* and it is measured in farads (F). [T5C01, T5C02] Capacitors used in radio circuits have values measured in picofarads (pF), nanofarads (nF) and microfarads (μF). As it stores and releases energy in a circuit, a capacitor smoothes out voltage changes. Because the electrodes are separated, a capacitor cannot pass dc current. AC current, however, can pass through a capacitor. In most capacitors, the electrodes and dielectric are sealed inside a protective coating as shown in **Figure 3.8**.

Inductors store energy in the *magnetic field* created by current flowing through a wire. [T6A06] This is called *inductance* and it is measured in henrys (H). [T5C03, T5C04] Inductors have values measured in nanohenrys (nH), microhenrys (μH), millihenrys (mH) and henrys (H). As it stores and releases energy in a circuit, an inductor smoothes out current changes. Inductors are made from wire wound in a coil, sometimes around a *core* of magnetic material that concentrates the magnetic energy. [T6A07] For dc currents, an inductor appears to be a short circuit, but it resists the flow of ac current. **Figure 3.9** shows several common types of inductors.

Many electronic components use markings, called a *color code*, to indicate their value. For example, a resistor with four colored stripes, red-violet-brown-gold, was made to have a resistance value of 270 Ω with an accepted variation in that value (called *tolerance*) of 5%. Other components have their value printed directly on their body, but may encode the value to save space. For example, a capacitor labeled with "683" has a value of 68×10^3 pF. You can find out how to read color and marking codes at the link on the *Ham Radio License Manual* web page.

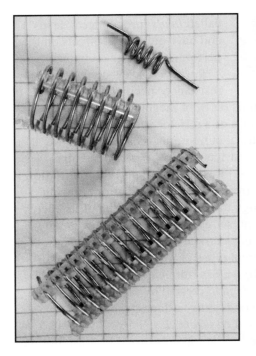

Ohm and Farad and Henry — Oh, My!

Electrical units of measure are almost all named for the scientists and experimenters that played an important role in understanding electricity and radio. For example, Georg Ohm (1787 –1854) discovered the relationship between current, voltage and resistance that now bears his name as Ohm's Law, as well as the unit of resistance, the ohm. **Table 3.1** lists units and the trailblazers for whom they are named. Complete biographical information on all of these electro-pioneers can be found online in the Wikipedia, **www.wikipedia.org**.

Table 3.1

Electrical Units and Their Namesakes

Unit	Measures	Named for
Ampere	Current	Andree Marie Ampere (1775 – 1836)
Coulomb	Charge	Charles Augustin Coulomb (1736 – 1806)
Farad	Capacitance	Michael Faraday (1791 – 1867)
Henry	Inductance	Joseph Henry (1797 – 1878)
Hertz	Frequency	Heinrich Hertz (1857 – 1894)
Ohm	Resistance	George Simon Ohm (1787 – 1854)
Watt	Power	James Watt (1736 – 1819)
Volt	Voltage	Alessandro Giuseppe Antonio Anastasio Volta (1745 – 1827)

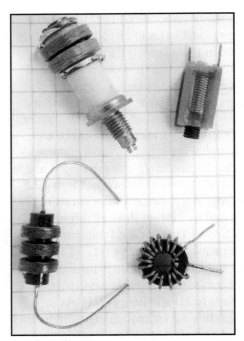

Figure 3.9 — Here are two types of inductors. On the top are air-core inductors. Those on the bottom use a magnetic material to concentrate the stored energy and increase inductance. At the lower right is an inductor wound on a toroid core.

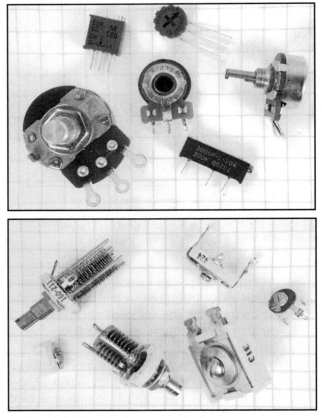

Figure 3.10 — Components with adjustable values are used to tune or calibrate circuits. This photograph shows variable resistors (top) and capacitors (bottom). Variable inductors can be seen at the top of the bottom photo in Figure 3.9.

All three types of basic components are also available as *adjustable* or *variable* models. A variable resistor is also called a *potentiometer* (poh-ten-chee-AH-meh-tur) or *pot* because it is frequently used to adjust voltage or potential, such as for a volume control. [T6A02, T6A03] Variable capacitors and inductors are used to tune radio circuits for a variety of purposes. **Figure 3.10** shows some examples of variable components.

Transformers are made from two or more inductors that share their stored energy. This allows energy to be transferred from one inductor to another while changing the combination of voltage and current. For example, a transformer is used to transfer energy from a home's 120 V ac outlets to a lower voltage for use in electronic equipment. [T6D06]

Before you go on, study these Technician exam questions from the question pool included at the back of this book or as a downloadable Study Guide version on the web:
T5C01 through T5C04
T6A01 through T6A07
T6D06
If you have difficulty with any question, review the preceding section.

REACTANCE AND IMPEDANCE

In a resistor, ac voltages and currents are exactly in step or *in phase*: As voltage increases, so does current and vice-versa. In capacitors and inductors, the relationship between ac voltage and current is altered so that there is an offset in time between changes in one and changes in the other as energy is stored, then released. That means voltage and current have a *phase difference*.

In a capacitor, the changes in current are a little ahead of, or *lead*, voltage changes because the capacitor's smoothing action works against changes in voltage. In an inductor, changes in the ac current *lag* a little behind changes in voltage because the inductor resists changes in current. The result is an opposition to ac current flow called *reactance*, represented by the capital letter X. Reactance is measured in ohms, as is resistance.

Reactance from a capacitor is called *capacitive reactance* and from an inductor, *inductive reactance*. The value of a component's reactance depends on the amount of capacitance or inductance and the frequency of the ac current.

The combination of resistance and reactance is called *impedance*, represented by the capital letter Z, and is also measured in ohms (Ω). [T5C13] Radio circuits almost always have both resistance and reactance, so impedance is often used as a general term to mean the circuit's opposition to ac current flow. [T5C12]

RESONANCE

The effect of capacitive and inductive reactance on ac current is different because of the opposite offset in time between voltage and current. In a circuit with both capacitive and inductive reactance, at some frequency the two types of reactance will be equal and cancel each other out. This brings the ac current and voltage exactly back in step with each other — a condition called *resonance*. The frequency at which resonance occurs is the *resonant frequency*.

Circuits that contain both capacitors and inductors will have at least one resonant frequency and are called *resonant circuits* or *tuned circuits*. [T6D08] If variable capacitors or inductors are used to create the tuned circuit, the resonant frequency can be varied, *tuning* the circuit. A tuned circuit acts as a filter, passing or rejecting signals at its resonant frequency. [T6D11] Tuned circuits are important in radio because they help radios generate or receive signals at a single frequency.

Before you go on, study this Technician exam question from the question pool included at the back of this book or as a downloadable Study Guide version on the web:
T5C12 T5C13
T6D08 T6D11
If you have difficulty with any question, review the preceding section.

DIODES, TRANSISTORS AND INTEGRATED CIRCUITS

Resistors, capacitors and inductors generally treat all values and polarities of voltage and current the same. There are other types of components whose response depends on the polarity and value of voltage and current. These are usually constructed from a *semiconductor* material.

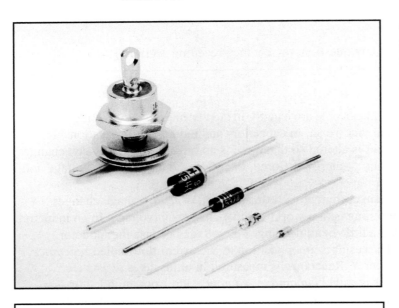

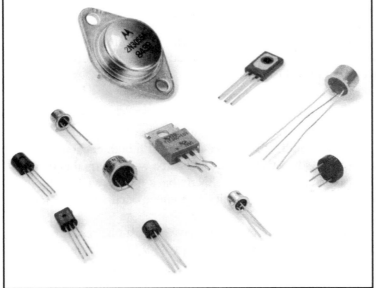

Figure 3.11 — Diodes (top) and transistors (bottom) come in a variety of body styles. The smaller types are used in low-power circuits to control very small signals. The larger types are used for controlling power and in transmitting circuits.

Some materials don't conduct electricity quite as well as a metallic conductor, nor are they a good insulator. These materials are called *semiconductors*. Some semiconductors, such as silicon, have the useful property that adding small amounts of certain impurities, called *doping*, changes their ability to conduct current. The impurities create *N-type* or *P-type* material, depending on the chemical properties of the impurity.

When N- and P-type material are placed in contact with each other, the result is a *PN junction* that conducts better in one direction than the other! This and other properties are used to create many useful electronic components, generally referred to as *semiconductors*. **Figure 3.11** shows several types of semiconductor components.

A semiconductor that only allows current flow in one direction is called a *diode*. [T6B02] Heavy duty diodes that can withstand large voltages and currents are called *rectifiers*. If an ac voltage is applied to a diode, the result is a unidirectional, pulsing dc current because current is blocked when the voltage tries to push electrons in the "wrong" direction. [T6D01] A diode has two electrodes, an *anode* and a *cathode*. [T6B09] As seen in Figure 3.11, on a diode component the cathode is usually identified by a stripe. [T6B06]

A special type of diode, the light-emitting diode or LED, gives off light when current flows through it. [T6B07] The material from which the LED is made determines the color of light emitted. LEDs are most often used as visual indicators. [T6D07] Requiring less power, LEDs are often used instead of incandescent indicator lamps.

Transistors are components made from patterns of N- and P-type material. The patterns form structures that allow the transistor to use small voltages and currents to control larger ones. [T6B01, T6D10] The transistor's electrodes are contacts made to a certain piece of the pattern. With the appropriate external circuit and a source of power, transistors can amplify or switch voltages and currents. [T6B03, T6B05] Using small signals to control larger signals is called *gain*. [T6B12]

There are two common types of transistors: *bipolar junction transistors* (*BJT*) and *field-effect transistors* (*FET*). [T6B08] The BJT is made from three alternating layers of N- and P-type material. [T6B04] The electrodes are called the *base*, *emitter* and *collector*. [T6B10] There are two types of BJT — NPN and PNP — referring to the arrangement of the layers.

The FET is constructed as a conducting path or *channel* of N- or P-type material. The ends of the channel form the *source* and *drain* electrodes. The *gate* electrode is used to control current flow through the channel. [T6B11]

An *integrated circuit* (*IC* or *chip*) is made of many components connected together as a useful circuit and packaged as a single component. [T6D09] ICs range from very simple circuits consisting of a few diodes all the way to complex microprocessors or signal-processing chips with many thousands of components.

Before you go on, study these Technician exam questions from the question pool included at the back of this book or as a downloadable Study Guide version on the web:
T6B01 through T6B12
T6D01 T6D07 T6D09 T6D10
If you have difficulty with any question, review the preceding section.

PROTECTIVE COMPONENTS

Protective components such as those in **Figure 3.12** are used to prevent equipment damage or safety hazards such as fire or electrical shock due to equipment malfunction. They are designed to have little or no effect on circuit behavior until the dangerous

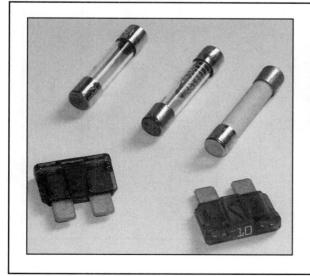

Figure 3.12 — Fuses (left) and circuit breakers (right) protect equipment by interrupting the current in case of an overload. Fuses "blow" by melting a metal wire or strip, seen in glass tube of the cartridge models. Circuit breakers can be reset once the problem creating the overload is removed.

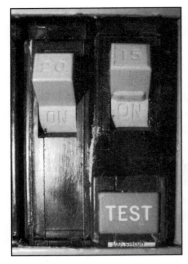

Figure 3.13 — The ground-fault circuit interrupter (GFCI) circuit breaker interrupts current flow when it senses imbalances between the hot and neutral circuits in ac wiring. The imbalance of currents indicates that a shock or other safety hazard exists in wiring supplied by the breaker. Electrical outlets with a built-in GFCI may be found in kitchens, bathrooms, basements, garages and other damp areas.

condition occurs. It is important to understand the different types of protective components and use them correctly.

Fuses interrupt current overloads by melting a short length of metal. [T6A09] When the metal melts or "blows," the current path is broken and power is removed from circuits supplied by the fuse. [T0A04] Fuses cannot be reused. Fuses are rated by the maximum current they can carry without blowing. "Slow-blow" fuses can withstand temporary overloads, but will blow if the overload is sustained.

Circuit breakers act like fuses by *tripping* when current overloads occur. Tripping opens or *breaks* the circuit. Unlike fuses, circuit breakers can be *reset* once the current overload is removed, closing the circuit and allowing current to flow again. Used in ac power wiring, a *ground-fault circuit interrupter* (GFCI) circuit breaker shown in **Figure 3.13** trips if an imbalance is sensed in the currents carried by the hot and neutral conductors. Current imbalances indicate the presence of an electrical shock hazard.

When replacing a fuse or circuit breaker, be sure to use the same model and current rating to avoid creating a safety hazard. Using a larger value, even temporarily, could allow the fault to permanently damage the equipment or start a fire. Resist the temptation to use a device with a higher current rating, even "just for a minute." [T0A05]

Surge protectors limit temporary voltage *transients* above normal voltages by turning into resistors when the voltage gets too high. They then dissipate as heat the energy that would otherwise be passed on to the equipment. Surge protectors are connected to a home's ac power circuits, often in power outlet strips, and to telephone lines. *Lightning arrestors* have a similar function, but are designed to handle the much higher voltages and currents of a lightning strike.

CIRCUIT GATEKEEPERS

Switches and *relays* are simple components that control current through a circuit by connecting and disconnecting the paths current can follow. [T6A08] Both can interrupt current — called *opening* a circuit — or allow it to flow — called *closing* a circuit. **Figure 3.14** shows different types of switches and a typical relay. A switch is operated manually while a relay is a switch controlled by an electromagnet. [T6D02]

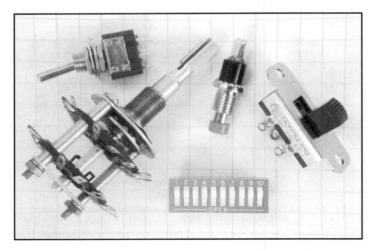

Figure 3.14 — Switches and relays control the path of current flow through a circuit. On the left are several types of switches. On the right is a typical relay, enclosed in a plastic case to reduce damage from dust or moisture.

Switches and relays are described by their number of *poles* and the number of *throws*. Each pole controls the path of one current. For example, a single-pole (SP) switch controls a single current flow and a double-pole (DP) switch controls two separate currents. Each throw refers to a different path for the current. A double-throw (DT) switch can route current through either of two paths while a single-throw (ST) switch can only open or close a single path. [T6D03]

The combination of poles and throws describes the switch. For example, the simplest switch that opens or closes a single current path is an SPST (single-pole single-throw) switch. SPDT, DPST and DPDT are other common configurations.

Indicators and displays are important components for radio equipment. An *indicator* is either ON or OFF, such as a power indicator or a label that appears when you are transmitting. A *meter* provides information as a value in the form of numbers or on a numeric scale. [T6D04] It is common for radios to combine several indicators and meters, such as a signal strength meter (S meter), into a single *display*. The most common type of display is the *liquid crystal display* or LCD found on the front panels of many radios.

Before you go on, study these Technician exam questions from the question pool included at the back of this book or as a downloadable Study Guide version on the web:
T6A08 T6A09
T6D02 T6D03 T6D04
T0A04 T0A05
If you have difficulty with any question, review the preceding section.

SCHEMATICS AND COMPONENT SYMBOLS

If a circuit contains more than two or three components, trying to describe it clearly in words becomes very difficult. To describe complicated circuits, engineers have developed the *schematic diagram* or simply *schematic*. Schematics create a visual description of a circuit by using standardized representations of electrical components called *circuit symbols*. [T6C01, T6C12] **Figure 3.15** shows the schematic symbols for a number of common components. Don't worry that you aren't familiar with all of them! Look for and identify the circuit symbols for a resistor, capacitor, inductor, diode and transistor. [T6C02 to T6C11]

Well-Grounded Symbols

Figure 3.15 includes four different symbols for "Grounds" which can be confusing. Remember from section 3.1 that "ground" meant a reference voltage at the Earth's surface. The upper right ground symbol in Figure 3.15 represents a connection directly to the Earth. Not all circuits require such a reference voltage. For some equipment, the reference voltage is obtained by connecting to the metal enclosure or *chassis*, the upper left-hand ground symbol. The chassis may or may not have its own connection to an earth ground.

Within a piece of equipment the triangular ground symbol on the bottom left generally indicates a connection for current to flow back to the power supply. This is usually referred to as *circuit common* or just *common*. If an A or D is added to the triangle, the circuit has digital (D) computing components (such as microprocessors) as well as circuits that handle analog signals (A). By keeping the return connection to the power supply separate for each type of circuitry, the analog signals are kept free of noise from the digital circuits.

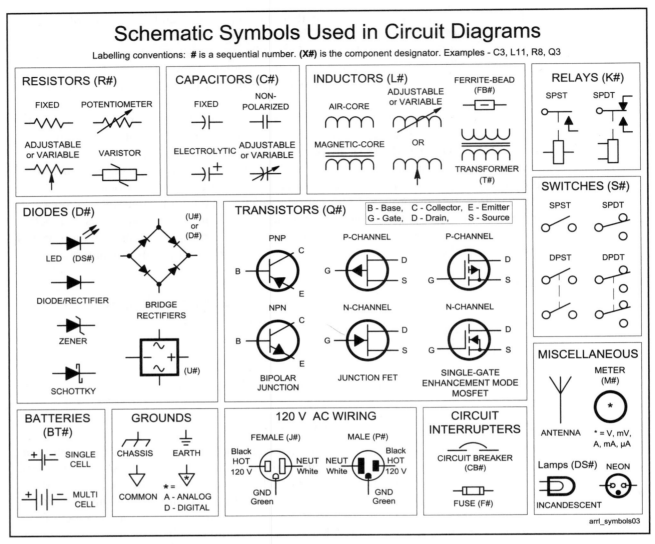

Schematic Symbols Used in Circuit Diagrams

Labelling conventions: **#** is a sequential number. **(X#)** is the component designator. Examples - C3, L11, R8, Q3

RESISTORS (R#)
FIXED POTENTIOMETER
ADJUSTABLE or VARIABLE VARISTOR

CAPACITORS (C#)
FIXED NON-POLARIZED
ELECTROLYTIC ADJUSTABLE or VARIABLE

INDUCTORS (L#)
AIR-CORE ADJUSTABLE or VARIABLE FERRITE-BEAD (FB#)
MAGNETIC-CORE OR TRANSFORMER (T#)

RELAYS (K#)
SPST SPDT

DIODES (D#)
LED (DS#) (U#) or (D#)
DIODE/RECTIFIER BRIDGE RECTIFIERS
ZENER
SCHOTTKY (U#)

TRANSISTORS (Q#)
B - Base, C - Collector, E - Emitter
G - Gate, D - Drain, S - Source

PNP P-CHANNEL P-CHANNEL
B C E G D S G D S
NPN N-CHANNEL N-CHANNEL
B C E G D S G D S
BIPOLAR JUNCTION JUNCTION FET SINGLE-GATE ENHANCEMENT MODE MOSFET

SWITCHES (S#)
SPST SPDT
DPST DPDT

MISCELLANEOUS
ANTENNA METER (M#) * = V, mV, A, mA, µA
Lamps (DS#) INCANDESCENT NEON

BATTERIES (BT#)
SINGLE CELL
MULTI CELL

GROUNDS
CHASSIS EARTH
COMMON * = A - ANALOG D - DIGITAL

120 V AC WIRING
FEMALE (J#) MALE (P#)
Black HOT 120 V NEUT White NEUT White Black HOT 120 V
GND Green GND Green

CIRCUIT INTERRUPTERS
CIRCUIT BREAKER (CB#)
FUSE (F#)

arrl_symbols03

Figure 3.15 — Symbols are used when drawing a circuit because there are so many types of components. Radio and electrical designers use them as a convenient way of describing a circuit.

Figure 3.16A shows the schematic for a simple transistor circuit. (These schematics are also used on the Technician exam.) Each component is assigned a unique *designator* within the circuit or a text label. Examples are BT1, R1, Q1 or "Input On" for the lamp's function. Resistors are designated with an R, capacitors with a C, inductors with an L, diodes with a D, transistors with a Q and so forth.

A schematic does not illustrate the actual physical layout of a circuit. (A *pictorial diagram* is used for that purpose.) It only shows how the components are connected electrically. The lines drawn from component to component, such as between R1 and Q1, represent those electrical connections. Each line does not necessarily correspond to a physical wire — it just indicates that an electrical connection exists between whatever is at each end of the line. [T6C13]

Shared connections are shown as solid, black dots where two lines intersect as in the power supply circuit, Figure 3.16B. If two lines cross without a dot, there is no connection. No dots are used at the connection to a component. In Figure 3.16B, the two dots between D1, C1, R1 and R2 show that these components are connected together. In Figure 3.16C,

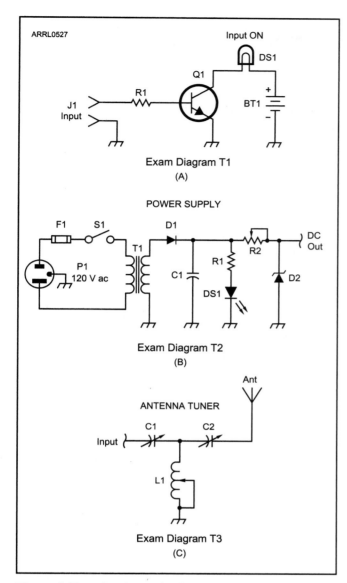

ARRL0527

Input ON
DS1

Q1

J1
Input

R1

BT1
+
−

Exam Diagram T1
(A)

POWER SUPPLY

F1   S1   D1   DC
                              R2   Out
P1   T1
120 V ac   C1   R1
                      DS1   D2

Exam Diagram T2
(B)

Ant

ANTENNA TUNER
        C1   C2
Input

        L1

Exam Diagram T3
(C)

Figure 3.16 — A schematic diagram describes complex circuits using symbols representing each type of component. Lines and dots show electrical connections between the components, but may not correspond to actual wires. These are diagrams T1 (A), T2 (B) and T3 (C) that are used on the Technician exam.

an antenna tuner schematic, the dot above L1 shows that C1, C2 and L1 are all connected together.

Take a moment to study the schematics of Figure 3.16A, B and C, making sure you can identify the type of component indicated by each symbol. In Figure 3.16A, the connector at the left, labeled "Input" also has the designator J1, with J indicating a *jack* type of connector. In Figure 3.16B, the ac *plug* symbol, P1, is labeled "120 V ac" to identify it as a power source. The schematic shows the important safety information that ac line voltage is connected to the fuse, F1, the switch S1 (assuming the fuse is not blown), and the transformer, T1.

On a well-constructed schematic, inputs to the circuit are located toward the left side of the schematic and outputs are toward the right. Positive power supply voltages are located toward the top of the schematic and ground or negative supply voltages are at the bottom. Components that work together performing a single function are usually drawn close together. Labels are often added to indicate circuit function. You can see this in Figure 3.16B as power from the ac line flows through the input components at left, through the transformer, and through the rectifier and regulator circuit components at the right. DC output voltage is available from the connection point at the right of the schematic labeled, "DC Out."

Remember that a schematic may have little resemblance to the actual physical layout of the circuit. It is just a convenient way to describe how the circuit is constructed electrically. The "First Steps in Radio" link on the *Ham Radio License Manual* web page will take you to a good article on reading schematics.

Before you go on, study these Technician exam questions from the question pool included at the back of this book or as a downloadable Study Guide version on the web:
T6C01 through T6C13
If you have difficulty with any question, review the preceding section.

3.3 Types of Radios and Radio Circuits

In Chapter 2, you learned about some of the terms that describe the functions of a radio signal used for communication — frequency, phase, modulation, bandwidth, sideband and so forth. In this section, we'll introduce some of the circuits that make those functions happen. The block diagrams that describe simple radios show how successive circuits, called *stages*, are arranged to construct basic radio transmitters and receivers.

TRANSMIT-RECEIVE SWITCHES

In the early days of radio, amateurs used separate transmitters and receivers. A switch was used to connect an antenna to either the transmitter or the receiver. As you might imagine, the circuit was called a *transmit-receive switch* or just a *TR switch*. The TR switch is still present in today's transceivers as an internal circuit. Figure 2.10 (in Chapter 2) shows the location of the TR switch between the transmitter and the receiver, with the antenna connected only to the switch.

OSCILLATORS AND AMPLIFIERS

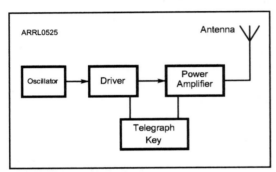

Figure 3.17 — A simple CW transmitter consists of an oscillator to generate a low-power signal and two amplifiers to increase the signal strength to a useful level. A telegraph key is used to turn the amplifiers on and off, creating the CW signal that is transmitted. The oscillator runs continuously so that its frequency remains stable.

An *oscillator* produces a steady signal at one frequency. [T7A05] Oscillators are used in both receivers and transmitters to determine the operating frequency. In a transmitter, the output signal from an oscillator is modulated and amplified before actually being applied to an antenna.

The output signal from the oscillator is not strong enough for reliable communication over long distances. An *amplifier* circuit called a *driver* allows the oscillator to operate continuously at low power so that its frequency remains stable. The output of the driver is then applied to a *power amplifier* which has an output strong enough for reliable communication with other stations. In order to turn the output signal on and off as Morse code, the driver and power amplifier stages are *keyed* (switched on and off) by a telegraph key. **Figure 3.17** illustrates how these stages are combined into a simple Morse code transmitter.

FILTERS

Filters are circuits that perform very important functions in radio. Just like a filter you might use to remove dust or impurities from air or water, radio filters are circuits that reject unwanted signals. A *passive filter* is made from resistors, inductors and capacitors. Tuned circuits made of inductors and capacitors are common examples of passive filters. Electronic circuits containing amplifiers can also act as filters and are called *active filters*.

Reducing a signal's strength is called *attenuation*. A filter circuit rejects the unwanted signals at its input by attenuating them. The filter passes the desired signals on to the output with little or no attenuation.

The name of a filter describes how it acts on signals of different frequencies. **Figure 3.18** shows the filtering action, called the filter's *response graph*, for common filter types. A filter that attenuates signals below a specified *cutoff* frequency is a *high-pass filter* (HPF). If the filter removes signals above the cutoff frequency, it is a *low-pass filter* (LPF).

Band-pass filters (BPF) reject signals outside the frequency band between the upper and

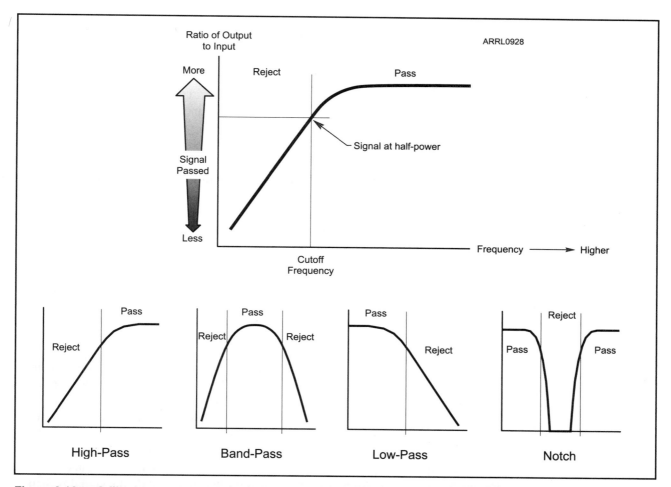

Figure 3.18 — A filter's response graph shows how it affects signals at different frequencies. The vertical axis shows the ratio of the signal at the filter's output to that at the input. The horizontal axis shows frequency. The point at which the output signal's strength is reduced to half of the input power is the cutoff frequency.

lower cutoff frequencies. They act like a low-pass filter above the upper cutoff frequency and a high-pass filter below the lower cutoff frequency. A *notch* or *band-reject* filter is the opposite of a band-pass filter — a band of frequencies is attenuated while all others pass. Band-pass and notch filters are called *wide* or *narrow* depending on whether the frequency range of interest is large or small.

Filters are used in receivers to help reject unwanted signals. They also reduce spurious signals from transmitters that might cause interference. Filters can also be used to remove noise or prevent transmitted signals from interfering with non-radio equipment.

MODULATORS

The process of combining data or voice signals with an RF signal is modulation, as you learned earlier. A circuit that performs the modulation function is therefore called a *modulator*. [T7A08] The function of the modulator is to add the data or voice signal to an RF signal or carrier. The result is an RF signal that can be communicated by radio. A *demodulator* circuit extracts the information from a modulated signal

A modulator can be as simple as an on-off switch (a telegraph key, for example) or it can be very complex in the case of a high-speed data transmission. You can learn more about modulator circuits in the *ARRL Handbook* or when studying for your General class license.

MIXERS

A *mixer* is closely related to a modulator. Both types of circuits combine signals with the intent of producing an output signal with a different frequency. The mixer however, is designed to combine two RF signals and shift one of them to a third frequency. [T7A03] It does not combine an RF signal with a data or voice signal like a modulator does. Mixers are used in both transmitters and receivers to shift signal frequencies for various purposes. (This circuit should not be confused with an *audio mixer* that adds audio signals together for recording or live entertainment.)

Before you go on, study these Technician exam questions from the question pool included at the back of this book or as a downloadable Study Guide version on the web:
T7A03 T7A05 T7A08
If you have difficulty with any question, review the preceding section.

RECEIVERS

If we include mobile phones, the most widely used type of receiver today is called a *direct-conversion receiver*. Direct-conversion receivers use mixers to convert a modulated signal directly back to speech or data. If the modulating signal is digital data, the output of the mixer is converted back to data using *digital signal processing* (*DSP*) techniques. *Software-defined radio (SDR)* receivers convert the incoming RF signal directly to digital data and a microprocessor performs all functions using software.

The *superheterodyne receiver* uses mixers to shift incoming signals of any frequency to a single fixed frequency, called the *intermediate frequency* (*IF*) where unwanted signals are rejected by filters. After amplification and filtering at the IF, the signal is then demodulated.

Superheterodyne techniques are also used to construct transmitters. Essentially, the process of mix-filter-amplify-demodulate is run "in reverse" to create a transmitter. This is just how many modern transceivers are constructed, using clever switching schemes to reuse many of the same circuits on both receive and transmit.

Selectivity and Sensitivity

Receivers are compared on the basis of two primary characteristics: *sensitivity* and *selectivity*. A receiver's sensitivity determines its ability to detect signals. [T7A01] Higher sensitivity means a receiver can detect weaker signals. Sensitivity is specified as a minimum detectable signal level, usually in microvolts (μV). If a receiver is not sensitive enough, a *preamplifier* or "preamp" can be used. The preamp is connected between the antenna and receiver. [T7A11] *Selectivity* is the ability of a receiver to discriminate between signals, retrieving only the information from the desired signal in the presence of unwanted signals. [T7A04] High selectivity means that a receiver can operate properly even in the presence of strong signals on nearby frequencies. Receivers use filters to reject the unwanted signals.

TRANSVERTERS

The superheterodyne technique of shifting signal frequencies can also be applied at the equipment level to convert an entire transceiver to operate on a band other than it was designed for. Instead of a mixer, a piece of equipment called a *transverter* is used. Low-power transmitter output signals on one band are shifted to the new output frequency where they are amplified for transmission. A *receiving converter* (basically a mixer) shifts incoming signals to the desired band where they are received as regular signals by the transceiver. Transverters are used by hams to allow one main transceiver to be used on one or more new bands. For example, with few transceivers available for CW and SSB operation on 222 MHz, a transverter is used to convert 222 MHz signals to and from the 28 MHz band available on all HF gear. [T7A06]

Before you go on, study these Technician exam questions from the question pool included at the back of this book or as a downloadable Study Guide version on the web:
T7A01 T7A04 T7A06 T7A11

If you have difficulty with any question, review the preceding section.

Chapter 4

Propagation, Antennas and Feed Lines

In this chapter, you'll learn about:

- **How radio signals travel from place to place**
- **Basic concepts of antennas**
- **How feed lines are constructed and used**
- **What SWR is and what it means to you**
- **Practical antenna system construction**

When you see the mouse, you'll find more information at www.arrl.org/ham-radio-license-manual

No piece of equipment has as great an effect on the performance of a radio station, whether handheld or home-based, as the antenna. Experimenting with antennas has been a favorite of hams from the very beginning, contributing greatly to the development of antennas for all radio services. To choose and use an antenna effectively, it's important to understand some basics of *propagation* — how radio waves get from one place to another. For these reasons, knowledge of antennas and propagation is very important for amateurs.

4.1 Propagation

Radio waves spread out from an antenna in straight lines unless reflected or diffracted along the way, just like light. Light waves are just a very, very, very high frequency form of radio waves! Like light, the strength of a radio wave decreases as it travels farther from the transmitting antenna. Eventually the wave becomes too weak to be received because it has spread out too much or something along its path absorbed or scattered it. The distance over which a radio transmission can be received is called *range*. The curvature of the Earth sets an effective range limit for many signals, creating a *radio horizon*. [T3C10]

Line-of-sight propagation takes place between transmitting and receiving antennas that are within direct sight of each other. Most propagation at VHF and higher frequencies is line-of-sight. Increasing antenna height or transmitter power also increases the range of line-of-sight propagation. Radio waves at HF and lower frequencies can also travel along the surface of the Earth as *ground wave* propagation.

Radio waves can be reflected by any sudden change in the media through which they are traveling, such as a building, hill, or even weather-related changes in the atmosphere. Obstructions such as buildings and hills create radio shadows, especially at VHF and UHF frequencies. **Figure 4.1** shows how radio waves can also be *diffracted* as they travel past sharp edges of these objects. This is called *knife-edge* propagation. [T3C05] *Refraction* is another type of propagation; a gradual bending of VHF and UHF radio waves in the atmosphere. By bending signals slightly back towards the ground, refraction counteracts the curvature of the Earth and allows signals at these frequencies to be received at distances somewhat beyond the visual line-of-sight horizon. [T3C11]

Radio waves can also penetrate openings in otherwise solid objects as long as at least one side of the opening is longer than about one-half wavelength. For this reason, the

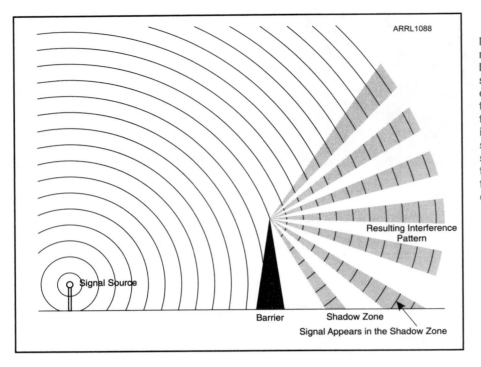

ARRL1088

Signal Source

Barrier   Shadow Zone

Resulting Interference Pattern

Signal Appears in the Shadow Zone

Figure 4.1 — VHF and UHF radio waves are diffracted by the edge of a solid object, such as a building, hill or other obstruction, bending them in different ways around the obstruction. The resulting interference pattern creates shadowed areas where little signal is present. If the edge of the object is small compared to a wavelength, the result is called "knife-edge diffraction."

shorter wavelengths of UHF signals make them more effective at propagating into and out of buildings in urban areas. [T3A02]

Radio signals arriving at a receiver after taking different paths from the transmitter can interfere with each other if they are out of phase, even canceling completely! This phenomenon is known as *multipath* and can cause a signal to become weak and distorted. Moving your antenna just a few feet may avoid the location at which cancellation is occurring. [T3A01] Multipath propagation of signals from distant stations results in irregular fading, even when reception is generally good. [T3A08] VHF or UHF signals from a mobile station moving through an area where multipath is present have a characteristic rapid variation in strength known as *mobile flutter* or *picket-fencing*. [T3A06] Variations in signal strength from multipath can also cause digital data signals to be received with a higher error rate, particularly at VHF and UHF. [T3A10]

Propagation at and above VHF frequencies assisted by atmospheric phenomena such as weather fronts or temperature inversions is called *tropospheric propagation* or just "tropo." [T3C08] Layers of air with different characteristics can also form structures called *ducts* that can guide even microwave signals for long distances. Tropo is regularly used by amateurs to make VHF and UHF contacts that would otherwise be impossible by line-of-sight propagation. Tropo contacts over 300-mile paths are not uncommon. [T3C06]

Radio signals are also reflected by conductive things in the atmosphere. The size and nature of the reflecting surface determines which frequencies are reflected best. For example, airplanes can reflect 2 meter and 70 cm signals over hundreds of miles. Yet there is one more conductive "thing" floating around "up there" that hams use to communicate around the world every day.

Before you go on, study these Technician exam questions from the question pool included at the back of this book or as a downloadable Study Guide version on the web:

T3A01 T3A02 T3A06 T3A08 T3A10
T3C05 T3C06 T3C08 T3C10 T3C11

If you have difficulty with any question, review the preceding section.

THE IONOSPHERE

Above the lower atmosphere where the air is relatively dense and below outer space where there isn't any air at all lies the *ionosphere*. In this region, from 30 to 260 miles above the Earth, atoms of oxygen and nitrogen gas are exposed to the intense and energetic *ultraviolet* (UV) rays of the sun. These rays have enough energy to create positively charged *ions* from the gas atoms by knocking off some of their negatively charged electrons. The resulting ions and electrons create a weakly conducting region high above the Earth.

The ionosphere forms in layers shown in **Figure 4.2**, called the D, E, F1 and F2 layers, with the D layer being the lowest. Depending on whether it is night or day and on the intensity of the solar radiation, these layers can refract (E, F1 and F2 layers) or absorb (D and E layers) radio waves.

Radio waves at HF (and sometimes VHF) can be completely bent back toward the Earth by refraction in the ionosphere's E and F layers as if they were reflected. This is called *sky wave* propagation or *skip*. Since the Earth's surface is also conductive, it can also reflect radio waves. This means that a radio wave can be reflected between the ionosphere and ground multiple times. Each reflection from the ionosphere is called a *hop* and allows radio waves to be received hundreds or thousands of miles away. [T3A11] This is the most common way for hams to make long-distance contacts on the HF bands.

The ability of the ionosphere to refract or bend radio waves also depends on the frequency of the radio wave. Higher frequency waves are bent less than those of lower frequencies. At VHF and higher frequencies, the waves usually pass through the ionosphere with only a little bending and are lost to space. This is why UHF signals from stations beyond the radio horizon are rarely heard without being relayed by a repeater. [T3C01]

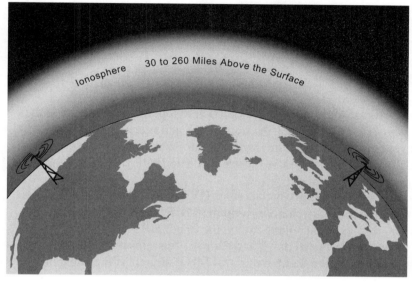

Figure 4.2 — The ionosphere is formed by solar ultraviolet (UV) radiation. The UV rays knock electrons loose from air molecules, creating weakly charged layers at different heights. These layers can absorb or refract radio signals, sometimes bending them back to the Earth.

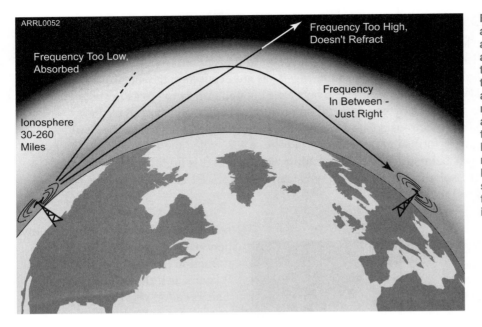

Figure 4.3 — Signals that are too low in frequency are absorbed by the ionosphere and lost. Signals that are too high in frequency pass through the ionosphere and are also lost. Signals in the right range of frequencies are refracted back toward the Earth and are received hundreds or thousands of miles away. Paths of this length generally require several ionospheric hops — the figure is intended only to illustrate MUF and LUF.

The highest frequency signal that can be reflected back to a point on the Earth is the *maximum usable frequency* (MUF) between the transmitter and receiver. The lowest frequency that can travel between those points without being absorbed is the *lowest usable frequency* (LUF). This is illustrated in **Figure 4.3**. The MUF rises as the sun illuminates the ionosphere.

As sunspot activity increases, the more intense solar UV rays become. This increases the peak level of ionization in the ionosphere, raising the MUF along the path between stations. During the years of maximum solar activity, the upper HF bands, such as 10 meters, are likely to be open between dawn and shortly after sunset. [T3C09] Occasionally, the F layers can even reflect 6 meter (50 MHz) signals at the sunspot cycle's peak. [T3C12]

When sky-wave propagation on an amateur band is possible between two points, the band is said to be *open*. If not, the band is *closed*. Because the ionosphere depends on solar radiation to form, areas in daylight have a different ionosphere above them than do those in nighttime areas. That means radio propagation may be supported in some directions, but not others, opening and closing to different locations as the Earth rotates and the seasons change. This makes pursuing long-distance contacts (DXing) very interesting!

VHF and UHF enthusiasts also experience exciting ionospheric propagation. When solar radiation becomes sufficiently intense, such as during the peak of the 11-year sunspot cycle, the F layers of the ionosphere can bend even VHF signals back to Earth. When those ham bands open, they support long-distance communication not possible under normal conditions. In addition, at all points in the solar cycle, patches of the ionosphere's E layer can become sufficiently ionized to reflect VHF and UHF signals back to Earth. [T3C02] This is called *sporadic E* or E$_s$ (or E-skip) propagation and it is most common during early summer and mid-winter months on 10, 6 and 2 meters. [T3C04]

Along with sporadic E propagation, the ionosphere is home to other radio wave reflectors. The aurora (northern lights) is the glow from thin sheets of charged particles flowing down through the lower layers of the ionosphere. Those thin sheets 50 miles or more above the Earth's surface reflect VHF and UHF signals. Because the aurora is constantly changing, the reflected signals change strength quickly and are often distorted. [T3C03]

This region of the ionosphere is also home to meteor trails. A meteoroid burning up in the upper atmosphere results in a *meteor* with a *meteor trail* of ionized gas lasting up to several seconds that can reflect radio signals. Bouncing signals off of these ionized trails is called *meteor scatter* propagation. The best band for meteor scatter is 6 meters, and contacts can be made at distances up to 1200 to 1500 miles. [T3C07]

Before you go on, study these Technician exam questions from the question pool included at the back of this book or as a downloadable Study Guide version on the web:
T3A11
T3C01 through T3C04 T3C07 T3C09 T3C12
If you have difficulty with any question, review the preceding section.

4.2 Antenna Fundamentals

An antenna is not necessarily something that you must purchase or that "looks like" a commercial product. Any electrical conductor can act as an antenna for radio signals — a wire, a pipe, a car body or even the proverbial bedsprings. However, for an antenna to radiate and receive radio signals efficiently, its dimensions must be an appreciable fraction of the signal's wavelength. **Figure 4.4** shows two examples of antennas commonly used by Technician class hams.

A *feed line* is used to deliver the radio signals to or from the antenna. The connection of antenna and feed line is called the *feed point* of the antenna. Just like Ohm's Law in Chapter 2, the ratio of radio frequency voltage to current at an antenna's feed point is the antenna's *feed point impedance*. An antenna is *resonant* when its feed point impedance is all resistance with no reactance.

An antenna's feed point impedance at a specific frequency depends on how its physical dimensions compare to the wavelength at that frequency. Feed point impedance changes with frequency because wavelength changes but the physical dimensions don't. An antenna's feed point impedance is also affected by nearby conductors and its height above ground.

The conducting portions of an antenna by which the radio signals are transmitted or received are called *elements*. An antenna with more than one element is called an *array*. The element connected to the feed line is called the *driven element*. If all of the elements are connected to a feed line, that's a *driven array*. Elements that are not directly connected to a feed line, but that influence the antenna performance, are called *parasitic elements*.

RF current in an antenna element creates radio waves that travel away from the antenna. The radio wave is composed of both electrical and magnetic energy supplied

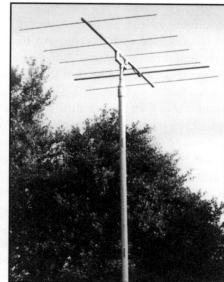

Figure 4.4 — The ground-plane antenna shown at the left radiates a signal from the vertical wire attached to the base. The vehicle's metal surface acts as an electrical mirror, creating the effect of another wire opposite to the one above the surface. On the right, the Yagi beam antenna uses parasitic elements to direct the signal in one direction and reject signals in the opposite direction.

by the electrons moving back and forth in the antenna. The wave is a combination of an electric and a magnetic field, just like the fields in a capacitor and inductor but spreading out into space like ripples traveling across the surface of water. [T3B03] Because a radio wave is made up of both types of fields, it is called an *electromagnetic wave*. [T5C07, T3A07] The wave's electric and magnetic fields oscillate at the same frequency as the RF current in the antenna.

If you could feel what an electron in an antenna element feels as a radio wave passes by, you would feel both an electric force and a magnetic force. The forces are caused by the radio wave's electric field and magnetic field and oscillate at the frequency of the radio wave's signal. These forces cause electrons in the receiving antenna element to move back and forth, creating an RF current that can be detected by a receiver.

Polarization refers to the orientation of the radio wave's electric field. [T3B02] A *horizontally polarized* antenna radiates a radio wave whose electric field is oriented horizontally. A *vertically polarized* antenna has an electric field perpendicular to the surface of the Earth. [T9A02]

When the electric field of the radio wave and the element of the antenna have the same polarization, the maximum amount of signal is created in the antenna by the wave. Primarily, the electric field is what causes the electrons to move back and forth along the antenna element, creating current and the signal for the receiver to detect. That is why it is important to hold your handheld radio so that its antenna is aligned with the antenna of the receiving station.

When the polarizations of transmit and receive antennas are misaligned, the received signal can be dramatically reduced — as much as by 100 times! Because the polarization of the radio wave doesn't match that of the receiving antenna, less current is created in the antenna. [T3A04] As a radio wave travels through the ionosphere its polarization changes from vertical or horizontal to a combination of the two, called *elliptical polarization*. As a result, a receiving antenna of any polarization will respond to the incoming wave at least partially. This means both vertical and horizontal antennas are effective for receiving and transmitting on the HF bands where skip propagation is common. [T3A09]

ANTENNA GAIN

The concentration of radio signals in a specific direction is called *gain*. (Antenna gain should not be confused with the gain of a transistor.) Antenna gain increases signal strength in a specified direction when compared to a reference antenna. [T9A11] Gain aids communication in the preferred direction by increasing transmitted and received signal strengths.

An antenna can create gain by radiating radio waves that add together in the preferred direction and cancel in others. Gain can also be created by reflecting radio waves so that they are focused in one direction. Gain only focuses power — it does not create power.

An *isotropic* antenna has no gain because it radiates equally in every possible direction. Isotropic antennas do not exist in the real world — they are only used as imaginary references. An *omnidirectional* antenna radiates a signal equally in every horizontal direction. An antenna with gain in a single direction is called a *beam* or *directional* antenna. Omnidirectional antennas are useful for communicating over a wide region while beam antennas are used when communication is desired in one direction.

An antenna's gain is specified in decibels (dB) with respect to some type of reference antenna. For example, the abbreviation dBi means gain in decibels with respect to an isotropic antenna. The abbreviation dBd means gain with respect to a dipole antenna's peak gain (discussed below). Gain, like voltage, is a relative measurement between an antenna and some reference antenna, most often an isotropic or dipole antenna.

Decibels — Bringing Large and Small Together

Radio signals vary dramatically in strength. At the input to a receiver, signals are frequently smaller than one ten-billionth of a watt. When they come out of a transmitter, they're often measured in kilowatts! Antennas, propagation and electronic circuits change signal strengths by many factors of ten. These big differences in value make it difficult to compare signal sizes. Enter the *decibel*, abbreviated dB and pronounced "dee-bee." The decibel measures the ratio of two quantities as a power of 10. The formula for computing decibels is:

dB = 10 log (power ratio)

dB = 20 log (voltage ratio).

Positive values of dB mean the ratio is greater than 1 and negative values of dB indicate a ratio of less than 1. For example, if an amplifier turns a 5-watt signal into a 25-watt signal, that's a change of 10 log (25 / 5) = 10 log (5) = 7 dB. On the other hand, if by adjusting a receiver's volume control the audio output signal voltage is reduced from 2 volts to 0.1 volt, that's a change of 20 log (0.1 / 2) = 20 log (0.05) = −26 dB.

A complete discussion of the decibel, its history and examples of how to work out the answers to exam questions about decibels are available on the *Ham Radio License Manual* web page. Look for the section on Chapter 4. [T5B09, T5B10, T5B11]

Before you go on, study these Technician exam questions from the question pool included at the back of this book or as a downloadable Study Guide version on the web:

T5B09   T5B10   T5B11

If you have difficulty with any question, review the preceding section.

RADIATION PATTERNS

The easiest way to describe how an antenna distributes its signals is a graph showing the antenna's gain in any direction around the antenna. That graph is called a *radiation pattern*. An antenna transmits and receives with the same pattern.

The most common type of radiation pattern is an *azimuthal* pattern that shows the antenna's gain in horizontal directions around the antenna. An azimuthal pattern can be imagined as looking down on the antenna from above as in **Figure 4.5**. An *elevation*

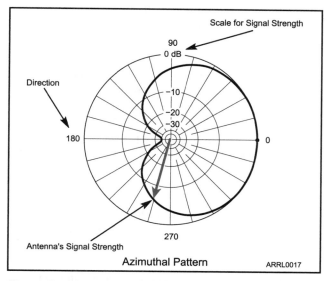

Figure 4.5 — As if looking down on the antenna from above, the azimuth radiation pattern shows how well the antenna transmits or receives in all horizontal directions. The distance from the center of the graph to the solid line is a measure of the antenna's ability to receive or transmit in that direction.

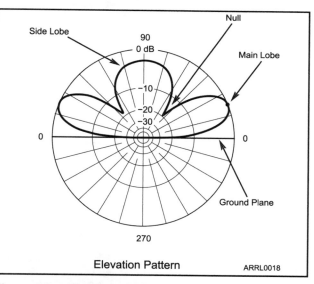

Figure 4.6 — The elevation pattern looks at the antenna from the side to see how well it receives and transmits at different angles above a horizontal plane.

pattern shows the strength of the radiated energy in vertical directions as if the antenna is viewed from the side, as shown in **Figure 4.6**. An antenna's radiation pattern may change as frequency changes for the same reasons that feed point impedance changes with frequency — changes in frequency change the wavelength, while the physical dimensions remain fixed, causing the antenna to behave differently electrically.

The region of the radiation pattern in which the antenna's gain is greatest is called the *main lobe*. Regions of lower gain are called *side lobes* and *nulls* where gain is a minimum. The ratio of gain in the preferred or *forward* direction to that in the opposite direction is called the *front-to-back ratio*. The ratio of gain in the forward direction to that at right angles called the *front-to-side ratio*. Antennas with high front-to-back and front-to-side ratios are useful in rejecting interference and noise from unwanted directions.

Before you go on, study these Technician exam questions from the question pool included at the back of this book or as a downloadable Study Guide version on the web:

T3A04   T3A07   T3A09   T3B02   T3B03
T5C07
T9A02   T9A11

If you have difficulty with any question, review the preceding section.

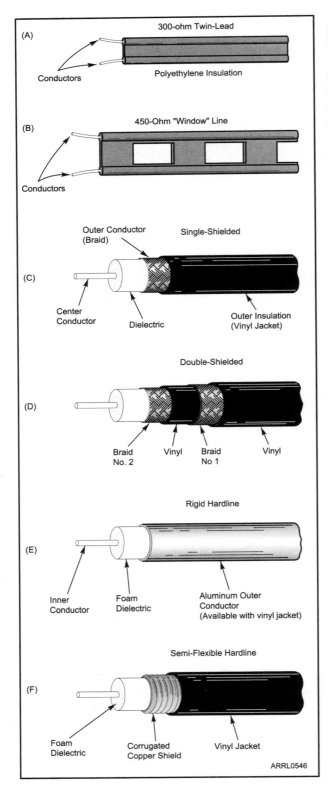

Figure 4.7 — This drawing illustrates some common types of open-wire and coaxial cables used by amateurs. Open-wire line (A and B) has two parallel conductors separated by insulation. "Coax" (C-F) has a center conductor surrounded by insulation. The second conductor, called the *shield*, covers the insulation and is in turn covered by the plastic outer *jacket*.

4.3 Feed Lines and SWR

Feed lines are used to connect a radio to an antenna. They are also used when an RF signal must be conducted from one piece of equipment to another. Feed lines are made from two conductors separated by an insulating material such as plastic. The radio signal is carried on the conductors and in the space between them. Feed lines used at radio frequencies use special materials and construction methods to minimize power being dissipated as heat by *feed line loss* and to avoid signals leaking in or out. [T7C07] Feed line loss increases with frequency for all types of feed lines. [T9B05]

COAXIAL CABLE

The most popular feed line used by amateurs to connect radios and antennas is *coaxial cable* or *coax*.

[T7C12] It is easy to use and requires few special installation considerations. [T9B03] **Figure 4.7** shows how coaxial cable is constructed. A wire *center conductor* is surrounded by insulation (the *center insulator* or *dielectric*). The insulation is covered with a tubular *shield* of braided wire or foil. Finally, the cable is covered with a plastic sheath called the *jacket*. The name "coaxial" comes from the shared central axis of the center conductor and the shield.

Coaxial cable carries the radio signal between the center conductor and the inside surface of the shield. That means it can be placed next to other cables or conducting surfaces such as conduit or antenna support masts without affecting the signal inside.

A special type of coaxial feed line is called *hardline* because its shield is made from a semi-flexible solid tube of aluminum or copper. This limits the amount of bending the cable can do, but hardline has the lowest loss of any type of coaxial feed line. [T9B11]

OPEN-WIRE FEED LINE

A feed line of two parallel wires separated by insulating material is called *open-wire, ladder line, window line* or *twin-lead*. This type of feed line, also shown in Figure 4.7 (parts A and B), has less insulating material and greater spacing between its conductors, so it has less loss than coaxial cable. Since the radio energy is not shielded by an outer tube, the signals in parallel conductor feed lines can be affected by nearby conductors. Open-wire feed lines cannot be buried or installed in metal conduits and must be kept clear of nearby conducting surfaces.

CHARACTERISTIC IMPEDANCE

Feed lines have a *characteristic impedance* which is denoted by Z_0, a measurement of how energy is carried by the feed line. This is not the same as the resistance of the conductors if measured from end to end of the feed line. To understand characteristic impedance, try this experiment: Obtain thin and fat drinking straws. Blow a single puff of air through each, feeling the resistance to air flow through each. You will feel more back-pressure in the thin straw because to the flow of air its characteristic impedance is higher.

The dimensions of feed line conductors, the spacing between them, and the insulating material determine characteristic impedance. Most coaxial cable used in ham radio has a characteristic impedance of 50 ohms. [T9B02] Coaxial cables used for video and cable television have a Z_0 of 75 ohms. Open-wire feed lines have a Z_0 of 300 to 450 ohms.

Before you go on, study these Technician exam questions from the question pool included at the back of this book or as a downloadable Study Guide version on the web:
T7C07   T7C12
T9B02   T9B03   T9B05   T9B11
If you have difficulty with any question, review the preceding section.

STANDING WAVE RATIO — SWR

The power carried by a feed line is transferred completely to a load, such as an antenna, when the load and feed line impedances are identical or *matched*. If the feed line and load impedances do not match, some of the power is *reflected* by the load. Power traveling toward the load is *forward power*. Power reflected by the load is *reflected power*. The greater the difference between the feed line and load impedances, the more power is reflected by the load. In the worst case where the feed line is connected to or *terminated* in an open or short circuit, all of the forward power is reflected.

Reflected and forward power traveling in opposite directions create a stationary

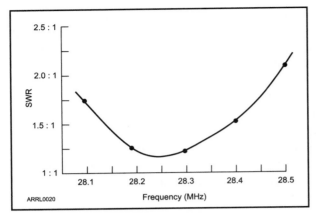

Figure 4.8 — Since an antenna's feed point impedance changes with frequency, so does SWR. The graph shows how the SWR of a typical antenna changes across the 10 meter amateur band.

wave-like interference pattern in the feed line called a *standing wave*. The ratio of the maximum value to minimum value of the interference pattern is called the *standing wave ratio* or *SWR*. SWR is the same everywhere along a feed line, but it is most commonly and conveniently measured at the transmitter's connection to the feed line.

Because SWR is determined by the proportions of forward and reflected power, SWR in an antenna system is also a measure of the how well the antenna and feed line impedances are matched. [T7C03] In fact, SWR is equal to the ratio of antenna-to-feed line or feed line-to-antenna impedances, whichever ratio is greater than 1.

When there is no reflected power there is no interference pattern and the SWR is 1:1. This condition is called a *perfect match*. [T7C04] As more power is reflected, SWR increases. SWR is always greater than or equal to 1:1. SWR greater than 1:1 is called an *impedance mismatch* or just *mismatch*. [T7C06] Since an antenna's feed point impedance changes with frequency while that of the feed line does not, SWR also changes with frequency as shown in **Figure 4.8**.

Why does SWR matter? Low SWR indicates the efficient transfer of power from the feed line because less power is reflected by the antenna. Low SWR also reduces losses in the feed line resulting from reflected power in the feed line traveling back and forth between the antenna and transmitter. [T9B01] With each pass through the feed line, some of the power is transferred to the antenna, but some is also lost as heat. As SWR increases, more power is reflected and more power is lost.

Another effect of high SWR is that the interference pattern causes voltages to increase in the feed line and at the transmitter's output where the feed line is connected. The higher voltages can damage a transmitter's output circuits. Most amateur transmitting equipment is designed to work at full power with an SWR of 2:1 or lower. SWR greater than 2:1 may cause the transmitter's protection circuits to reduce power automatically to avoid damage to the output transistors. [T7C05]

What causes high SWR? Antennas that are much too short or too long for the frequency being used often have extreme feed point impedances, causing high SWR. A faulty feed line or feed line connectors can also raise SWR. Erratic SWR usually indicates a loose connection in the feed line or antenna. [T9B09] To correct high SWR not caused by a fault, an impedance matching device is used. This device is called a *transmatch*, an *impedance matcher* or an *antenna tuner*. The antenna tuner is typically connected at the output of the transmitter.

Before you go on, study these Technician exam questions from the question pool included at the back of this book or as a downloadable Study Guide version on the web:
T7C03 through T7C06
T9B01   T9B09
If you have difficulty with any question, review the preceding section.

4.4 Practical Antenna Systems

DIPOLES AND GROUND-PLANES

The simplest type of antenna is a *dipole*, meaning essentially "two electrical parts." Dipoles are made from a straight conductor of wire or tubing one-half wavelength (½-λ) long with a feed point somewhere along the antenna, usually in the middle. Dipoles are easy to make, easy to use, and work quite well in a variety of environments. Most are oriented horizontally, particularly on the lower frequency bands, and radiate a horizontally polarized signal. [T9A03] Dipoles can also be installed vertically, sloping or even drooping from a single support in the middle (the *inverted-vee* or *inverted-V*). **Figure 4.9** shows the central portion of a typical wire dipole.

A dipole radiates strongest broadside to the axis of the dipole and weakest off the ends. The radiation pattern for a dipole isolated in space looks like a donut as seen in **Figure 4.10**. [T9A10] The figure shows both two- and three-dimensional patterns. The two-dimensional pattern shows what the three-dimensional pattern would look like if cut through the plane of the dipole.

Another popular antenna is the *ground-plane* antenna. The most common type of ground-plane antenna is one-half of a dipole (¼-λ long) with the missing portion made up by an electrical mirror, called the ground plane, which is made from sheet metal or a screen of wires called *radials*. The ground-plane is generally one-quarter wavelength long with the feed point at the base of the antenna, as shown in **Figure 4.11**. One conductor of the feed line is connected to the antenna and the other to the ground plane. The length of a ground-plane is half that of a dipole and is often estimated as: length (in feet) = 234 / frequency (in MHz). [T9A08, T9A09] At HF, a longer length is often required as discussed in the sidebar on dipoles.

Figure 4.9 — The center of a basic dipole antenna with the balanced feed line connected to a center insulator and the legs of the dipole extending to the left and right.

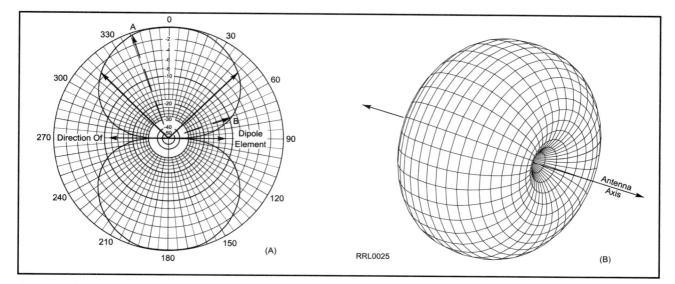

Figure 4.10 — The radiation pattern of a dipole far from ground (in free-space). At (A) the pattern is shown in a plane containing the dipole. The lengths of the arrows indicate the relative strength of the radiated power in that direction. The dipole radiates best broadside to its length. At (B) the 3-D pattern shows radiated strength in all directions.

How Long is That Dipole?

If a resonant dipole antenna is ½-λ long, how long is that in feet? This traditional formula for thin wire dipoles that are lower than one-half wavelength above ground is often used, even though the estimate is usually too short at HF:

Length (in feet) = 468 / frequency (in MHz)

or

Length = 468 / f

Example: At 50.1 MHz (in the 6 meter band), dipole length is calculated as 468 / 50.1 = 112 inches long (9.33 feet).

The value of the constant used in the formula accounts for effects that cause an antenna to act like it is a little longer electrically than it is physically. The actual resonant length is affected by height above ground, its electrical properties, and nearby conductive objects.

Make the dipole a few percent longer at first (use 490 instead of 468), then use an SWR meter or antenna analyzer to determine the resonant frequency. Shorten the dipole until it is resonant at the desired frequency. [T9A05] This allows you to compensate for the effects of ground or nearby conductors that might affect the antenna. For dipoles made out of wire, be sure to add a little extra wire to fasten it to insulators and supports.

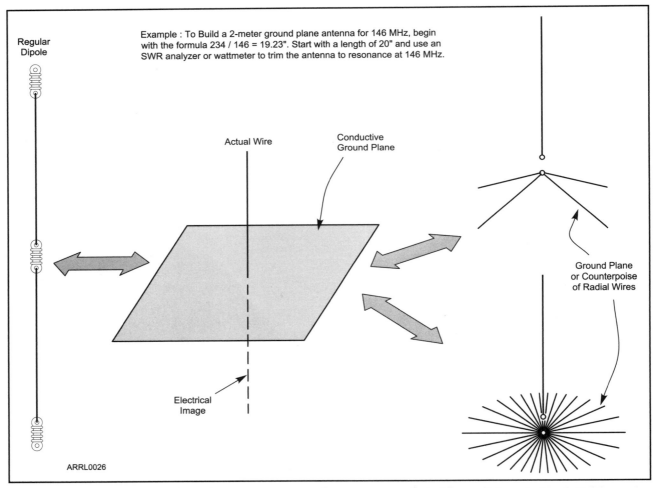

Regular Dipole

Example : To Build a 2-meter ground plane antenna for 146 MHz, begin with the formula 234 / 146 = 19.23". Start with a length of 20" and use an SWR analyzer or wattmeter to trim the antenna to resonance at 146 MHz.

Actual Wire

Conductive Ground Plane

Electrical Image

Ground Plane or Counterpoise of Radial Wires

ARRL0026

Figure 4.11 — A ground-plane makes up an electrical mirror that creates an image of the missing half of a ground-plane antenna. The result is an antenna that acts very much like a dipole. The ground plane can be made up of a screen of wires (often used at HF) or a metal surface at VHF and UHF. For VHF and UHF antennas mounted on masts, a counterpoise of a few wires serves the same purpose.

In Figure 4.11, what is meant by "electrical mirror"? Have you ever watched as someone put their face at the end of a mirror, the reflection recreating the face's missing half? A ground-plane works in much the same way. The more complete the ground-plane plane, the better the reflection and more dipole-like the result. A good ground-plane should extend at least ¼-λ from the base of the antenna in all directions. As you might expect, since a ground-plane antenna emulates a dipole, it radiates strongest perpendicular to the axis of the antenna and weakest off its end.

Ground-plane antennas are often called *verticals* because it is easiest to mount them so they are perpendicular to the ground, with the ground-plane on or parallel to the ground. When mounted this way, the radiation pattern of the single-element ground-plane is *omnidirectional*, transmitting and receiving equally well in all directions. This type of pattern is used when there is no preferred direction of communications.

Good examples of practical vertical antennas are the small whips placed on the roofs or trunks of cars. The metallic body of the car makes up the ground plane. If mounted on the vehicle roof in the clear, the result is a relatively uniform radiation pattern in all directions. [T9A13] On the ham bands at 50 MHz and above, these antennas give great performance at low cost.

The ⅝-λ vertical antenna offers some improvement over the ¼-λ vertical. Due to its extended length, the ⅝-λ vertical focuses a bit more energy toward the horizon, improving range. [T9A12]

On the HF bands below 24 MHz, the increasing wavelength makes a ¼-λ vertical less practical for portable and mobile operation. To reduce the physical size of the antenna, it is often constructed with some of the radiating conductor wound into a coil or a separate inductor inserted in the antenna. This technique is called *inductive loading* and it makes the antenna longer electrically than it is physically. [T9A14]

Before you go on, study these Technician exam questions from the question pool included at the back of this book or as a downloadable Study Guide version on the web:
T9A03   T9A04   T9A05   T9A07   T9A08   T9A09
T9A10   T9A12   T9A13   T9A14
If you have difficulty with any question, review the preceding section.

Antennas for Handheld Radios

The flexible antenna used with most handheld radios is called a *rubber duck*. It's a ground-plane antenna shortened by coiling the conductor inside a plastic coating. The body of the radio and the operator form the antenna's ground plane. The rubber duck is conveniently sized, but doesn't transmit or receive as well as a full-sized ground-plane antenna. [T9A04] While using a rubber duck antenna, for best performance hold the transceiver so that the antenna is vertical as shown in Figure 4.A. This aligns the handheld antenna with those of repeaters and most other handhelds so that signal strength is maximized.

When using a handheld transceiver inside a vehicle, a rubber duck may not be an effective antenna. The vehicle's metal roof and doors act like shields, trapping the radio waves inside. Some of the signal gets out through the windows (unless they're tinted with a thin metal coating), but it's as much as 10 to 20 times weaker than an external mobile antenna. [T9A07]

A handheld radio can also be connected to a full-size or external base-style antenna for better performance, replacing the rubber duck. The radio usually has a standard RF connector that allows a mobile antenna to be used in a vehicle or with base station-style antennas at home.

Figure 4.A — To get the best range from a handheld radio, hold it so that the antenna is vertical. This aligns your antenna with those of a repeater or another handheld user. Also, turn the microphone slightly away from your face when talking so that your breath does not blow directly into it.

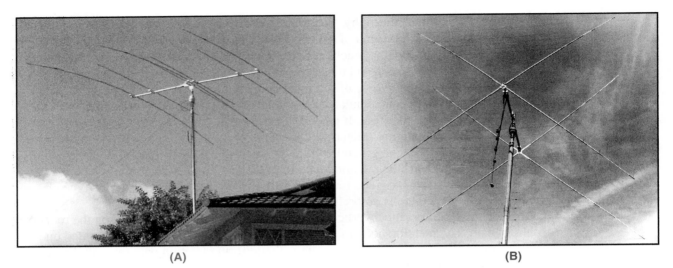

Figure 4.12 — Yagi (A) and quad (B) antennas are beam antennas that focus radiated power along their axis. Yagi elements are tubes or rods and can be oriented horizontally (as shown here) or vertically. Quad elements are one-wavelength loops. By feeding the driven element of a quad at different points, the radiated signal can be made either horizontally or vertically polarized.

DIRECTIONAL ANTENNAS

Simple dipoles, ground-planes and loops work well, but they have little gain. That is, their radiation patterns don't have strongly preferred directions. In many situations, it is desired to focus transmitted power (and to optimize reception) in one direction, so a directional beam antenna is used. [T9A01] Beams can be used to increase signal level at a distant station or to reject interference or noise. On VHF and UHF, if a direct signal path is blocked by building or other obstruction, a beam antenna can be used to aim the signal at a reflecting surface to bypass the obstruction. [T3A05]

Beams are created from arrays of multiple elements. (Dipoles and ground-planes are *single-element* antennas.) The two most widely used types of beam antennas used by hams are *Yagis* and *quads*, shown in **Figure 4.12**. The radiation pattern for a typical beam is shown in **Figure 4.13**.

The Yagi is named after one of its two inventors, Dr Yagi and Dr Uda. Both Yagis and quads are arrays, constructed from two or more dipoles (the Yagi) or loops (the quad) mounted on a central support called a *boom*. Only the driven element of the Yagi or quad is connected to a feed line. The remaining parasitic elements determine the radiation pattern of the antenna. The length of the parasitic elements and their arrangement along the boom are chosen to cause the antenna to focus energy along the boom in one direction. Yagis and quads are thus known as *parasitic arrays*.

Careful inspection of parasitic elements shows that those located along the boom in the

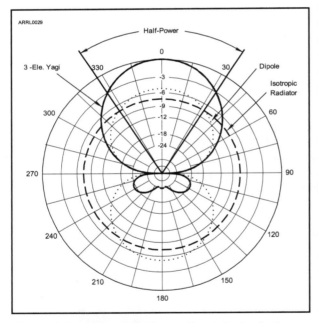

Figure 4.13 — The radiation pattern of a typical, three-element Yagi antenna with a driven element, reflector, and director shows that most of the antenna's energy is focused in one direction along the boom of the antenna. (along the 0-180 axis of the graph.) Smaller amounts are radiated toward the side and back. This antenna also rejects noise and interference from the side and back. The round pattern of the isotropic antenna and the figure-eight pattern of a dipole are included for reference.

Figure 4.14 — The 10-meter diameter dish shown on the left was used on 1296 MHz to bounce signals off the moon and for long-distance terrestrial contacts. N7CFO's much smaller dish (right) operates on 10 GHz and is portable enough to be taken on contest outings.

direction of maximum gain are slightly shorter than the driven element. These elements are called *directors*. Parasitic elements in the direction of minimum gain are called *reflectors* and are slightly longer than the driven element.

Horizontally polarized Yagis and quads are usually used for long-distance communications, especially for weak signal SSB and CW contacts on the VHF and UHF bands. Horizontal polarization is preferred because it results in lower *ground losses* when the wave reflects from or travels along the ground. [T3A03]

As frequency increases and the size of Yagi and quad elements become smaller, it becomes more difficult to construct practical antennas. At frequencies above 1 GHz, a different style of antenna becomes practical — the *dish*. [T9A06] Amateur dish antennas work very much like the satellite TV dishes often seen on homes and apartment railings. **Figure 4.14** shows two examples of amateur dish antennas.

A dish antenna has much more gain than a Yagi — by a factor of 10 or more! On the 33 cm and 23 cm bands, dishes of a few feet in diameter are common and really large dishes are not unknown. At higher frequencies of 10 GHz or more, dish size shrinks below ½-meter in diameter, while still providing plenty of gain.

Before you go on, study these Technician exam questions from the question pool included at the back of this book or as a downloadable Study Guide version on the web:

T3A03   T3A05

T9A01   T9A06

If you have difficulty with any question, review the preceding section.

PRACTICAL FEED LINES AND ASSOCIATED EQUIPMENT

Feed Line Selection and Maintenance

Popular types of coax used by amateurs are shown in **Table 4.1**. Next to characteristic impedance, most important characteristic of coax is feed line loss. Loss is specified in dB per 100 feet of cable at a specific frequency. Table 4.1 gives cable loss at 30 MHz (close to the 10 meter band) and at 150 MHz (close to the 2 meter band). In general, a larger diameter cable such as RG-8 will have less loss than a small cable such as RG-58. [T9B10]

Coaxial cables must be protected. The performance of coaxial cable depends on the integrity of its outer jacket. Nicks, cuts and scrapes can all breach the jacket allowing moisture contamination, the most common cause of coaxial cable failure. [T7C09] Prolonged exposure to the ultraviolet (UV) in sunlight will also cause the plastic in the jacket to degrade, causing small cracks that allow water into the cable. [T7C10] To protect the cable against UV damage the jacket usually contains a pigment that absorbs and blocks the UV. Coax should also not be bent sharply, lest the center conductor be forced gradually through the soft center insulation, eventually causing a short circuit to the outer braid.

When would you use open-wire line? Open-wire lines have much lower loss than coaxial cable. If you needed a really long run of feed line it would be a good choice as long as it can be kept clear of nearby conductors. For example, at 30 MHz, 450-Ω "window" line has a loss of only 0.15 dB/100 feet while RG-8 loses 1.1 dB in the same length. At 150 MHz, RG-8 is losing 2.5 dB/100 feet and the window line a bit less than 0.4 dB/100 feet.

Open-wire feed lines also need care. These feed lines are often constructed using solid wire. Prolonged flexing in the wind will eventually crack and break the conductors if no strain relief is provided. Moss, vines or lichen growing on the cable will also increase loss, as will a coating of ice or standing water. Tree limbs rubbing on the line will eventually break the wires. Protect splices from weather damage with good-quality electrical tape or a paint-on coating.

Coaxial Feed Line Connectors

Connectors for coaxial cable ("coax connectors") are required to make connections to radios, accessory equipment and most antennas. "Pigtail" style connections, where the braid and center conductor are separated and attached to screw terminals, are generally

---

Table 4.1
Common Types of Coaxial Cable

Type	Impedance Ω	Loss per 100 feet (in dB) at 30 MHz	Loss per 100 feet (in dB) at 150 MHz
RG-6	75	1.4	3.3
RG-8	50	1.1	2.5
RG-8X	50	2.0	4.5
RG-58	50	2.5	5.6
RG-59	75	1.8	4.1
RG-174	50	4.6	10.3
RG-213	50	1.1	2.5
LMR-400	50	0.7	1.5

Values in this table were calculated using the online calculator at the Times-Microwave website.

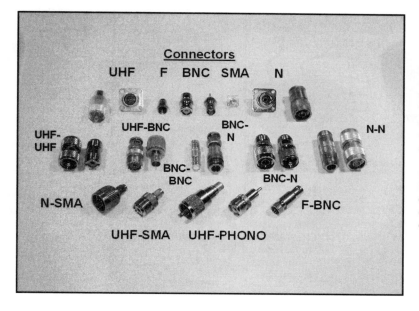

Figure 4.15 — The photo shows a variety of common coaxial connectors that hams use. The larger connectors are used for higher power transmitters and antennas. The most common are the UHF and N styles. Special *adapters* are used to make connections between cables and equipment that have different styles of connectors.

unsuitable at frequencies above HF. Pigtails are also difficult to seal and expose the inside of the cable to water which increases feed line loss.

Figure 4.15 shows several common types of coaxial connectors. The figure also shows *adapters* that make connections from one type of connector to another. Which connector to use depends on the frequency of the signals being used. The UHF series of connectors — also called PL259s — are the most widely-used for HF equipment. (UHF does not stand for "ultra-high frequency" in this case.) [T9B07] Above 400 MHz, the Type N connectors are used. [T9B06] You'll find both UHF and N connectors on 6, 2 and 1¼ meter equipment.

Coax connectors exposed to the weather must be carefully waterproofed. Water in coaxial cable degrades the effectiveness of the braided shield and dramatically increases losses. [T9B08] If you use low-loss air-core or "open-cell foam" coax, you will need to pay extra attention to waterproofing the connectors because special techniques are required to prevent water absorption by this cable. [T7C11]

Complete information on common coax connectors, including assembly instructions, is available in any edition of the *ARRL Handbook* and *ARRL Antenna Book*.

Melting Metal — In a Good Way

As you assemble your antenna system, if you don't know how already this is the time to learn the basics of soldering. Some coax connectors can be installed without soldering by crimping or compression fittings, but many cannot. Learning how to install your own coax connectors not only saves money but allows you to make repairs at home and under emergency conditions! Start by reading the ARRL Technical Information Service's online article "The Art of Soldering." (**www.arrl.org/circuit-construction**) Follow up with "Connectors For All Occasions, Parts 1 and 2". You'll learn what kind of solder to use for electronics (rosin-core), what a "cold" solder joint looks like (it has a grainy or dull surface), and many other useful tips to get you melting metal like a pro! [T7D08, T7D09]

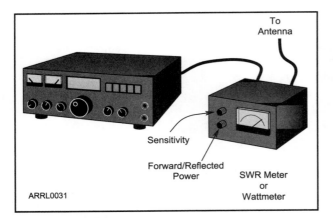

ARRL0031

Sensitivity

Forward/Reflected
Power

SWR Meter
or
Wattmeter

To
Antenna

Figure 4.16 — The SWR meter measures power flowing toward the antenna (forward) and toward the transmitter (reflected or reverse). The meter is calibrated to show SWR or power. A wattmeter does not measure SWR, just power, and the SWR can be calculated from the power readings.

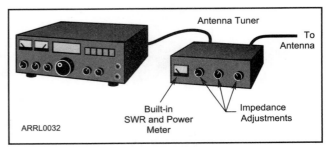

ARRL0032

Antenna Tuner

To
Antenna

Built-in
SWR and Power
Meter

Impedance
Adjustments

Figure 4.17 — An antenna tuner acts like an electrical version of a mechanical gearbox. By adjusting the tuner's controls, the impedance present at the end of the feed line can be converted to the impedance that best suits the transceiver's output circuits, usually 50 ohms.

SWR Meters and Wattmeters

To measure the SWR in a feed line the *SWR meter* is used. This device is placed in the feed line, usually right at the output of the radio as shown in **Figure 4.16**, and senses the power flowing in each direction to compute SWR. [T4A05] This makes it easy to see exactly what SWR is being presented to the radio by the antenna system and to keep an eye out for high SWR due to mistuned antennas. High SWR can also be caused by loose connections or connectors, faulty feed line, or even a faulty antenna. Having an SWR meter "in the line" will help you discover and fix those problems. Many radios include a built-in SWR meter.

Instead of SWR meters, many amateurs prefer a *wattmeter* and better yet, a *directional wattmeter*. Wattmeters measure power in a feed line and can be placed in the line to read power flowing in either direction. Directional wattmeters can measure power flowing toward the antenna and power reflected from the antenna by rotating a sensing element or turning a switch. The operator can then convert the forward and reflected power readings to SWR by using a table or formula. [T7C08]

Antenna Tuners

If the SWR at the end of the feed line is too high for the radio to operate properly, devices called *impedance matchers* or *transmatches* or *antenna tuners* are connected at the output of the transmitter as shown in **Figure 4.17**. An antenna tuner is adjusted until the SWR measured at the transmitter output is acceptably close to 1:1. This means the antenna system's impedance has been matched to that of the transmitter output. [T9B04] Think of the antenna tuner as an electrical gearbox that lets the engine (the transmitter) run at the speed it likes no matter how fast the tires are turning (feed point impedance).

Note that the antenna is not really tuned — it's the impedance at the output of the feed line that is converted to some other value. This allows the transmitter to deliver full power output without damage from the high SWR. For convenience, most tuners combine the functions of impedance matcher, directional wattmeter and antenna switch. There are also *automatic tuners* that sense when SWR is high and make the necessary adjustments under the control of a microprocessor to match the impedances.

Figure 4.18 — The popular MFJ series of antenna analyzers are used to adjust and troubleshoot antenna systems. The instrument contains a low-power signal source with an adjustable frequency and an SWR meter. The LCD display shows the operating frequency and information about the antenna impedance. The meters show SWR and feed point impedance.

Antenna Analyzers

Figure 4.18 shows an *antenna analyzer* consisting of a very low-power signal source with an adjustable frequency and one or more meters to show the impedance and SWR. It is used to measure an antenna system without using a transmitter whose signal might cause interference. [T7C02] It can be connected to a feed line in place of an SWR meter or directly to an antenna. Antenna analyzers are very handy and can be used for many types of antenna and feed line measurements, such as determining the frequency at which an antenna is resonant.

Before you go on, study these Technician exam questions from the question pool included at the back of this book or as a downloadable Study Guide version on the web:
T4A05   T7C02   T7C08 through T7C11   T7D08   T7D09
T9B04   T9B06   T9B07   T9B08   T9B10
If you have difficulty with any question, review the preceding section.

Chapter 5

Amateur Radio Equipment

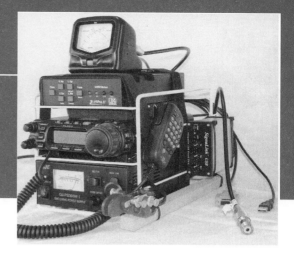

In this chapter, you'll learn about:

- **Basic operation of transmitters and receivers**
- **Special features of handheld transceivers**
- **Power supplies and batteries**
- **Digital mode basics and setup**
- **RF interference — symptoms and cures**
- **RF grounding**

When you see the mouse, you'll find more information at **www.arrl.org/ham-radio-license-manual**

Before we begin this chapter, it's important to mention the one piece of equipment that applies to just about every part of Amateur Radio — the computer! You may use a computer as much or as little as you choose, of course. However, just about any function or capability of the ham radio experience has the potential to involve a computer: bookkeeping chores such as logging contacts, operating on the digital modes, and even sending and receiving CW! [T4A02]

5.1 Transmitters and Receivers

In the previous chapters you have become acquainted with the basic equipment used by hams and the fundamentals of electronics and radio ideas that make the equipment go. We are now ready to start learning about Real Ham Radio, where knobs and dials get turned, meters jump and signals crackle back and forth over the airwaves!

By the end of this chapter, you'll know your way around common radio controls and their functions. Figures and photos will show examples of controls used for these functions. Your radio will be different, of course, so rely on the owner's manual for complete information. You can learn more about radios from this book's supplement "Choosing a Ham Radio" by or downloading user's manuals from the manufacturers' websites.

You'll get a look at how antennas, feed lines and power supplies are used by the typical amateur. We'll also cover the basics of dealing with interference. Knowing these practical techniques will get you started on the right foot as a ham. It also simplifies learning about operating conventions, rules and regulations that follow in upcoming chapters. Ready? Let's find that power switch and get to it!

Figure 5.1 — The frequency and mode controls are shown for a pair of HF radios, the Icom IC-718 (left) and the Ten-Tec Eagle (right). Frequency within a band can be adjusted with the large tuning knob or by the numeric keypad. If a computer is connected to the radio, frequency may be changed by the computer software. Pressing the MODE key switches between CW, USB and LSB and other available modes.

SELECTING BAND, FREQUENCY AND MODE

Regardless of whether you use a transceiver (most likely) or a separate receiver and transmitter, there are two functions that are the same for all three — control of frequency and mode. Amateur Radio is somewhat unique among the other radio services. Amateurs can tune anywhere in their assigned bands and are not required to use channels pre-assigned by the FCC, except in the 60 meter band. Repeaters operate on fixed channels, but hams developed that system on their own.

Amateurs can also use many different modes, a few of which we covered in a previous chapter. Most other radio services are restricted to a single mode. **Figure 5.1** shows the typical frequency and mode controls of two common transceivers. If the radio is a transceiver, frequency and mode settings will apply to both transmit and receive. With the Amateur service's flexibility and freedom comes the obligation to know how to choose and set a frequency and select a mode. It is the operator that controls the frequency of the transmitted and received signals.

Start by selecting the band if your radio operates on multiple bands. For example, you might have a *dual-band* handheld transceiver for the 2 meter and 70 cm bands or a *multiband* HF rig that covers 160 through 10 meters. The band selection controls may be labeled in terms of frequency or wavelength. **Table 5.1** shows how the labels on your radio may indicate the band.

Then a frequency is selected within the band — this is called *tuning*. The control you use for tuning is the *VFO*, an abbreviation for *variable frequency oscillator*. In older radios, the VFO knob changes the resonant frequency of an oscillator circuit that in turn determines the radio's operating frequency. In modern radios, the VFO control is read by a microprocessor that controls the radio's frequency.

Remember that it's not legal to use radios sold for Amateur Radio to transmit on non-amateur frequencies. Your radio may have the ability to listen outside the ham bands, but aside from the special Military Auxiliary Radio System (MARS) frequencies, you can't transmit except in an emergency.

Some radios also have a keypad that you can use to enter frequencies directly. [T4B02] The keypad will be on the radio's front panel or on the microphone. Numeric keys are used to enter the exact frequency. The radio's manual will show you how to use the keypad.

Table 5.1
Band Selection Labels

Frequency (MHz)	Wavelength (meters or cm)
HF	
1.8	160 meters
3.5	80
5	60
7	40
10	30
14	20
18	17
21	15
24	12
28	10
VHF	
50	6 meters
144	2
222	1.25
UHF	
440	70 cm
902	33
1240	23
2300	13

On *multimode* radios, the operator also selects the signal mode:

- AM or SSB (USB or LSB)
- FM
- CW or Morse code
- Data (for RTTY or other digital modes)

Radios that cover HF and the all-band radios that cover HF/VHF/UHF are multimode. Most handheld and mobile transceivers designed to be used with repeaters are single-mode radios that use only FM.

Memories or *memory channels* are used to store frequencies and modes for later recall. Memories are provided so that you can quickly tune to frequently used or favorite frequencies. [T4B04] Depending on the type of radio, the memory channels may also store information such as power level and repeater access tones. Memories on VHF/UHF FM radios are often labeled as channels and store a pair of frequencies for use with repeaters. (Repeater operation is described in detail in the next chapter.) With dozens of memories available on modern radios, the frequencies or repeaters you use are easy to access.

Before you go on, study these Technician exam questions from the question pool included at the back of this book or as a downloadable Study Guide version on the web:

T4A02   T4B02   T4B04

If you have difficulty with any question, review the preceding section.

TRANSMITTER FUNCTIONS

In radios that operate using AM/SSB and CW, the transmitter output power is controlled by an RF power control. A typical RF power control, usually a knob, is shown in **Figure 5.3**. (This may be a menu option or a knob shared by several features.) FM transmitters, such as handheld and mobile radios, have the operator select from a set of fixed power levels.

Control and Function Keys

With all the amazing functions and flexibility of today's radios, manufacturers have had to strike a balance between having separate controls for each function and having a reasonable number of buttons and keys. This means many of the keys on your radio do double or even triple duty! If you look closely, you may see several labels in different colors or styles on a single key, each indicating a separate controlling action when the key is pressed.

To distinguish between the different actions, control or function keys are used, such as the MON-F key shown on the VX-7R handheld radio in **Figure 5.2**. (F stands for "function," referring to the key's alternate use). When the MON-F key is pressed, the numeric keys all take on different meanings. For example, pressing the MON-F key changes the key usually reserved for the number 0 to a key that activates the receiver squelch setting function.

Some radios display menus of choices on a screen. The menu system is often activated with a MENU key. Inside the menu system, the radio keys have new functions to navigate the menu and select or adjust the various parameters and settings. Both keys and menus can be confusing, but reading the operator manual and a little practice will soon make it second-nature to use the most common functions.

Figure 5.2 — Because front-panel space is limited, some buttons and keys can perform more than one function. A control or function key causes the alternate functions to be used. On the VX-7R handheld a single MON-F key controls both alternate functions and writing information into the radio's memories.

Figure 5.3 — The maximum output power of HF transceivers is set by an RF PWR knob. This control does not affect MIC GAIN or other audio settings. On mobile rigs, output power is varied between several preset levels with a button or key.

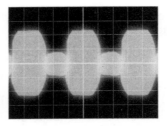

Figure 5.4 — The MIC GAIN settings can have a big effect on your signal. If the MIC GAIN is too low, the signal is undermodulated and will sound weak. Too much MIC GAIN results in the distorted signal shown here. Notice that the peaks are flattened or rounded. The signal will likely cause interference on adjacent frequencies.

A second control that affects transmitter output power on AM and SSB transmitters is *microphone gain* (MIC GAIN or MIC as seen in Figure 5.3). Microphone gain controls the level of speech audio that is applied to the modulator circuit of the transmitter.

For AM and SSB transmitters, more modulation generally means more power in the output signal sidebands and a louder received signal. Too much speech audio (**Figure 5.4**) causes signal distortion and interference to others. Microphone gain has no effect on CW transmissions. On FM transmitters, the microphone gain setting is fixed and output power doesn't change.

Most radios also have a *speech compressor* or *speech processor* control. These circuits increase the average power of the transmitted signal by applying more amplification to the weak parts of the speech signal. While this creates a small amount of distortion, it improves reception when conditions are noisy or the received signal is weak. Compressors and processors should be used with care. Too much *compression* or *processing* creates severe distortion and can create interference on nearby frequencies.

Peak envelope power (PEP) is the measure of an AM or SSB signal's power. PEP is the transmitter output power when speech into the microphone is the loudest. CW and FM transmissions have a constant power output, so PEP is equal to that constant level.

To avoid interfering with other stations while you're adjusting your transmitter or measuring its output power, it's a good idea to use a *dummy load*. A dummy load is a heavy-duty resistor that can absorb and dissipate the output power from a transmitter. [T7C01, T7C13]

Spurious Signals

Excessive modulation for all types of speech results in distortion of transmitted speech and unwanted or *spurious* transmitter outputs on adjacent frequencies where they cause interference. Those unwanted transmitter outputs have lots of names, but the most common is *splatter*. Generating those outputs is called *splattering*, as in, "You're splattering 10 kHz away!"

An overmodulated FM signal has excessive deviation and is said to be *overdeviating*. Overdeviation is usually caused by speaking too loudly into the microphone and may cause interference on adjacent channels. An FM transmitter can also be misadjusted internally to overdeviate at normal speech levels. To reduce overdeviation, speak more softly or move the microphone farther from your mouth. [T7B01]

Overmodulation of an AM or SSB signal as shown in Figure 5.4 is caused by speaking too loudly or by setting the microphone gain or speech compression too high, possibly resulting in distortion of the transmitted signal. [T4B01] To eliminate overmodulation, speak more softly or reduce microphone gain or speech compression.

To help prevent overmodulating, the *automatic level control* (ALC) circuit automatically reduces output power as the transmitter's limit is reached. ALC is not a foolproof remedy, but will help keep your signal *clean* — that is, free of spurious outputs. When using digital modes, however, the ALC system should never activate. Your radio's operating manual

(A)

(B)

(C)

Figure 5.5 — Several common types of microphones are shown in the photos. A hand mike shown in (A) is usually provided with the radio (in this case a TM-281A by Kenwood). Desk mikes shown in (B) are popular at home stations with operators that like to have extended conversations, or ragchews. Contest and public service communications operators prefer headsets as shown in (C) because they allow hands-free operation.

will show you how to operate your radio's controls to avoid overmodulating.

A signal that sounds overmodulated could also be the victim of *RF feedback*. In other words, signals from your own transceiver could be picked up by the microphone input circuit and distort the speech signal.

CW transmissions can cause interference by generating *key clicks*. Key clicks are sharp, brief sounds heard on frequencies near the offending signal. They are generated by a transmitter turning its output signal on and off too rapidly as it forms the dots and dashes. Key clicks can often be reduced by adjusting the transmitter's settings that control how it generates CW signals.

Microphones and Keys

The choice of microphone or key is very personal. When you buy your radio, a perfectly good microphone will usually be included. As you become experienced, you'll develop preferences for a certain style of operating and that includes the microphone and key. **Figures 5.5** and **5.6** show just a few of the many styles of microphones and keys.

Hand mikes, held in the hand during use, are included with all mobile and many home station transceivers. Hand mikes may also

Figure 5.6 — Most operators begin sending Morse with a straight key (top). The key is attached to a table or board so that it doesn't move as the operator "pounds brass." When you've become more skilled and your sending speed is higher, you may want to switch to a paddle (bottom) used with a keyer. The paddle is like two keys back-to-back; one causes the keyer to send dots and the other dashes. (Top photo courtesy of Wiley Publishing, *Ham Radio for Dummies*)

allow a handheld radio to be used while clipped to a belt or pocket. This keeps the antenna and the radiated signal away from your face and head — convenient and safe. Hand mikes may also feature operating and tuning controls. *Desk mikes* sit on the operating table so that the operator does not have to hold them.

Switching between receive and transmit on voice can be performed manually with a *push-to-talk* (PTT) switch or an automatic *voice-operated transmitter* control circuit (VOX). [T7A07] *Headsets* or *boomsets* combine a microphone with headphones so the operator can use the transmitter with hands free. A *boom mike* is held in front of the face by a thin support (the *boom*) attached to the headphones. Boom mikes are particularly useful during contests or other long periods of operating. Boom mikes make it easy to use VOX, but many operators use a *footswitch* to control the PTT function instead.

The microphone connector of a transceiver is likely to include push-to-talk connections and also supply voltage for powering electret-style microphones. [T4A01] The wiring of the connector varies for the different manufacturers, so check the connections before trying a new microphone.

For sending Morse code, a *key* is used to turn the transmitter output signal on and off. Morse code's dots and dashes are known as the *elements* of the code. When using a *straight key*, the operator generates the dots and dashes manually. This is called *hand keying*. The military surplus J-38 key is a common type of straight key and is popular with beginners.

Once you are skilled at "the code," you'll want to go faster by using a *keyer*. This electronic device turns contact closures by a Morse *paddle* into a stream of Morse code elements. A keyer may be a standalone device or it can be built into a transceiver. A *paddle* is a pair of levers mounted side by side, each having its own set of contacts, one for dots and one for dashes. Keyers and paddles can generate Morse code much faster than by using a straight key. Computers software can send CW by using a *keying interface* connected to the key input of a radio. [T8D10]

Before you go on, study these Technician exam questions from the question pool included at the back of this book or as a downloadable Study Guide version on the web:
T4A01   T4B01
T7A07   T7B01   T7C01   T7C13
T8D10
If you have difficulty with any question, review the preceding section.

RECEIVER FUNCTIONS

The receiver has a difficult task — picking just the one signal you want out of all the other signals. Nevertheless, modern receivers do a great job as long as they're adjusted properly. Knowing how a receiver's controls work makes a big difference, as you will see when you start tuning the bands! **Figure 5.7** shows where you might find receiver controls on a typical transceiver.

The most familiar control is the *AF gain* or *volume* control. Just like on a home or car radio, this sets the speaker or headphone listening level. On an HF rig, the *RF gain* control will be nearby. RF gain adjusts the sensitivity of the receiver to incoming signals. *Attenuators* are used to reduce the strength of signals at the receiver input. They are used when excessively strong signals overload the receiver. Using the RF gain and attenuator controls can remove a lot of distortion when the band is full of strong signals — give them a try!

A receiver's *automatic gain control (AGC)* circuitry constantly adjusts the receiver's sensitivity to keep the output volume relatively constant for both weak and strong signals.

ATT

AF

CLAR

SHIFT

Figure 5.7 — The front panel of this Yaesu FT-840 transceiver has most of the receiver controls you'll use every day. AF GAIN (labeled AF) controls volume. CLAR (RIT) and SHIFT vary the receiver frequency without changing your transmit frequency. ATT adds attenuation to reduce the strength of very strong signals.

[T4B12] AGC can respond quickly or slowly to changes in signal strength. A *fast AGC* response is used for CW and data. *Slow AGC* response is used for AM and SSB signals. FM receivers don't use AGC.

To keep from having to listen to continuous noise when no signal is present, the *squelch* circuit was invented. The squelch circuit (sometimes called *carrier squelch*) mutes the receiver's audio output when no signal is present. [T2B03, T4B03] The *squelch threshold*, controlled by the squelch control, is the signal level at which muting is turned off and the signal becomes audible. If the receiver's output is not muted, squelch is *open*. If the signal is muted, squelch is *closed*. Raising the squelch threshold is called *tightening the squelch*.

A receiver rejects unwanted signals through the use of *band-pass filters*. Every receiver passes radio signals through a sequence of increasingly narrow filters. Right at the receiver input, a band-pass filter passes only signals from the selected band. Further into the receiver, the signal passes through *intermediate frequency* or *IF filters* narrow enough to reject signals on the same band as the desired signal. IF filters are specified by bandwidth in Hz or kHz. Wide filters (around 2.4 kHz) are used for SSB reception on phone. [T4B09] Narrow filters (around 500 Hz) are used for Morse code and data mode reception. [T4B10] Having multiple filters available allows you to reduce noise or interference by selecting a filter with just enough bandwidth to pass the desired signal. [T4B08]

Receivers have other weapons against noise and interference. A *notch filter* removes a very narrow range of frequencies from a receiver's audio output. This is useful when an interfering tone is encountered. The tone is removed, leaving the desired speech, code, or data relatively unaffected. Receivers may also feature *noise reduction* that removes audio noise by using digital signal processing.

Receiver incremental tuning (RIT) is a fine-tuning control used for SSB or CW operation. RIT allows the operator to adjust the receiver frequency without changing the transmitter frequency. [T4B07] This allows you to tune in a station that is slightly off frequency or to adjust the pitch of an operator's voice that seems too high or low. [T4B06] On some radios, the RIT is called a *clarifier* which is labeled CLAR.

There are lots of types of noise in the radio spectrum. Special circuits are employed to get rid of noises or at least limit their effect. A *noise blanker* senses the sharp buzz of pulses from arcing power lines, motors, or vehicle ignition systems and temporarily mutes the receiver during the pulse. [T4B05] A *noise limiter* doesn't try to remove the noise, it just *clamps* the audio signal at a controllable level to protect your ears. This is particularly useful on HF when storms are nearby, generating powerful static crashes.

Receivers usually have a way to indicate signal strength, such as a meter with a moving needle or a variable-length bar at the side or bottom of the display. On the rig in Figure 5.7, the meter shows signal strength in *S units*. Although not strictly calibrated, a change of one S unit corresponds to about a factor of four in signal strength. S units are numbered

Digital Signal Processing (DSP) — Radios with Software

DSP is the technique of using a special-purpose microprocessor in a radio to perform filtering and other functions on the received signal. Common DSP functions include:

- Noise reduction
- Variable signal filtering
- Automatic notch filtering
- Audio response tailoring on both receive and transmit

DSP operates on a received signal converted to digital form. A microprocessor performs the filtering and other functions on the digital data. Finally, the digitally modified data is converted back into an IF or audio signal and output through the speaker or headphones. DSP is making rapid advances and is a key part of most newer radios.

from S-1 to S-9, with S-9 being the strongest. The strength of signals stronger than S-9 is reported as the number of dB (decibels) greater than S-9. For example, "Your signal is 20 dB over S-9!"

HANDHELD TRANSCEIVERS

Handheld transceivers or *handhelds* are incredibly popular and offer a variety of useful features. Because they can be carried by the operator and used while in motion, they're particularly useful for disaster or public service communications. Even if you purchase a mobile rig, having a handheld radio for use away from the car is a good idea.

Dual-band handhelds usually cover 2 meters and 70 cm. Multiband handhelds add coverage of 6 or 1.25 meters or even 23 cm. Many handhelds also feature the ability to receive frequencies for other services, such as public safety, aircraft or broadcast stations. This is called *extended* or *wide-band receive*. Check the radio's specifications for the exact frequency coverage.

Using the handheld's built-in microphone and speaker requires the operator to hold the radio close to his or her mouth and ear. This may not be a problem, but occasionally it is inconvenient. One solution is to use a *speaker-mike* and use the belt clip supplied with the radio to hold the radio on a belt or in your pocket. The speaker-mike can then be clipped onto a shirt or jacket where you can hear the radio and use the microphone without also having to hold the whole radio. Headsets are also available for many rigs and provide completely hands-free operation. The headset will have to provide its own VOX circuit to activate the radio's PTT circuit, unless the radio has VOX built-in.

If you find that your handheld radio has difficulty accessing distant repeaters or making simplex contacts, an RF power amplifier can be used to increase the output power by a factor of five or more. [T7A10] To use an amplifier, you'll also need to replace the flexible rubber duck antenna with an external antenna rated for the higher power.

Before you go on, study these Technician exam questions from the question pool included at the back of this book or as a downloadable Study Guide version on the web:
T2B03
T4B03   T4B05 to T4B10   T4B12
T7A10
If you have difficulty with any question, review the preceding section.

5.2 Digital Communications

Amateurs have developed or adapted techniques for exchanging digital data by transforming the 1s and 0s of data into tones that are in the same frequency range as the human voice. Radios designed for voice transmission can then transmit and receive the tones as either AM or FM signals. *Digital* or *data modes* combine modulation (the addition of information to a radio signal) with a *protocol*, the rules by which data is packaged and exchanged. The protocol also controls how the transmitter and receiver coordinate the exchange of data. The method used to represent each character as digital 1s and 0s is called a *code* and yes, Morse is a digital code!

Why use digital modes, anyway? Voice and CW are quite effective but they don't have the ability of digital systems to automatically correct errors caused by noise and interference. Special codes and characters embedded in the stream of data allow the receiving modem to detect, and sometimes correct, errors. The result is that some digital modes offer error-free communications at speeds that adjust automatically to propagation and noise. Protocol design is an area in which amateur experimentation is definitely advancing the state of the radio art.

A *modem*, an abbreviation for "modulator-demodulator," changes data signals to and from audio signals. Modems may be standalone devices built into a radio or they can combine computer software and a sound card. If a microprocessor that performs the protocol rules is combined with a modem, the result is a *terminal node controller* (TNC). A TNC that can perform several protocols is called a *multiple protocol controller* (MPC). A modem, TNC or MPC is installed between a computer and the radio.

Software modems or software TNCs use a sound card to do the work of the TNC. The power of sound cards now found in every computer has enabled amateur protocol developers to experiment and develop many new modes, such as PSK31 and MFSK. [T8D01]

AMATEUR DIGITAL MODES

Different combinations of protocols, codes and modulation methods, such as SSB or FM, are used to create *digital modes* that work best in particular frequency ranges. The combination determines the speed and reliability of communications. Frequency range (HF or VHF/UHF) affects the characteristics of radio communication because propagation changes with frequency. On HF, where SSB modulation is the norm, popular digital modes include:

- Radioteletype (RTTY) that uses the 5-bit Baudot code
- Winlink 2000 using the PACTOR or WINMOR protocols
- Keyboard-to-keyboard modes such as PSK31 or MFSK

On VHF/UHF, FM is the standard and popular modes include:
- Packet radio that uses the AX.25 protocol
- Winlink 2000 using the B2F code so that e-mail can be exchanged via packet radio

Even CW is a type of digital mode that uses the Morse code and AM! [T8D09]

What is That Signal?

With so many different digital modes active on the air, it can be difficult to tell what type of signal you're listening to. After a while, you'll find that it's easy learn to recognize the most common modes by ear or even by eye on a band scope or waterfall display. In the meantime, a collection of sound samples from several popular modes is available online at **www.arrl.org/resources-for-license-instruction** under "Digital Modes: Audio Files."

Errors in Digital Data

Just changing the data to tones is not enough for reliable communications. A radio signal experiences many disruptions between the transmitter and receiver. Fading, interference and noise often introduce errors into the stream of data, measured by the *bit error rate* (BER). These disruptions would cause errors in the received information if no precautions were taken.

Some codes include extra data elements to allow the receiver to detect an error. For example, some codes include a *parity bit* can be used to detect simple errors in a single character of data. Some protocols add special data so that common errors can be detected and sometimes even corrected.

Packet and Packet Networks

On VHF and UHF, the most common digital mode is *packet radio*. Packet signals are often found on simplex channels from 145.01 to 145.09 MHz. Information on packet radio can be found in the *ARRL Handbook* and the manuals for TNC data interfaces.

Packets are transmitted in noise-like bursts. *Frequency-shift keying* (FSK) is used to transmit the individual characters as a series of rapidly alternating audio tones. A receiving modem and terminal node controller (TNC) then reassemble the data from the received packets. The characters in a packet are transmitted at 1200 or 9600 baud, so that the overall *throughput* of a packet system is about 400 or 3000 bits per second.

Each packet consists of a *header* and *data*. The header contains information about the packet and the call sign of the destination station. The header also includes a *checksum* that allows the receiver to detect errors. If an error is detected, the receiver automatically requests that the packet be retransmitted until the data is received properly. This is called ARQ for *automatic repeat request*. [T8D08, T8D11] The rules of the packet radio protocol, AX.25, ensure that data accepted by the receiver is error-free.

Individual packet stations can *connect* to each other directly, with the operators typing messages to each other, similar to text messaging on a mobile phone. *Node* stations act as routing centers for packet connections. Nodes are also connected to other nodes, forming networks that allow communications between individual stations through the network.

Stations can also use relays called *digipeaters* to connect to out-of-range stations. A digipeater stores the data in each packet and retransmits it. Packets can be forwarded by nodes or individual stations as well. If the operator has a list of packet stations and their locations, the packets can be forwarded from station to station over long distances.

Before you go on, study these Technician exam questions from the question pool included at the back of this book or as a downloadable Study Guide version on the web:

T8D01   T8D08   T8D09   T8D11

If you have difficulty with any question, review the preceding section.

Keyboard-to-Keyboard

Digital modes that are designed for real-time person-to-person communication are called *keyboard-to-keyboard* modes. Most popular on the HF bands, keyboard-to-keyboard mode signals are found at frequencies just above CW signals.

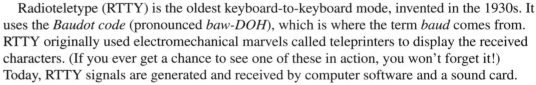

Radioteletype (RTTY) is the oldest keyboard-to-keyboard mode, invented in the 1930s. It uses the *Baudot code* (pronounced *baw-DOH*), which is where the term *baud* comes from. RTTY originally used electromechanical marvels called teleprinters to display the received characters. (If you ever get a chance to see one of these in action, you won't forget it!) Today, RTTY signals are generated and received by computer software and a sound card.

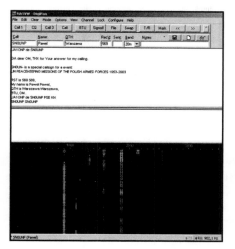

Figure 5.8 — PSK31 is a popular keyboard-to-keyboard mode on HF. Many amateurs use the *DigiPan* software shown here. At the top is a window that shows the text decoded by the modem software, the bottom shows a *waterfall display* that is a visual presentation of what the computer's sound card is receiving from the radio.

The most popular keyboard-to-keyboard mode today is PSK31, which stands for *phase shift keying, 31 baud*. [T8D06] PSK31 uses very precise signal timing to aid the receiving modem in recovering the signal from noise and interference. Although it sends data at a low rate, it works very well in noisy conditions. [T8D07] Best of all, the software to use PSK31 is free and runs on most computers. The popular *DigiPan* PSK31 software is shown in **Figure 5.8** and software packages such as *FLDIGI* by W1HKJ support not only PSK31 but many other modes, as well.

APRS

The *Automatic Packet Reporting System* (APRS) uses packet radio to transmit the position information from a moving or portable station to a system of APRS digipeaters. [T8D02] These relay points forward the position information and call sign to a system of server computers via the Internet. Once the information is stored on the servers, websites can access the data and show the position of the station on maps in various ways. A typical APRS tracking map is shown in **Figure 5.9**. A common public service application of APRS is to provide maps of station locations while they are providing real-time tactical communications. [T8D05]

A portable APRS station is basically a packet radio station with a *Global Positioning System* (GPS) receiver attached to the TNC. [T8D03] The GPS receiver outputs a stream of position data in standard format which is interpreted by the suitably-equipped TNC and transmitted in packet form. Along with position information, some stations also transmit weather information.

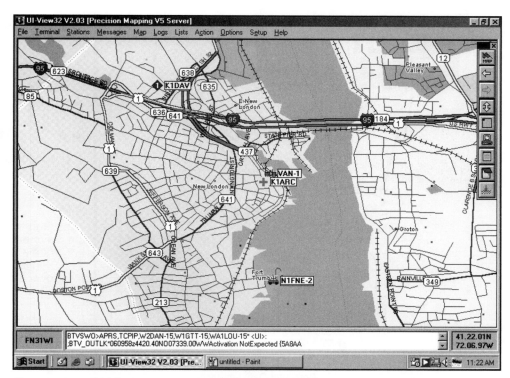

Figure 5.9 — APRS, the Automatic Packet Reporting System, uses packet radio signals to display the positions of fixed and mobile stations on a computer-generated map.

Winlink

Another Amateur Radio innovation gaining considerable popularity for public service communications and traveling hams is the Winlink system that sends e-mail over the air waves. Winlink combines *RMS Express*, an e-mail program, with a digital mode such as PACTOR III or WINMOR on HF and packet radio on VHF/UHF.

An overview diagram of the Winlink system is shown in **Figure 5.10**. Traveling hams connect to a gateway called a *radio message server* (RMS) station that is in turn connected to e-mail servers via the Internet. When the traveler connects to an RMS, any e-mail waiting on the server can be retrieved, no matter where the traveler happens to be. It is common for hams at sea to connect to RMS stations thousands of miles away.

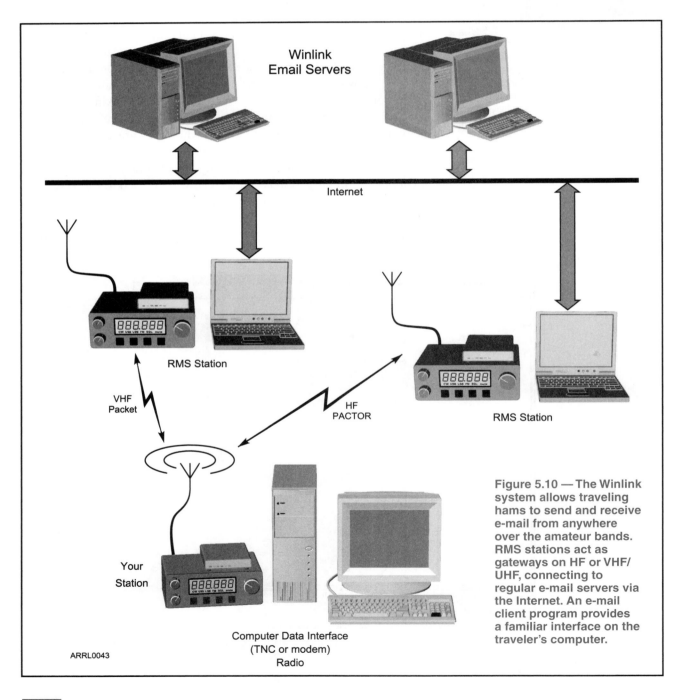

Figure 5.10 — The Winlink system allows traveling hams to send and receive e-mail from anywhere over the amateur bands. RMS stations act as gateways on HF or VHF/UHF, connecting to regular e-mail servers via the Internet. An e-mail client program provides a familiar interface on the traveler's computer.

ARRL0043

SETTING UP FOR DIGITAL MODES

Figure 5.11 shows a generic example of how a station is configured to use digital modes. [T4A06] A TNC is connected to one of the computer's digital data ports via an RS-232 (COM port) or USB interface. The TNC is then connected to the radio's microphone input and speaker or headphone output. If a sound card is used, its output is connected to the radio's microphone input and the speaker or headphone output is connected to the sound card input. [T4A07] If you use a sound card, you will also need a digital communications interface to electrically *isolate* the radio and computer. This prevents hum or RF feedback from interfering with the data signals.

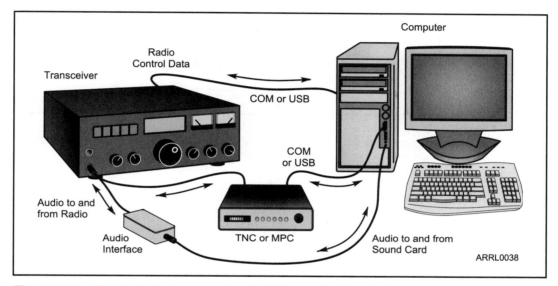

Figure 5.11 — Data interfaces are connected between the transceiver's audio inputs and outputs and the computer's data connections (USB or COM ports) or sound card jacks. A TNC or MPC (multi-protocol controller) converts between data and audio. An audio interface isolates the computer sound card from the radio to prevent hum.

To make sure your digital signal is transmitted and received correctly, you'll need to adjust or configure all of the following:
*   Transmit audio level — all of the same cautions apply as with voice operation about overmodulation caused by excessive signal strength at the transmitter input.
*   Receive audio level — the output from the receiver must be at the proper volume level for the data interface to turn it back into data. Levels that are too high distort the data tones and if too low, allow noise to be added. Both cause errors in the tone-to-data conversion.
*   Digital interface — if you are using a computer, you may need to configure the connection to the data interface so that the proper control signals are connected.
*   Transceiver control — turning the transmitter on and off at the right time may require a connection to the PTT (push-to-talk) input of the radio.

The operating manual supplied with your data interface or software will show you how to make all the necessary adjustments on the radio or computer. Once the adjustments are set properly, record the position of your audio and microphone gain controls and save the computer sound card settings. You can then return to digital operation quickly.

When using data modes, the microphone is not required because the transmitted audio is provided by the sound card or TNC. To return to voice operation, the data connection is replaced by the microphone either by physically swapping the connectors or using a microphone/data switch.

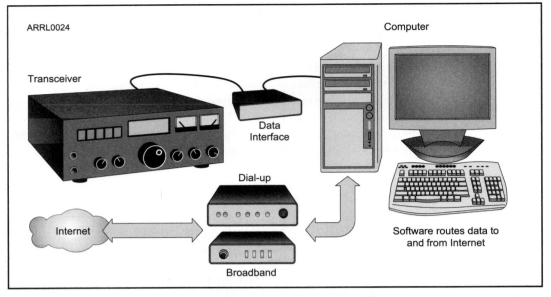

Figure 5.12 — An Internet gateway station acts like a regular digital mode station except that the computer runs software that relays data between the radio and the Internet. The most common example of gateway stations is a Winlink RMS station.

GATEWAYS

An *Internet gateway* shown in **Figure 5.12** is a special kind of digital station that provides a connection to the Internet for data transmitted by Amateur Radio from other stations. [T8C11] Most gateways are set up to *forward* messages. The most common examples are packet radio *bulletin board systems* (BBS) and the Winlink RMS stations described above. Messages with a recognized Internet e-mail address can be sent and retrieved over these systems.

Another type of gateway provides direct Internet connectivity so that a computer running standard web browser software can connect to any Internet address. The radio has a built-in data interface with an Ethernet connection to which the computer connects, just like a home network. This type of gateway is provided by radios using the D-STAR system.

Caution! All of the rules and regulations about commercial and business-related messages and communications apply to Internet gateways. For example, it is definitely not okay to exchange e-mails for your employer or to access websites with third-party advertising. Because so much of the Internet is associated with commercial activity, take extra care to comply with Amateur Radio operating rules.

Before you go on, study these Technician exam questions from the question pool included at the back of this book or as a downloadable Study Guide version on the web:
T4A06   T4A07
T8C11   T8D02   T8D03   T8D05   T8D06   T8D07
If you have difficulty with any question, review the preceding section.

5.3 Power Supplies and Batteries

The most exciting part of radio is, of course, radio signals, but a solid power source is just as important. You'll find that a proper source of power makes for a clean, noise-free signal and better reception. Using a power source whose output voltage is too high or too low or that cannot supply sufficient current can damage radio equipment or make it operate improperly.

POWER SUPPLIES

Power supplies that operate from household ac power are the most common source of power for radios. One is shown in **Figure 5.13**. They convert the ac input power to a smooth dc current that the radio can use. A power supply has two main ratings: its output voltage and the amount of current it can supply continuously. For example, a power supply for a typical mobile radio might be rated as a "12 V, 20 Amp" supply. Radios that operate from a "12 V" supply may actually work best at the slightly higher voltage of 13.8 V typical of vehicle power systems with the engine running. Check the equipment manual and be sure your power supply can generate the correct voltage. [T5A06]

Figure 5.13 — The traditional or "linear" power supply is a reliable, if heavy and bulky, performer. The modern "switching" supply shown here uses sophisticated electronics to avoid large iron transformers, delivering equivalent power at a fraction of the weight.

A supply's output voltage changes with the amount of output current. The percentage of voltage change between zero current (*no load*) and maximum current (*full load*) is the *regulation* of the supply. To achieve "tight" regulation, meaning little variation as current changes, requires a *regulator* circuit in the supply. [T6D05] These supplies are *regulated supplies* and have regulation of a few percent, the output voltage varying only a small amount with load. Regulated supplies are the best power source for radios because they prevent voltage fluctuations from affecting the functions of the radio's sensitive circuits. [T4A03]

The current rating of a supply must be at least as much as the sum of the maximum current needs of everything hooked up to the supply. The equipment's manual will tell you how much current is required. A single power supply can be shared between two or more pieces of equipment if it can supply enough current.

MOBILE POWER WIRING

Mobile installations have special requirements for obtaining power safely. Remember that a vehicle battery stores a lot of energy! Accidental short circuits can not only damage your radio equipment, they can start fires and do a lot of expensive damage to your car. General guidelines include:

- Fuse both the positive and negative leads of your radio near the power connection.
- Connect the radio's negative lead to the negative battery terminal or where the battery ground lead is connected to the vehicle body. [T4A11]
- Use grommets or sleeves to protect wiring from chafing or rubbing on exposed metal, especially where it passes through a bulkhead or firewall.
- Don't assume all metal is connected to battery ground — vehicle bodies are often a mix of plastic and metal.

Vehicle power wiring often carries a significant amount of noise that can affect your radio's operation. *Alternator whine* is caused by noise on the dc power system inside your own vehicle. You might hear it with the received audio but more likely it will be heard by others as a high-pitched whine on your transmitted audio that varies with your engine speed. It can be removed by a dc power filter at your radio. [T4A10, T4A12]

GENERATORS AND INVERTERS

For portable and disaster or public service operation, ac power from the utility grid might not be available. In situations like this, generators and inverters are used. A *generator* turns mechanical energy from an engine into electrical ac or dc power. An *inverter* turns dc power into ac power. The ac power from either can then be used to run a regular ac power supply or charge batteries.

Generators and inverters are rated in watts, the amount of power they can supply while still maintaining output voltage within the specified range. Rarely can either be used at full power continuously, so it is best to be conservative when choosing or using them.

Electronic equipment that runs from ac power generally expects its input current to be a relatively undistorted sine wave. A distorted output from an inverter can cause the equipment to overheat or even malfunction. If possible, use "sine wave output" inverters to run electronic equipment.

Voltage regulation is also important. The voltage from generators and inverters varies a lot more than voltage from the ac utility grid. Generators made for running tools or motors are usually less expensive, but are poorly regulated. For radio use, check the specifications to be sure the output voltage is acceptable both at no load and full load.

Before you go on, study these Technician exam questions from the question pool included at the back of this book or as a downloadable Study Guide version on the web:
T4A03   T4A10   T4A11   T4A12
T5A06
T6D05
If you have difficulty with any question, review the preceding section.

BATTERIES

Batteries supply dc power in place of power supplies for portable radios, emergency power, and other uses where ac power isn't available or practical. Batteries are made up of one or more *cells*. A cell is an individual package that contains chemicals to produce current from a chemical reaction. (The general term "battery" is often used to refer to single cells, as well.) The cell is constructed so that the chemical reaction can't occur until there is a path or circuit for electrons to flow between the cell terminals. When the chemicals are "used up" and the reaction stops, so does the current. The types of chemicals also determine the voltage of the cell.

The cells of a multiple-cell battery are connected in series so that the voltages from each cell add together. The common "9 volt" alkaline battery is actually six smaller 1.5-volt cells connected in series. An automobile's 12-volt battery is made of six cells, each a separate compartment holding the necessary chemicals to produce two volts.

There are many different types of batteries, but they fall into three basic groups:
- Disposable or Primary — the chemicals can only react once, then the battery must be discarded
- Rechargeable or Secondary — the chemical reaction can be reversed, recharging the battery

- Storage Batteries — rechargeable batteries used for long-term energy storage

The most common types and sizes of disposable and rechargeable batteries used by hams are listed in **Table 5.2**. [T6A10] The column labeled "Chemistry Type" describes the chemicals used in the battery and the "Fully-Charged Voltage" column represents the output of a fresh or recently-charged battery. [T6A11] The battery's *"Energy Rating"* in ampere-hours (Ah) or milliampere hours (mAh) specifies its ability to deliver current while still maintaining a steady output voltage. **Figure 5.14** shows several common types of batteries and their relative sizes.

Table 5.2
Battery Types and Characteristics

Battery Style	Chemistry Type	Fully-Charged Voltage	Energy Rating (average)
AAA	Alkaline — Disposable	1.5 V	1100 mAh
AA	Alkaline — Disposable	1.5 V	2600 – 3200 mAh
AA	Carbon-Zinc — Disposable	1.5 V	600 mAh
AA	Nickel-Cadmium (NiCd) — Rechargeable	1.2 V	700 mAh
AA	Nickel-Metal Hydride (NiMH) — Rechargeable	1.2 V	1500 – 2200 mAh
C	Alkaline — Disposable	1.5 V	7500 mAh
D	Alkaline — Disposable	1.5 V	14000 mAh
9V	Alkaline — Disposable	9 V	580 mAh
9V	Nickel-Cadmium (NiCd) — Rechargeable	9 V	110 mAh
9V	Nickel-Metal Hydride — Rechargeable	9 V	150 mAh
Coin Cells	Lithium — Disposable	3 – 3.3 V	25 – 1000 mAh
Packs	Lithium ion (Li-ion) — Rechargeable	3.3 – 3.6 V per cell	Varies

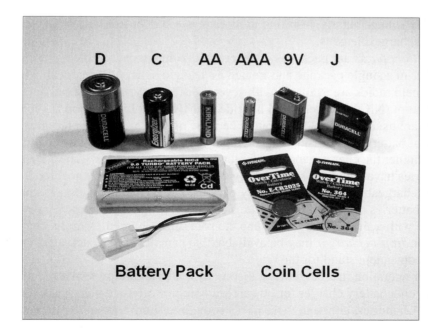

Figure 5.14 — The photo shows several common sizes of batteries. Coin cells are usually used in radios as a source of backup power for the microprocessor circuitry. Battery packs are packages of several cells in a single enclosure or case. The photo shows a battery pack used for remote control vehicles. (Courtesy Wiley Publishing, *Two-Way Radios and Scanners for Dummies*)

To get the most energy from a battery, limit the amount of current drawn from it. A low discharge rate keeps the battery cool inside and minimizes losses due to the battery's natural internal resistance. To maximize battery life and capacity, store them in a cool, dry place. You may refrigerate batteries, but never freeze them since the resulting ice may expand enough to crack the battery. Heat accelerates the battery chemical's tendency to *self-discharge* so that they can no longer deliver as much charge. Moisture allows charge to leak slowly between the battery's external terminals. Regularly inspect batteries for damage or leakage and perform an occasional maintenance charge as part of your battery plan.

When recharging batteries, be sure to use a *charger* designed for that particular type of battery. Each different battery chemistry and size requires a certain method of charging. Too much recharging current may damage the battery. Too little current may keep the battery from reaching full charge.

Storage batteries, such as deep-cycle marine or RV batteries, are often used as an emergency power source in place of a power supply operating from ac power. They can store hundreds of times the energy in a small battery. Storage batteries are often left connected to a charger that can keep them fully charged with a small current, called *trickle* or *float charging*. Be sure that the charger will switch to this lower current automatically or it can overcharge and ruin these expensive batteries. In an emergency, with no commercial power available, a 12-V lead-acid storage battery can be recharged by connecting it to a vehicle's battery and running the engine. [T2C02] Care must be taken not to overcharge a small battery or overload the charging system.

Storage batteries hold a lot of energy and must be treated with respect. They contain strong acids that can be hazardous if spilled or allowed to leak. Storage batteries can also release or *vent* flammable hydrogen gas, that can cause an explosion. [T0A09, T0A10] Be sure to store and charge these batteries in a well-ventilated place. Accidentally short-circuiting a storage battery with a tool or faulty wiring can easily cause a fire and damage the battery.

When you are done with a battery, don't just throw it in the garbage. The materials in most batteries are mildly toxic, so check at hardware stores or with your local government to see if there is a battery recycling program available.

BATTERIES FOR HANDHELD RADIOS

Handhelds use battery packs — groups of individual cells connected together to form a higher voltage battery, resulting in more power output from the radio. Most battery packs use rechargeable cells.

Battery packs are packages of several individual rechargeable batteries connected together in a single package and treated as a single battery. Rechargeable battery packs for handheld radios are available with several different types of internal batteries: nickel-cadmium (NiCd), nickel-metal hydride (NiMH), and Lithium-ion (Li-ion). (Lithium-ion cells are usually not sold separately because of special charging requirements.) For a given size of battery pack, Li-ion has the highest energy capacity, followed by NiMH and NiCd. The higher the energy capacity, the longer the battery pack will last. Disposable alkaline batteries have about the same energy capacity as NiMH rechargeables.

Rechargeable batteries or battery packs are convenient and less expensive than disposable batteries over time, but require a charger that operates from ac power. The radio will come with a simple charger to keep the pack fully charged. A more sophisticated *fast charger* or *drop-in charger* may be available that can charge the pack quicker and usually acts as a convenient stand for the radio.

For operation in emergencies, disposable batteries are preferred because they do not depend on a battery charger for power. Many radios offer a battery pack for disposable AAA or AA batteries, although these packs sometimes limit transmitter output power. If your interests include disaster communications or public service, be sure the radio you select offers this option. In any case, you should have at least one spare battery pack.

Before you go on, study these Technician exam questions from the question pool included at the back of this book or as a downloadable Study Guide version on the web:
T2C02
T6A10 T6A11
T0A09 T0A10
If you have difficulty with any question, review the preceding section.

5.4 RF Interference (RFI)

As more and more electronic devices and electrical appliances are put in use every day, interference between them and radios, called *radio frequency interference* (RFI), becomes more commonplace. RFI can occur in either direction — to or from the Amateur Radio equipment. Interference becomes more severe with higher power or closer spacing to the signal source. In this section, we'll cover the basic causes of RFI so that you know what to expect when you encounter it, how to react and some of the techniques to eliminate it. [T7B03, T7B07] The ARRL's Technical Information Service website provides information on all kinds of RFI and the means to correct it.

FILTERS

Filters are an important part of radio and nowhere are they as important as in preventing and eliminating RFI. Filters are used both to prevent unwanted signals from being radiated in the first place and to keep unwanted signals from being received. To select the correct type of filter, you have to know the nature of the interfering signals. **Figure 5.15** shows

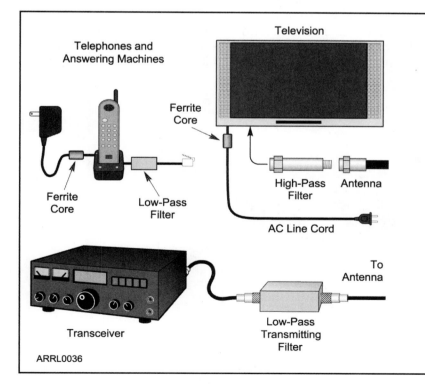

Figure 5.15 — Filters are used in a number of different ways to reduce interference. Telephone or answering machine interference is often cured with a ferrite core placed on the power supply cable and a low-pass filter on the phone line. Both prevent RF signals from getting to the phone's electronics. A high-pass filter in the antenna feed line often prevents strong amateur signals from entering the TV receiver and causing overload. Ferrite cores on the power cable keep RF from entering the TV by that route. A low-pass filter on the output of a transceiver prevents weak harmonics from being picked up by a radio or TV as interference. In difficult cases you may need a notch filter (not shown) tuned to the offending signal's frequency.

ARRL0036

how filters are applied for several common types of RFI.

AC power line filters are low-pass filters that keep RF signals from passing into or out of equipment via the ac power connection. They reject all signals with frequencies greater than a few kHz. Low-pass and high-pass *feed line filters* are installed in antenna feed lines to reject interfering RF signals above or below the desired signal, respectively.

RF choke or *common-mode* filters made of *ferrite* material are used to reduce RF currents flowing on unshielded wires such as speaker cables, ac power cords, and telephone modular cords. *Ferrite chokes* are also used to reduce RF current flowing on the outside of shielded audio, microphone, and computer cables. [T4A09]

DIRECT DETECTION

A device need not be a receiver to experience interference. Telephones, music players, touch-sensitive lamps and other electronics can be affected by a strong RF signal. If RF signals can gain entry to an electronic device, components such as transistors, diodes and ICs turn them into voltages and currents that affect the device's function, possibly upsetting its operation or distorting an audio signal. This is called *direct detection*. The symptoms of direct detection are thumps or pulses when a transmitter is turned on and off. A garbled voice might be heard during AM or SSB transmissions. Strong FM signals are

Ferrite — the RFI Buster

One of the most useful materials in dealing with RFI is the *ferrite core*. Ferrite is a ceramic magnetic material — you may have used ferrite magnets. The type of ferrite used for RFI suppression is specially designed to absorb RF energy over a broad frequency range, such as HF or VHF. Ferrite is available in many different *mixes* of slightly different composition that absorbs best in a particular range.

One popular form of ferrite is the snap-core shown in the figure. The actual ferrite is a rectangular block with a large hole in it, sawn in half. A plastic case with a snap holds the two pieces together. This allows cords or cables to be wound on the core even if they already have connectors attached, such as power cords or video cables.

Ferrite is available as round *cores* (toroids), rods and beads shown in **Figure 5.16**. Wires or cables are then wound on or passed through the ferrite forms. Beads are made large enough that they can be slipped over coaxial cables and secured with tape or a locking wire-tie. A wire or cable wound on such a ferrite core forms a choke filter.

Figure 5.16 — Ferrite is a ceramic magnetic material used to make choke filters for RFI suppression. It is available in many different forms: rings (toroids), rods and beads. Cables can be passed through or wound on these cores to prevent RF signals from flowing along their outside surfaces. (Courtesy Wiley Publishing, *Two-Way Radios and Scanners for Dummies*)

usually detected as hum. To eliminate RFI caused by direct detection, the RF signal must be prevented from entering the equipment.

Direct detection is the most common form of interference to telephones, since they are rarely designed to reject RF signals. Low-pass RF filters connected at the telephone's input connector are the best way to reduce RFI from direct detection. [T7B04]

OVERLOAD

Very strong signals may overwhelm a receiver's ability to reject them. This is called *fundamental overload*. Overload usually results in severe interference to all channels of a broadcast TV, AM, or FM receiver. Consumer equipment is often unable to reject strong signals outside the bands it is intended to receive. [T7B02] Similarly, an amateur may hear noise across an entire band when the strong signal is present. If adding attenuation (either by turning on a receiver's attenuator or removing an antenna) causes the interference to disappear, it's probably caused by overload.

A high-pass RF feed line filter can be connected at the antenna input of FM and TV receivers as shown in Figure 5.15 to reject strong lower-frequency signals from amateur HF and Citizen's Band stations. [T7B05] (Do not use feed line filters of any sort in a cable TV feed line.) A filter at the amateur's transmitter will not solve overload problems — the problem lies within the receiver. It is the receiver owner's responsibility to solve overload problems from a properly functioning transmitter.

If the interfering signal's frequency is close to the desired signal's frequency, it may not be possible to prevent overload with a high-pass or low-pass filter because the desired signal will be attenuated, as well. For example, when a TV receiver is overloaded by a nearby 2 meter transmitter, a notch or band-reject filter is required to attenuate the 2 meter signal without also filtering out over-the-air broadcast channels 2 through 13. [T7B12]

Both consumer and amateur receivers can experience overload from nearby broadcast stations. *Broadcast-reject* filters attenuate signals from nearby AM, FM or TV broadcast stations. The type of signal to be rejected must be specified when purchasing the filter since those stations transmit on different frequencies.

HARMONICS AND SPURIOUS EMISSIONS

Due to minor imperfections, every transmitter's RF output signal contains weak harmonics of the desired output signal and other spurious emissions that can cause interference to nearby equipment. In extreme cases, a misadjusted or defective transmitter can generate strong interfering signals. To prevent harmonics from being radiated by your station, a low-pass or band-pass filter must be installed at the transmitter's connection to the antenna feed line as shown in Figure 5.15. [T4A04]

As a matter of good practice, amateur HF stations often use a low-pass filter to keep any VHF harmonics generated by the transmitter from reaching the antenna. Even if the transmitter is completely within FCC rules, a nearby TV receiver could still pick up the VHF signal and experience interference. Harmonics that cause interference cannot be filtered out at the receiver because they are on the same frequency as the desired signal. This is called *in-band interference*. Remember that filters to be used in a feed line carrying signals from a transmitter must be rated to carry the full transmitter output power.

NOISE SOURCES

Interference to amateur stations is not usually caused by a transmitter. It's far more common for amateurs to receive interfering noises from *unintentional radiators*. These signals are either leaked from electronic circuitry or generated as a byproduct from electrical equipment. The following is a list of the most common "offenders."

- Electrical arcs from motors, thermostats, electric fences, neon signs and so forth generate raspy or clicking noises over a wide range of frequencies. The noise is strongest on the lower HF bands and gets weaker as frequency increases. AC power line filters on the offending equipment sometimes work. The on-and-off pattern or rhythm of the noise is often a clue to what causes it. For example, a noise generated by a furnace fan motor will appear for a few minutes at a time and then stop.
- Cracked or dirty insulators, loose connections, wires rubbing together — all these can cause power line noise from arcs. Each peak of the 60 Hz sine wave creates an arc, so power line noise has a characteristic 120 Hz buzz. If you can track down the noise to a particular power pole or piece of equipment by using a portable broadcast or vehicle AM receiver, record the pole's identification numbers and call the power company. Do not attempt to shake or bump the pole — loose hardware can fail and drop a live wire right on you!
- Ignition noise caused by motor vehicles is a whine or buzz that varies with engine speed. It often only lasts while the vehicle travels past your location.
- Alternator whine is a type of noise caused by noise on the dc power system inside your own vehicle. You might hear it with the received audio but more likely it will be heard by others as a high-pitched whine on your transmitted signal that varies with your engine speed. It can be removed by a dc power filter at your radio.
- Switching power supplies used by computers and consumer electronics generate mostly HF noise heard as unsteady tones at evenly spaced frequencies. Once the source has been identified, an ac power line filter is often effective.
- Computer and networking electronics may also directly generate signals at a single frequency that are steady or that vary in patterns. These can be very difficult to eliminate if the equipment is not shielded properly. Choke filters on all input and output cables are a good technique to start.

UNKNOWN SIGNALS

Sometimes amateurs will experience interference from a transmitter that is mistakenly transmitting on an amateur frequency or from a ham's transceiver that is transmitting unintentionally. For example, sitting on one's microphone push-to-talk switch while driving is not unknown! Rarely, cases of intentional interference or jamming occur. In either case, it's important to be able to locate the source of the interference. This is done by *radio direction finding* or *RDF*. By using directional antennas and maps, it's often possible to quickly find the offending transmitter.

GUIDELINES

Dealing with interference is just a fact of life for hams. Regardless of the source, you can reduce or eliminate much interference by making sure your own station follows good amateur practices for grounding and filtering.
- Start by making sure your station is in good working order with appropriate grounding, filtering and good quality connections, particularly for the RF signals.
- Use shielded wire and shielded cables to prevent coupling with unwanted signals and undesired radiation. [T6D12] Be sure to connect the shield properly, such as to the outside of metal equipment enclosures.
- Eliminate interference to your own home appliances first. Demonstrating that you aren't interfering with your own devices is a good start. Eliminating interference at home is considered good practice! [T7B06]
- Eliminate sources of interference in your own home, such as worn out motor brushes, poorly filtered power supplies and so forth. Not only will it make operating more pleasant, it will be much easier to determine whether noise is caused elsewhere.

RFI AND NEIGHBORS

You may eventually encounter a situation where your signals are causing interference to a neighbor, or a device the neighbor owns is causing interference to you. Diplomacy is often required, even though your transmissions may not be at fault. Techniques for dealing with RFI to or from others is discussed on the ARRL's RFI Resources website.

Remember these simple suggestions:

- Start by making sure it's really your transmissions that are causing the problem. It's not unknown for the mere presence of an antenna to generate a report of interference, deserved or not!
- Offer to help determine the nature of interference — detection, overload or harmonics. Knowing the cause leads to solutions.
- If you've determined that the noise is caused by a neighbor's equipment, offer to help determine the source of interference. Severe noise often indicates defective equipment that could be a safety or fire hazard. Again, determining the source leads to solutions.
- You may have to politely explain to the neighbor that FCC rules prohibit them from using a device that causes harmful interference.
- Consult the ARRL RFI Resources website and printed material.

Be diplomatic in dealing with your neighbors, even though it may be their responsibility to deal with interference to or from their devices. They are probably unaware of FCC rules! Remember that the FCC is a last resort for everyone. Before getting involved, the FCC will require that all parties to take all reasonable steps to identify and mitigate the effects of the interference. [T7B08]

PART 15 RULES

Part 15 of the FCC's rules governs the responsibilities of owners of unlicensed devices that use low-power RF communications or radiate low-power signals on frequencies used by licensed services, such as Amateur Radio. Examples include cordless phones or wireless data transceivers and power lines, electric fences or computers. These are called *Part 15 devices*. [T7B09]

Reducing Part 15 to its basic principles:

- An unlicensed device permitted under Part 15 or an unintentional radiator may not cause interference to a licensed communications station, such as to an Amateur Radio station. Its owner must prevent it from causing such interference or stop operating it.
- An unlicensed device permitted under Part 15 must accept interference caused by a properly operating licensed communications station, such as from an Amateur Radio station.

What this means is that as long as your station is operating properly under the FCC's rules, then your operation is protected against interference by and complaints of interference to unlicensed equipment. If your signals are interfering with a television or telephone, it is the TV or telephone owner's responsibility to eliminate it, even though you may offer to assist them. Similarly, it is the owner's responsibility to eliminate interference caused by their device, even with assistance from you. In such cases, you can see the advantages of being federally licensed! These rules are printed in the owner's manual for all unlicensed devices and are available on the FCC website.

Before you go on, study these Technician exam questions from the question pool included at the back of this book or as a downloadable Study Guide version on the web:

T4A04   T4A09

T6D12

T7B02 through T7B09   T7B12

If you have difficulty with any question, review the preceding section.

5.5 RF Grounding

In amateur stations, it's necessary to consider *RF grounding* in addition to the ac safety ground. (See the **Safety** chapter for information on ac safety grounding.) That's why there are different symbols for ground as you found in the section on schematic diagrams. The type of grounding makes a difference in radio systems.

The ARRL Technical Information Service dedicates an entire page to RF grounding with several *QST* magazine articles and web references. Each station is a little different, so you may have to experiment to get the results you expect. Low power VHF/UHF stations usually have few RF grounding problems.

"RF grounding" doesn't really keep your equipment exactly at earth ground potential. It refers to keeping all of the radio equipment at the *same* RF voltage. Your station's feed lines and control cables act as antennas for your transmitted signal. The feed lines and cables are in turn connected to your equipment enclosures and the connections between them. By keeping all of the equipment at the same RF voltage, current will not flow between them.

RF current in your station can cause audio distortion, erratic operation of computer equipment, and occasionally RF "burns" where the RF voltage happens to be high. (RF burns rarely cause injury.) It is far more likely for RF current flowing in sensitive audio cables or data cables to interfere with your station's normal function, just as your strong transmitter signal might be picked up and detected by a neighbor's telephone or audio system. "RF feedback" via a microphone cable can cause distorted transmitted audio, for example. [T7B11]

Every station is different and so will have to create a unique RF ground, but **Figure 5.17** shows the general idea. Here are some guidelines for RF grounding:

- An RF ground rod should *never* be substituted for a properly wired ac safety ground. Your local building codes specify how ac safety grounds should be wired and bonded together.
- Bond all metal equipment enclosures to a common ground *bus*.
- Connect the ground bus to a ground rod or grounded pipe with a short, wide conductor such as copper flashing or strap or heavy solid wire (#8 AWG or larger). Solid strap is

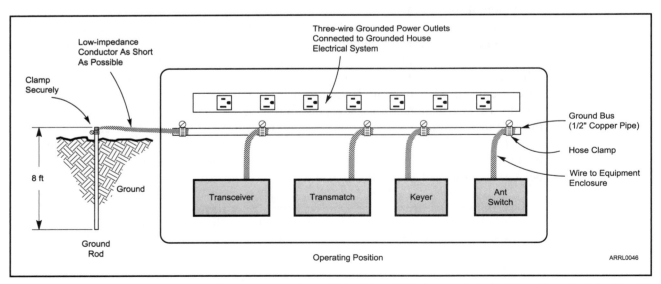

Figure 5.17 — An effective station RF ground consists of keeping all equipment bonded together electrically with short, low-impedance connections. The bus is made of material such as copper pipe, flashing, or even heavy copper wire. The bus is then connected to a nearby ground rod with strap or wire.

best because it presents the lowest impedance to RF current. [T4A08]

- Keep all connections, straps and wires as short and direct as possible. Connection lengths longer than 8 to 10 feet start to act like antennas on the higher-frequency HF bands. At VHF and UHF, ground rods are ineffective because of the length of the connection.
- Braid removed from coaxial cable should not be used for RF grounding because the cable jacket no longer presses the individual strands of wire braid together to ensure good contact. Wire braid from coaxial cable also begins to oxidize when no longer protected by the cable jacket.
- If your shack is located above the ground floor, the length of connection from the equipment would probably be too long to act as an RF ground on any band. In such cases, just use a ground bus between your equipment to keep it all at the same RF voltage.
- In difficult situations, a piece of wide flashing or screen can be placed under the equipment and connected to the ground bus.
- Basement or ground floor stations may be able to use a cold water pipe for a ground, since they travel underground for many feet, connecting to the water main. (Hot water pipes have to go to a water heater and so do not generally make good RF grounds.) Before you decide to use a pipe, make sure it takes a short, direct route to exit the house. It doesn't do much good if the pipe runs all the way across the house before entering the ground! Also, make sure the pipe is metallic — plastic pipe is worthless as an RF ground.

Before you go on, study these Technician exam questions from the question pool included at the back of this book or as a downloadable Study Guide version on the web:
T4A08
T7B11
If you have difficulty with any question, review the preceding section.

Chapter 6

Communicating With Other Hams

When you see the mouse, you'll find more information at www.arrl.org/ham-radio-license-manual

Having learned all that interesting material about rigs and electronics and radio waves, you know a lot about the technology of radio. In this section we turn to operating — how are contacts made and what does a contact consist of? We begin with the elements common to nearly every casual ham radio contact. Once you know these, you'll learn how the ham bands are organized so that you know where to tune. You'll then discover how ham radio is conducted using repeaters and in the organized activities called "nets," especially during emergencies. This chapter concludes with coverage of a few of ham radio's many specialty activities. Clear your throat and get ready for that first contact!

6.1 Contact Basics

This part of the book is about "good amateur practices." There aren't many FCC rules about operating procedures, so amateurs have developed their own rules of etiquette. They work pretty well, so to be a successful and effective communicator it's a good idea to learn and practice them. The FCC (and other amateurs) expects you to follow established practices fairly closely, just as you expect other people to do in daily life.

CONTACT CONTENTS

The simplest possible ham-to-ham contact is an exchange of call signs — you can't get much simpler than that! Most contacts have a bit more to them — from casual conversations to coordinating a public event to formal radio messages. Yet all have some common elements. Some of the conventions and procedures radio operators use may be unexpected, but they are necessary for two or more people to communicate by radio.

General Principles

Having a radio contact is different from a face-to-face meeting. When talking by radio, you can't see the other person so you can't see the hand and facial movements we use to communicate in person. Except for your voice (or characters that are displayed on the computer screen), you're invisible.

The most important thing is to *identify* regularly on the air. Your radio name is your call

Fun with Phonetics

"My name is Chris, Charlie Hotel Radio India Sierra." Why do hams (and radio operators in general) use those strange words instead of just spelling with letters? Remember that you only hear what comes through the radio, so it's often difficult to tell the letters apart. For example, the letters E, T, B, C, D and so forth can sound a lot alike. Not only that, but non-English speakers don't pronounce letters in the same way at all! The solution to both of these problems is *phonetics*.

Created in the early days of international radio, the International Telecommunication Union (ITU) developed a standardized *phonetic alphabet* that operators of all languages could use to exchange precise information. The alphabet is shown in Table 6.1. Each word was chosen because it was easy to understand over the radio. Hams should learn these standard words and use them whenever there is a need for precise, exact spelling — for example, your call sign!

You may also be familiar with the US military version of phonetics and these are used, too. Hams that specialize in DX contacts overseas sometimes use the names of countries or cities. For example, you might hear "Norway" instead of "November" or "Santiago" instead of "Sierra." There is no FCC rule about which set is required. Use the set that best suits the need.

Use caution in making up your own set of funny or cute words, such as "Wanted, One Aged Whiskey" for W1AW. Those may be fine in your local club since they're just nicknames, really, but they can be very confusing on the air, particularly to a foreign ham.

Table 6.1
ITU Phonetic Alphabet

Letter	Word	Pronunciation
A	Alfa	**AL** FAH
B	Bravo	**BRAH** VOH
C	Charlie	**CHAR** LEE
D	Delta	**DELL** TAH
E	Echo	**ECK** OH
F	Foxtrot	**FOKS** TROT
G	Golf	GOLF
H	Hotel	HOH **TELL**
I	India	**IN** DEE AH
J	Juliet	**JEW** LEE ETT
K	Kilo	**KEY** LOH
L	Lima	**LEE** MAH
M	Mike	MIKE
N	November	NO **VEM** BER
O	Oscar	**OSS** CAH
P	Papa	PAH **PAH**
Q	Quebec	KEH **BECK**
R	Romeo	**ROW** ME OH
S	Sierra	SEE **AIR** RAH
T	Tango	TANG GO
U	Uniform	**YOU** NEE FORM
V	Victor	**VIK** TAH
W	Whiskey	**WISS** KEY
X	X-Ray	**ECKS** RAY
Y	Yankee	**YANG** KEY
Z	Zulu	**ZOO** LOO

Note: The **boldfaced** syllables are emphasized. The pronunciations shown in this table were designed for those who speak any of the international languages. The pronunciations given for "Oscar" and "Victor" may seem awkward to English-speaking people in the US.

sign, so use it whenever you announce your presence or participate in a conversation. It's not enough to just identify yourself to the person you're in contact with — others listening need to know who you are, too! They may not know who "Bob" or "Carol" is. Give your call sign at the beginning of and during the contact. Then identify once more at the contact's end. The FCC rules encourage the use of a *phonetic alphabet* when identifying on phone so that your call sign is clear. **Table 6.1** shows the ITU phonetic alphabet, the most widely used.

Only one of you can talk at a time. As you converse, you should also make it clear when you're done transmitting. Remember, the listeners can't see you, so they can't tell if you're done talking or just thinking. That's why phone operators say, "Over" and a CW operator sends "K" — so the other party knows it's their turn. When the contact is completely done, "Clear" is used to let others know that the frequency is available for a contact. These abbreviations are *procedural signals* and have been part of communicating since the days of the telegraph!

Take pains to speak clearly and a little slower than in normal conversation. Remember, all you can hear comes through the radio speaker — no eye contact! If you mumble or rush or slur during voice contacts, you will be hard to understand. On the digital modes, efforts to spell and punctuate correctly are appreciated. The same goes for Morse code — send carefully and take pride in forming your characters clearly by sending at a comfortable speed.

You'll also find that hams use a lot of jargon such as abbreviations, procedural signals, acronyms and slang. These can be hard to figure out on your own but are

Table 6.2
The RST System

READABILITY
1 — Unreadable.
2 — Barely readable, occasional words distinguishable.
3 — Readable with considerable difficulty.
4 — Readable with practically no difficulty.
5 — Perfectly readable.

SIGNAL STRENGTH
1 — Faint signals barely perceptible.
2 — Very weak signals.
3 — Weak signals.
4 — Fair signals.
5 — Fairly good signals.
6 — Good signals.
7 — Moderately strong signals.
8 — Strong signals.
9 — Extremely strong signals.

TONE (CW and Digital)
1 — Sixty-cycle ac or less, very rough and broad.
2 — Very rough ac, very harsh and broad.
3 — Rough ac tone, rectified but not filtered.
4 — Rough note, some trace of filtering.
5 — Filtered rectified ac but strongly ripple- modulated.
6 — Filtered tone, definite trace of ripple modulation.
7 — Near pure tone, trace of ripple modulation.
8 — Near perfect tone, slight trace of modulation.
9 — Perfect tone, no trace of ripple or modulation of any kind.

The "tone" report refers only to the purity of the signal. It has no connection with its stability or freedom from clicks or chirps. Most of the signals you hear will be a T-9. Other tone reports occur mainly if the power supply filter capacitors are not doing a thorough job. If so, some trace of ac ripple finds its way onto the transmitted signal. If it has a chirp (either on "make" or "break") add C (for example, 469C). If it has clicks or noticeable other keying transients, add K (for example, 469K). Of course a signal could have both chirps and clicks, in which case both C and K could be used (for example, RST 469CK).

important to learn because they save time, maximize efficiency, and help you fit in. Your Elmer will help you understand them and it's also a good idea to have a reference book handy, such as those listed in the first chapter. With a little practice you'll soon sound like an OT (Old Timer).

Using a Frequency

After you've been on the air for a while, you'll notice that hams tend to use the same frequencies on a regular basis. For example, NØAX might regularly "monitor" the 443.50 repeater or use CW on 7.035 MHz. No matter how much NØAX stays on that frequency, though, no station has exclusive rights to any particular frequency. Even groups that meet on the same frequency every day at the same time have no priority right to use it. If another ham is using the frequency, they should be prepared to move to another frequency or wait. While it's easy to see how this could lead to difficulties, most hams realize the need to be flexible and are accommodating of the needs of other hams. The key is to be polite and always have a "Plan B" for making your contacts.

Signal Reports

Immediately after establishing contact, it is customary to let the other station know how well their signal is being received. This is important because the other station can then compensate for poor conditions by sending or speaking more slowly or repeating items. Conversely, if signals are strong, unnecessary repetitions can be avoided. This information is called a *signal report*.

The "RST" format in **Table 6.2** is used during SSB, digital and CW contacts. The letters stand for Readability, Strength and Tone. RST is sent as three numbers, such as 339, 599 or 457. Readability has a value of 1 to 5 and rates how well the signal can be understood. Strength is rated on a relative scale of 1 to 9. Tone is used for CW and digital transmissions and refers to the quality of the signal. Clear, pure tones that have a steady frequency or with no audible hum or distortion rate a 9 — as do most of the signals produced by modern radios.

Voice contacts also use the "Q" or "Quality" system. A number from 1 (meaning barely understandable) to 5 (meaning perfectly readable) follows the Q. Signal reports for repeater contacts are discussed later in the sections on using a repeater.

Power Level

It's also nice to know how much power the other station is "running." If a QRP (low power) station's signal is worthy of an RS of 59, that's an accomplishment! Power is not generally of that much interest during repeater contacts, since the signals are relayed.

The FCC rules are quite clear that hams should use the minimum amount of power needed to make the contact. That doesn't mean you should reduce power until your signal can just barely be heard. The intent of the rule is to use an appropriate power level that does not deny others the use of the frequency farther away — more a concern on SSB, digital and CW modes than it is on FM repeaters where the repeater output power can't be adjusted. In general, if your signal is perfectly understandable while running "barefoot" (meaning without an external power amplifier), don't turn the amplifier on. As always, be sure to comply with the legal power limits in the rules!

Locators

You have exchanged signal reports and power, so it is now time to find out where the other station is. The exact way to identify location varies, from the local mobile station's "I'm on the Beltway, just south of the toll road" to "Location is latitude 43 North and longitude 39 West" of a radio officer on a freighter in the Atlantic. Location can always, however, be denoted or queried with the *Q-signal* QTH. "My QTH is…" means, "My location is…" and "QTH?" means, "What is your location?" (See the sidebar on Q-signals and **Table 6.3** for more on these abbreviations.)

An increasingly popular method of identifying location is the Maidenhead Locator System, better known as *grid squares*. In this system, named for the town outside London, England where the method was first created, the Earth's surface is divided into a system of rectangles based on latitude and longitude. Each rectangle is identified with a combination of letters and numbers. [T8C05] A four-digit code of two letters followed by two numbers identifies a unique rectangle of 1° latitude by 2° longitude. For example, ARRL Headquarters in Newington, Connecticut is located in grid square FN31. A further two letters can be added for greater precision, such as FN31pq for the precise location of the ARRL station.

Appropriate Topics

As you listen to amateurs discussing everything from the weather to sporting events and public service, you'll wonder, "What don't hams talk about?" Hams pride themselves on professional, high-quality procedures and conduct, but there are really very few restrictions about what you can talk about over the air.

No matter what the topic, indecent and obscene language is flatly prohibited. No cussin' allowed! Just don't! If you find yourself getting upset for any reason, it's best to just turn off the radio for a while. While there is no official list of "the dirty words you can't say on the radio," use good sense and an extra helping of manners to avoid offending other hams. That especially includes racial and ethnic references. Everything you say travels a long way and you never know who might be listening!

Similarly to the use of language, hams also try to stay clear of provocative subjects that often create strong feelings, such as politics, religion and sexual topics. There are plenty of forums for these discussions, such as the Internet and e-mail. Ham radio is for developing communications expertise and goodwill. If the discussion doesn't further those aims, it's best taken elsewhere.

Signing Off

When your contact is complete, concluding it is simple, but as in real life, it's rude to just "hang up." Hams have their own way of saying goodbye that can get to be somewhat involved. The best way to learn (like many other things in ham radio) is to listen. Here are a few of the terms you'll hear as hams go their separate ways:

Table 6.3

Q-Signals

These Q-signals are the ones used most often on the air. (Q abbreviations take the form of questions only when they are sent followed by a question mark.)

QRG   Your exact frequency (or that of _____) is _____ kHz.
       Will you tell me my exact frequency (or that of _____)?
QRL   I am busy (or I am busy with _____). Are you busy? Usually used to see if a frequency is busy.
QRM   Your transmission is being interfered with _____
       (1. Nil; 2. Slightly; 3. Moderately; 4. Severely; 5. Extremely.)
       Is my transmission being interfered with?
QRN   I am troubled by static _____ . (1 to 5 as under QRM.)
       Are you troubled by static?
QRO   Increase power. Shall I increase power?
QRP   Decrease power. Shall I decrease power?
QRQ   Send faster (_____ wpm). Shall I send faster?
QRS   Send more slowly (_____ wpm). Shall I send more slowly?
QRT   Stop sending. Shall I stop sending?
QRU   I have nothing for you. Have you anything for me?
QRV   I am ready. Are you ready?
QRX   I will call you again at _____ hours (on _____ kHz).
       When will you call me again? Minutes are usually implied rather than hours.
QRZ   You are being called by _____ (on _____ kHz).
       Who is calling me?
QSB   Your signals are fading. Are my signals fading?
QSK   I can hear you between signals; break in on my transmission.
       Can you hear me between your signals and if so can I break in on your transmission?
QSL   I am acknowledging receipt.
       Can you acknowledge receipt (of a message or transmission)?
QSO   I can communicate with _____ direct (or relay through _____).
       Can you communicate with _____ direct or by relay?
QSP   I will relay to _____ . Will you relay to _____ ?
QST   General call preceding a message addressed to all amateurs and ARRL members.
       This is in effect "CQ ARRL."
QSX   I am listening to _____ on _____ kHz. Will you listen to _____ on _____ kHz?
QSY   Change to transmission on another frequency (or on _____ kHz).
       Shall I change to transmission on another frequency (or on _____ kHz)?
QTC   I have _____ messages for you (or for _____). How many messages have you to send?
QTH   My location is _____ . What is your location?
QTR   The time is _____ . What is the correct time?

Q-Signals

Q-signals are a system of radio shorthand as old as wireless and developed from even older telegraphy codes. Q-signals are a set of abbreviations for common information that save time and allow communication between operators who don't speak a common language. Modern ham radio uses them extensively. Table 6.3 lists the most common Q-signals used by hams. While Q-signals were developed for use by Morse operators, their use is common on phone, as well. You will often hear, "QRZed?" as someone asks "Who is calling me?" or "I'm getting a little QRM" from an operator receiving some interference or "Let's QSY to 146.55" as two operators change from a repeater frequency to a nearby simplex communications frequency. [T2B10, T2B11]

- Final — the last transmission, as in "I will be clear on your final."
- QRU — the Q-signal that means, "I have nothing more for you"
- Down the log — "I'll see you later"
- 73 — "Best Regards," an almost universal closing motif
- 88 — added for a female operator, originally "Love and Kisses"

As you complete your final transmission of a contact and do not intend to respond, it is customary to add "Clear" to let everyone know you are through. Occasionally you may hear "Out" as in "Over and out," but that is mostly a relic of the movies. "Clear" or "Off and clear" is unambiguous.

ADVISING AND ASSISTING

The longest tradition of Amateur Radio is that of mutual assistance. A century ago there were no radio texts or handbooks so all hams were self-taught, relying on others to help them get on the air and learn how to operate. The same ethic is alive and well today. While as a new ham you may be in need of assistance today, you'll be glad to help another ham tomorrow.

Methods and Procedure

It's important to start by noting that every ham was once at the same level of learning that you are today. Crack operators and technical whizzes are not born, they're made. Don't be embarrassed because you don't know everything about ham radio or make a mistake! There are plenty of ways to learn. Others will help (particularly if you ask questions), and you'll learn rapidly. Accept criticism in the helpful spirit of its offering and be sure to extend the same spirit to others.

The most common mistakes made by any ham are ones of technique — transmitting too soon or too late, using the wrong procedure, misunderstanding instructions and so forth. While on the air, be sure to take a helpful tone in correcting the operator. Tell them what you expected and give guidance as to what to do next time. For example, if a station makes a call on a repeater in the middle of a public service exercise, perhaps they just didn't listen long enough before transmitting. You may be annoyed, but don't let that color your response. An appropriate response would be something like, "W1AW, this is K1ZZ, net control for the Podunk Valley Parade. The repeater is being used to coordinate the parade until about 1 o'clock. You could use the 145.35 repeater, OK?" This avoids prejudging W1AW's error, lets them know what's happening and makes a helpful suggestion. This will generally correct the situation promptly and avoids hurt feelings or unnecessary embarrassment.

With all the features of modern radios, it's also easy to get confused and transmit in a way you didn't expect, such as on the wrong mode or frequency. (Hint — that's why you might want a simple radio to start with instead of one with all the bells and whistles.) Even experienced hams goof now and then and you will, too! If you hear someone having trouble operating their radio, such as transmitting off frequency or using FM in the SSB part of the band, use the same technique to correct them as in the preceding example. Contact them and let them know what the problem is. If you can, offer a solution; "Your voice is distorted, try speaking more softly." A helpful, friendly voice when you're having trouble is appreciated!

Radio and Antenna Checks

There will be times when you are not sure your radio is working properly. When you need an on-the-air evaluation of your signal, that's called a *radio check*. Before asking for help, make sure you can clearly describe the problem and what kind of evaluation you need. If you can arrange for a friend to meet you on the air that's the best method of troubleshooting.

If a friend's not available, you may need to make a general call. The technique is to find an ongoing contact and break in. When the stations acknowledge your call sign, you respond with, "This is W1AW and I need a radio check, please." If the stations can help you, they'll take it from there. Be sure to identify yourself with your call sign and be polite. If they are busy and can't help, move to another frequency. You can also make the same call on an unused frequency where someone may hear you and respond. Avoid using a busy repeater to do the tests.

If you are asked to give a radio check, listen carefully and respond with detailed information. Distortion of voice signals could be microphone or RF feedback problems. A signal abruptly cutting in and out indicates a broken or intermittent connection. Hum on any type of signal might indicate power supply or battery problems.

Antenna comparisons are the next most common testing need. Hams are always experimenting or replacing antennas in search of a better signal. The same procedure applies for radio checks; pay attention to details and don't hesitate to relay your observations to the testing station. This is the time for an accurate and precise signal report! If you are conducting the test, be sure to give the evaluating station enough time to observe the antenna's performance with a *five-count* where you count on the air, "One-two-three-four-five" or on CW send "V" several times or a few seconds of steady key-down signal. Always be sure to properly identify your station!

Noting Violations

Hams have a long-standing tradition of self-policing to help each other play by the rules. Part of that tradition is notifying each other of apparent infractions. You may also come across a station violating the Amateur service rules, such as transmitting on voice in a CW-only band segment. Don't commit an infraction yourself! Contact the station by e-mail, mail or phone and let them know. Be diplomatic and keep your tone helpful.

The ARRL has also established the Official Observer (OO) program that relies on technically skilled hams. The OOs keep an ear on the bands, often finding problems before they become a problem for other hams.

LOGGING AND CONFIRMING CONTACTS

Hams are no longer required by the rules to keep a logbook, but it is an excellent idea to do so. There are a variety of paper logbooks and software logging programs available from most ham radio vendors. Enter *radio logging software* into an Internet search engine and

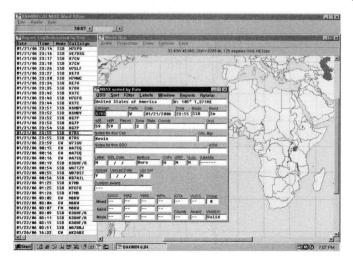

you'll find plenty of choices. **Figure 6.1** shows the data entry window of *DX4WIN*, a typical logging program. Why keep a log? It's useful to have a record of contacts with the hams you meet on the air. Keeping a log can also help you identify good times to operate or help identify

Figure 6.1 — Software logging programs can store and cross-reference thousands of contacts. For use away from a computer, the paper logbooks are a reliable standby and the information can be transcribed into a computer later. Many programs will help you upload your logbook to the ARRL's electronic *Logbook of the World* (LoTW) to confirm contacts automatically for ARRL awards.

sources of interference, both to and from your station. It's your record of how you make use of your license!

Most hams log random contacts made on HF or VHF/UHF bands while at home or operating in a portable location, such as during a vacation or even a radio *DXpedition* to an unusual spot. Casual and regular contacts made with friends aren't usually logged, nor are most local repeater contacts. Contacts made during mobile operation, which requires extra attention to the hazards of driving, are rarely logged.

When keeping a log, record the time and date of the contact, the band and mode, and the call sign of the station you contacted. These are the minimum needed to confirm a contact later. You might want to record the exact frequency, your power level, signal reports exchanged, and information about the other station such as location and operator name. Some logging software can connect directly to your radio and record much of that information for you!

Many hams exchange cards after a contact, particularly on the HF bands or when using SSB and CW. (Repeater and local contacts are rarely acknowledged in this way.) Nearly every ham has a personalized card like those in **Figure 6.2**, called a *QSL* after the Q-signal that means

Figure 6.2 — Exchanging QSLs after a contact is a cherished part of ham radio for many hams. The cards can be simple or ornate, text-only or beautiful photos. With the availability of excellent home color printers, many hams design and print their own cards.

"received and understood." Sending the cards is called *QSLing*. Hams collect the cards just for personal interest or to submit as proof-of-contacts for award programs.

You can create your own card with a home computer or order them from several vendors. If you decide to make your own, get a few cards from friends and study them to be sure you include all the necessary information as described above. Make sure to include an accurate description of your location, such as county and grid square, since many award programs require this information. There are also electronic QSL systems such as the ARRL's *Logbook of the World*. Most logging software can automatically transfer your contact information to these electronic systems, saving postage and time. Nevertheless, a paper QSL arriving by mail has a special meaning for many hams.

Before you go on, study these Technician exam questions from the question pool included at the back of this book or as a downloadable Study Guide version on the web:
T2B10   T2B11
T8C05
If you have difficulty with any question, review the preceding section.

6.2 Band Plans

The amateur bands may not appear to be very big on a chart of the radio spectrum, but once you start tuning around, you'll find they are quite large! How can you tell where to tune so that you can find your preferred activity? What keeps the different types of operating from being spread out randomly all over the band?

DEFINITIONS AND FINDING BAND PLANS

The easiest place to find the band plans for US amateurs is via the ARRL's home page or on the *Ham Radio License Manual* website. You should take a few minutes to browse the bands plans, particularly those for the 2 meter and 70 cm bands where many Technician licensees begin operating. **Table 6.4** shows the 2 meter band plan. You'll probably be surprised at all the different types of activity!

On the HF bands, band plans tend to be simpler because there are no repeaters except on the 10 meter band. On the VHF and UHF bands where repeater operation is common, the band plans show where repeaters listen as well as transmit. Here are some definitions of the band plan terms:

- DX Window — Because bands are often full of local or regional contacts, DX windows give the weaker signals traveling a long distance a relatively clear space to be heard. DX contacts can certainly be made elsewhere, but the windows are good places to start listening.

- Digital modes, RTTY (radioteletype), packet — Digital operation is less common than voice and CW, so keeping the signals together makes it easier to find other digital mode signals.

Table 6.4

2 meter (144-148 MHz) Band Plan

Frequency	Description
144.00-144.05	EME (CW)
144.05-144.10	General CW and weak signals
144.10-144.20	EME and weak-signal SSB
144.200	SSB calling frequency
144.200-144.275	General SSB operation
144.275-144.300	Propagation beacons
144.30-144.50	OSCAR subband
144.50-144.60	Linear translator inputs
144.60-144.90	FM repeater inputs
144.90-145.10	Weak signal and FM simplex (145.01, 03, 05, 07,09 are widely used for packet)
145.10-145.20	Linear translator outputs
145.20-145.50	FM repeater outputs
145.50-145.80	Miscellaneous and experimental modes
145.80-146.00	OSCAR subband
146.01-146.37	Repeater inputs
146.40-146.58	Simplex
146.52	National Simplex Calling Frequency
146.61-147.39	Repeater outputs
147.42-147.57	Simplex
147.60-147.99	Repeater inputs

- Beacons — Automated transmissions are used to tell when the band is "open" to the area of the world from which the beacon signal originates. The Northern California DX Foundation operates a series of beacons around the world on the HF bands and many individual hams operate a beacon on 10 meters or the VHF/UHF bands.

- Weak signal — Areas of the band are designated for making contacts with distant stations or using difficult propagation paths compared to the strong signals of local FM repeaters. Modes that work better at lower signal strengths are used here, such as CW, SSB, and some digital modes. Every amateur band from 50 MHz on up has some portion available to CW and SSB operation. [T2B13]

- Satellite uplinks and downlinks — These are segments of the bands where signals are sent to (*uplink*) and received from (*downlink*) satellites.

- Simplex — Transmitting and receiving on the same frequency [T2B01]

- Repeater inputs and outputs — This is where the usual repeater operations are found and where nearly all FM operation occurs.
- Control links — Repeaters and other stations that are linked together or controlled by a remote operator use radio links to carry audio and control signals.

CALLING FREQUENCIES AND BEACONS

Nearly every band has one or more *calling frequencies* listed. These are frequencies on which users of a specific mode or style of operation make contact. They then tune to a nearby frequency to continue the contact. A calling frequency allows stations to find each other quickly. Continuously monitoring the calling frequency and making an occasional CQ is useful for detecting long-distance propagation opportunities on the VHF and UHF bands. A sudden change in propagation will then be noticed by many amateurs instead of one lucky operator who just happened to be on the right frequency.

WHY BAND PLANS ARE NEEDED

You are probably beginning to understand band plans. These voluntary conventions simply make it easier for everyone to maximize their own success and share the sometimes-crowded ham bands. Sharing is particularly important because the different styles of operating are not always compatible. For example, a digital station and CW or voice station simply can't share a common frequency — they are too different. It's a lot easier for signals of the same type to operate close together than it is for a mix of modes.

The other major reason for having band plans at VHF and above is the effective use of repeater stations. As we mentioned in the previous section, repeaters have an input and an output frequency separated by the repeater offset frequency. Having the repeater input frequencies close together and the output frequencies close together serve two purposes. It's a lot easier to tune from repeater to repeater that way, and the repeater receiver's sensitivity is improved by keeping the input frequencies separate from the powerful output signals. Everything works a lot better when the band plan is followed!

WHO MAKES BAND PLANS

You might think that the ARRL created all these band plans, but that's not the case. The ARRL publishes what it understands to be the band plan based on how hams use the various frequencies. These voluntary "gentlemen's agreements" evolved over the years as the various groups of hams decided how best to coordinate their operating. For example, the international clubs of (QRP) low-power operators forged an agreement to congregate around a specific set of frequencies on all the bands and promoted them. The ARRL band plans simply reflect those agreements.

Repeater band plans at VHF and UHF were a bit more difficult to organize, since each region was isolated by limits to signal propagation. This often had a big impact on decisions about where to place input and output frequencies and the *channel spacing* between one repeater and the next. Fortunately, it was quickly realized that a certain amount of compatibility between the plans was needed. Hams have almost completely aligned the regional repeater band plans so that travelers can easily use repeaters anywhere.

RULES ABOUT BAND PLANS

As mentioned earlier, band plans are considered "good practice" by the FCC and so it expects hams to abide by them. The only time this really comes into question is when persistent interference is caused by a station operating in conflict with the band plan. For example, if someone decides to operate a repeater whose output frequency is the same as a nearby repeater's input frequency, interference will result. If the parties are unable to resolve the issue with the help of the regional *frequency coordinator*, the FCC will first look to see which repeater is operating according to the band plan.

Band plans by themselves cannot guarantee that no interference exists. For example, on 40 meters, the international RTTY calling frequency is the same as that of the QRP CW operators! The powerful RTTY and weak Morse code signals don't coexist very well. Luckily, it is relatively uncommon for interference between the two groups to last very long. Hams, being much more flexible or *frequency-agile* than any other radio service, just retune a bit and carry on.

It's important to keep in mind that band plans are voluntary agreements designed for normal conditions. They are not regulations. Except for repeaters, whose operating frequencies are fixed, stations are expected to operate flexibly. Amateur Radio is the only service that can tune freely and use multiple modes within their allocations, so we are expected to do so. You may find that a special event or contest crowds the bands or conditions may be unfavorable or public service communications may be underway. Utilize ham radio's flexibility and you will be happier and more successful on the air.

Before you go on, study these Technician exam questions from the question pool included at the back of this book or as a downloadable Study Guide version on the web:
T2B01   T2B13
If you have difficulty with any question, review the preceding section.

6.3 Making Contacts

Before you transmit, be sure you are authorized to use that frequency and mode! As a Technician licensee, you will have full amateur privileges on 50 MHz and up, but there are several mode-restricted band segments. On HF, your privileges are restricted as shown in the chapter on **Licensing Regulations**.

This section covers the basic procedures you'll use for making ham radio contacts. While these "radio manners" apply to just about all contacts, repeater operation has some special aspects that are covered in the next section.

REPEATER CONTACTS

Repeaters are frequently like social clubs; they have a regular clientele and usually there are manners everyone uses to share the facility. To "join the club" you should learn by listening and abide by those manners. Here are some repeater manners shared by all repeaters users:
* Listen so that you are aware of someone using the repeater
* Keep transmissions short
* Identify your station legally; identifying at the beginning of the contact is also a good idea
* Pause briefly between transmissions to listen for another station trying to break in
Remember, too, that using a repeater does not mean you are no longer responsible for operating legally. As the control operator of the originating station you are responsible, not the repeater owner, if your transmissions violate FCC rules.

Because repeaters usually have a strong signal on a known frequency, it's not necessary to make a long transmission to attract listeners tuning by. The easiest way is to just give your call sign. [T2A09] That announces you are listening and are available for a contact.

Many stations add a few words to make it clear that they would like to make a contact: "W1AW is monitoring." Or "W1AW is monitoring and standing by for a call." That announces to everyone listening that W1AW is available for contacts through the repeater. Customs vary from region to region and even from repeater to repeater, so listen a while first to see how others make contacts and follow their lead. (Hint — listening to learn is a sure path to success in ham radio!)

If you want to respond to a station asking for a call or want to contact a station whose call sign you already know, take advantage of the repeater's strong signal and keep it short. Just say the other station's call sign once, followed by "this is" or "from," then give your call sign. [T2A04] It sounds like this: "N0AX this is WA7KYI"

Giving signal reports on a repeater is also part of repeater etiquette. Because the repeater is retransmitting the signal, a signal report tells the transmitting station how well the repeater hears them! Along with the Q-system of signal reports, you'll often hear the following terms, from strongest to weakest:

- Full quieting — your signal is strong enough that no receiver noise is heard
- White noise — not as strong as full quieting, some noise is present
- Scratchy — weaker still, noise is almost as strong as your voice
- Mobile flutter or picket fencing — rapid fading due to moving through an area of multipath propagation or shadowing
- Dropping out — mostly audible, but frequent periods of no signal
- Broken or breaking up — short periods of audibility, but mostly unreadable

These give a good description of what the signal sounds like on the repeater's output. Keep your handheld radio's antenna vertical if your signal is less than full-quieting strength.

Repeaters often add a short *courtesy beep* to the retransmitted signal when the transmitting station's signal disappears. This useful feature becomes the "over" cue to other stations to start speaking, although saying "Over" is common on repeaters as well. Some radios have the capability to add their own courtesy beep, but it is not necessary (and often confusing) on a repeater with a courtesy beep enabled already.

What if someone else is using the repeater and you accidentally interrupt them? Don't panic, just say "Sorry, W1AW clear" and wait for their contact to end or tune to a different repeater.

What if you receive a report that your signal's audio is strong, but distorted? A station that is slightly off-frequency will be strong, but distorted. This sometimes happens when a radio control key gets bumped, changing frequency by a small amount. If accidentally pressing a control key on your radio is a frequent problem (smaller radios are particularly prone to it), try using the LOCK feature of your radio to disable unintended key presses. You could also be causing excessive deviation by speaking too loudly into the microphone. Either lower your voice or hold the microphone farther away from your mouth. Weak or *low* batteries can also cause distorted audio. [T7B10]

Another common symptom of a weak battery pack is a signal that sounds great for a couple of seconds at first, then suddenly fades. The batteries are too weak to sustain a transmission and the transmitter shuts off. Keep a fully charged spare battery pack handy!

What if you are sure that your signal is being received by the repeater but the stations don't respond? Don't feel slighted! Remember that many hams use repeaters as a kind of on-the-air meeting place for club members or acquaintances. They may not be looking for a random contact all the time. For best success on repeaters, wait until an existing contact is finishing and then call one of the stations. This works especially well if you can discuss a topic of their just-concluded contact — a real conversation starter! Participating regularly in nets (discussed later) or other events on repeaters is also a good way to break in with a group.

SSB, CW AND DIGITAL CONTACTS

Because contacts on these modes generally don't use repeaters on fixed channels, finding and attracting other stations is done differently. To attract the attention of other stations, you have to make a *call* long enough for someone to tune their radio to your frequency and determine your call sign. This done by *calling CQ*. CQ is a procedural signal that means "I am calling any station." [T2A08] The station calling CQ will actually send or say "CQ" several times followed by their call sign. It sounds like this:

W1AW: "CQ CQ CQ, this is W1AW Whiskey One Alfa Whiskey calling CQ and standing by."

On CW or a digital mode it looks like this:

W1AW: "CQ CQ CQ DE W1AW W1AW W1AW K"

DE means "from" and the procedural signal K means that your transmission is finished and you're ready to receive. K is used at the end of transmissions of all sorts, just as "Over" is used on phone.

Before you call CQ you should do three things [T2A12]:

* Be sure the frequency is one your license privileges authorize you to use!
* Listen to be sure the frequency is not already in use. If you don't hear any signals in five to ten seconds the frequency may be available.
* Make a short transmission asking if the frequency is in use — an ongoing contact or activity may have paused or you may not be able to hear the station currently transmitting. Simply asking "Is the frequency in use?" followed by your call sign is sufficient.

As with repeater contacts, if you accidentally interfere with or "step on" another contact, simply apologize, give your call sign and tune to another frequency or wait until the frequency is clear.

If you hear a station calling CQ, it's easy to respond. Give the CQing station's call sign once (they already know their own call!) then yours once (if they are strong and clear) or twice. Give your call clearly and distinctly so that they can understand it if there is noise or interference. [T2A05] Your response should sound like this:

N6ZFO: "W1AW this is November Six Zulu Foxtrot Oscar, November Six Zulu Foxtrot Oscar, over."

Use phonetics (see Table 6.1) when using phone, so the other station gets your call sign correct.

TAKING TURNS AND BREAKING IN

Once you've successfully joined or initiated a contact, you may find yourself in a friendly extended conversation, called a *ragchew*. If you're sharing the frequency with several other stations, that's a *roundtable*. Stations in a rare location may only exchange signal reports and call signs so that many stations may contact them. Organized on-the-air meetings are called *nets*. These are only a few of the activities that you'll find on the air.

Once again recalling that radio communications can't rely on visual cues, there are certain methods that hams use to control the flow of a conversation. The most common is the word "Over" or the procedural signal K for CW or digital. Hams learn not to begin speaking until transmitting station stops or "turns it over." That's why you should say "Over" (or send K) when you are ready for the other station to transmit. It lets the other station know that you are ready for a reply and are not just thinking or pausing. If you are having a conversation with good signal quality and are familiar with the other operators, the need to use "Over" every time is relaxed, but it is still very useful in keeping the conversation flowing.

Even the best procedure can't guarantee that two stations won't accidentally start transmitting at the same time. Because ham transceivers can't receive while transmitting, there's no way to know of the simultaneous transmissions until one station stops.

Two stations transmitting at the same time is called *doubling*. When that happens, someone listening in might say, "You doubled!" At that point, one station asks for a repeat transmission and waits or yields the frequency to the other station.

If two stations are having a contact and you need to enter the conversation for some reason, it is necessary to get their attention to get an opportunity to speak. This is called *breaking in* and requires a little finesse in technique. The best way to break in is to wait for one station to stop transmitting and then quickly say your call sign. It is good manners during your contacts to pause briefly before transmitting to give a *breaking* station the chance to transmit. If you hear someone breaking in, the appropriate response is to pause and ask the breaking station to go ahead. For emergency break-ins, refer to this chapter's section on emergency communications for a discussion of how to go about it. *Never* use an emergency signal when you don't need to!

SIMPLEX CHANNELS

Repeaters provide such good coverage, why would anyone not use a repeater for a contact? A repeater's wide coverage and signal strength are precisely why it's not always appropriate to use them. Since only so many repeaters can share a band in a region, hams must use them wisely. It's easy to become so used to using repeaters that direct or simplex communication isn't considered. In fact, it's often quite easy to make contact directly.

If you are close to the station with whom you're in contact, why not give simplex a try? If you can hear the other station directly when listening on the repeater input frequency, you should consider using a simplex channel. [T2B12] Doing so avoids occupying a repeater and makes your conversation a lot less public than having it retransmitted over the repeater's entire coverage area! Here's how to move a contact to a simplex frequency:

W1AW: "NK7U this is W1AW, are you on the repeater this morning?"

NK7U: "W1AW this is NK7U. Yes, and you're strong on the input. Let's move to 146.55 simplex."

W1AW: "OK, I'll meet you on 146.55. W1AW clear."

NK7U: "NK7U clear"

You can tell if you are within range just by listening to the other station transmit on the repeater's input frequency. Many radios have a *reverse* function that swaps your transmit and receive frequencies, making it easy to listen for the other station.

Listening on the repeater's input is often helpful when a weak station is trying to access the repeater, but isn't quite strong enough. If the weak station is near you, it's likely that you'll hear their signal better than the repeater does!

Simplex channels are conveniently located between bands of repeater input and output channels. For example, the national simplex calling frequency on 2 meters is 146.52 MHz and on 70 cm it is 446.00 MHz. [T2A02] That means the antenna you use for repeater contacts will work just fine for simplex, too.

Before you go on, study these Technician exam questions from the question pool included at the back of this book or as a downloadable Study Guide version on the web:

T2A02 T2A04 T2A05 T2A08 T2A09 T2A12 T2B12
T7B10

If you have difficulty with any question, review the preceding section.

6.4 Using Repeaters

As you learned earlier, repeater stations are located in high spots — on towers, hills or buildings. The output of the repeater's receiver is simultaneously retransmitted at the same time on one or more different frequencies or channels. The strong output signal is received easily over a wide area by mobile and handheld radios. Transmitting on one frequency and receiving on another frequency is called *duplex* communication.

As a Technician licensee, you are likely to make contacts through a repeater. Because repeater signals can be heard by many amateurs, repeater contacts have some operating procedures that are different than for SSB, CW and digital contacts.

FINDING REPEATERS

First, how do you find a repeater that you can use? Start by looking at the band plan. Find the band segment filled with repeater output frequencies. You can tune through that part of the band or use your radio's *scanning* functions, but there is no guarantee that you'll find all of the repeaters because they don't transmit all the time.

If you hear a repeater, how can you tell which one it is? Listen for a while and you may hear an automated voice announcing a call sign and possibly some other information, such as location or time. This is the repeater's *ID* and it allows you to look up the call sign online and tell for sure where it is. The repeater may also send Morse code, a *CW ID*. This is very common and is a good reason to know Morse code, even if you don't use Morse to make contacts.

To find all of the repeaters in your area, you'll need a listing sorted by area, such as the *ARRL Repeater Directory* shown in **Figure 6.3**, or a source such as a club newsletter. Repeater frequencies are evenly spaced in *channels*, so you know exactly what frequency to use when you do find or select a repeater.

This would be a good opportunity to use your radio's scanning functions. By continually scanning the repeater *output frequencies*, you'll eventually find all of the active local repeaters.

144-148 MHz

Location	Output	Input	Notes	Call	Sponsor
ALABAMA					
Albertville	145.1100	–	O 107.2 (CA)eWXx	KI4RYX	MCEARS
Alexander City	146.9600	–	O 88.5/88.5 l	WA4KIK	96 Radio Club
Andalusia	147.2600	+	O 100.0e WX	WC4M	South AL R
Anniston/Cheaha Mt	147.0900	+	O 131.800a eWXz	WB4GNA	CCARA
Anniston/Oak Mt	146.7800	–	OaelRB WXz	KG4EUD	Calhoun EMA
Arab	146.9200	–	O 77.0/77.0 elRBz	KE4Y	BMARA
Argo/Trussville	145.2600	–	O	K4YNZ	KE4ADV / D-STA
Ashland	147.2550	+	O 131.8ae WXz	KI4PSG	KI4PSG
Athens	145.1500	–	O 100.0/OR OFF/ARES (CA)eWXz	N4SEV	Limestone
Auburn	147.0600	–	O	KA4Y	KA4Y
Auburn	147.2400	+	O 156.7ae	K4RY	Auburn Uni
Auburn	147.3000	+	O 123.0/123.0elRB	W4HOD	HODARS
B,Ham /Shades	147.1400	+	O	WA4CYA	WA4CYA

Figure 6.3 — Listings of repeaters, such as the ARRL's *Repeater Directory or TravelPlus for Repeaters* make it easy to find repeaters in your area or where you intend to travel. This is a typical section of a page in the *Repeater Directory*.

Once you have located an active repeater, to access it you will need to know three things: the repeater transmitter's output frequency, the repeater receiver's *input frequency* and the frequency of any access control tones.

REPEATER OFFSET OR SHIFT

Let's start with the repeater's output frequency. This is the frequency on which you hear the repeater's transmitted signal and it's the frequency by which repeaters are listed in a directory, such as the *ARRL Repeater Directory*. Hams will say, "Meet you on the 443.50 machine" or "Let's move to the 94 repeater." (*Machine* is slang for repeater.) "94" means 146.94 MHz, a standard repeater output channel frequency. To listen to the repeater, tune to its output frequency.

To send a signal through the repeater, you must transmit on the repeater's input frequency where the repeater receiver listens. It would be chaos if every repeater owner used a different separation of input and output frequencies, so hams have decided on a standard separation between input and output frequencies. The difference between repeater

Table 6.5

Standard Repeater Offsets by Band

Band	Offset
10 meters	−100 kHz
6 meters	Varies by region: −500 kHz, −1 MHz, −1.7 MHz
2 meters	+ or −600 kHz
1.25 meters	−1.6 MHz
70 cm	+ or −5 MHz
902 MHz	12 MHz
1296 MHz	12 MHz

input and output frequencies is called the repeater's *offset* or *shift*. [T4B11] The shift is the same for almost all repeaters on one band as shown in **Table 6.5**. [T2A01, T2A03] If the repeater's input frequency is higher than the output frequency, that is a *positive shift*. *Negative shifts* place the repeater's input frequency below the output frequency.

Instead of having to remember two frequencies, using a standard shift allows you to remember only the repeater output frequency. The SHIFT or OFFSET key or menu setting on your radio allows you to switch between positive and negative shifts, or no shift to use simplex communications. Your radio is probably already configured to use the standard shift on each band, usually referred to as *auto-repeat*. The operating manual will have complete instructions on changing the direction and amount of shift.

REPEATER ACCESS TONES

Most repeaters won't pass a signal from the receiver to the transmitter for retransmission unless it contains a special tone. The tone is one of 38 different frequencies, all below 300 Hz. Each repeater can have a different tone. The tone indicates to the repeater that your signal is intended for it and should be retransmitted.

Repeater access tones were invented by Motorola to allow different commercial users to share a repeater without having to listen to each other's conversations. These tones are known by various names: *Continuous Tone Coded Squelch System* (CTCSS), *PL* (for Private Line, the Motorola trade name) or *subaudible*. FRS/GMRS radio users know these tones as *privacy codes* or *privacy tones*. [T2B02]

Why wouldn't a signal on the proper frequency be intended for the repeater? Most repeater installations are close to other repeaters, paging transmitters and broadcast transmitters. The powerful signals from all these transmitters sometimes mix together and create false signals, called *intermod*, an abbreviation of *intermodulation*. Intermod can easily appear on a repeater's input frequency and would be retransmitted, disrupting normal communications. To prevent these signals from being retransmitted, the repeater receiver listens for the proper tone in the received signal. No tone or an improper tone indicates the signal is not intended for that repeater, so it won't be retransmitted.

Your radio will have a TONE key or menu selection that allows you to both select a tone and add it to the transmitted signal, if desired. You may also be able to set your radio's squelch to require a CTCSS tone to pass received audio to the speaker. This is called *tone squelch*. You should be aware that most repeaters filter out CTCSS tones before the received audio is retransmitted.

Picking a Tone

CTCSS tones are added to your signal to cue the repeater that it should relay your signal. Before you can transmit through a repeater that requires a CTCSS tone, you'll have to find out which of the 38 possible tones it could be. If you have an *ARRL Repeater Directory* or are using a club website or newsletter, the tone will be listed along with the output frequency of the repeater. The listing will look like this:

PODUNK VALLEY   146.94 (-)   103.5

You will hear the repeater output on 146.94 MHz and the input is 600 kHz (the standard 2 meter offset) below the output frequency, or 146.34 MHz. The (-) means that the repeater offset or shift is negative. The entry "103.5" is the frequency of the CTCSS or subaudible tone. Your radio's operating manual will explain how to select and activate the tone. There may be several tone options, such as tone squelch and *digital code squelch* (DCS). Leave them off for now.

Some radios also have the ability to determine the tone frequency from on-air signals. This is called *tone scan*. Radios with this feature can often be configured to automatically set their own CTCSS tone to the same frequency without any operator intervention. This is very handy when you are new to an area, just visiting or driving through.

Another method of squelch control is *digital code squelch* (DCS). A continuous sequence of subaudible tones must be received during a transmission to keep the output audio turned on. If the proper tone sequence is received, your receiver will open up the squelch and you can hear the calling station. Check the operating manual of your radio for information about how to configure DCS.

If you can hear a repeater's signal and you're sure you are using the right offset, but you can't access the repeater, then you probably don't have your radio set up to use the right type or frequency of access tone. [T2B04]

ACCESSING A REPEATER

First, you'll need to program your radio to use the correct offset and CTCSS tone. Offset is generally standardized on the different bands so that radios with an *autorepeater* function will automatically select the right amount and direction of frequency shift. Choose a subaudible tone as described in the radio's user manual.

Now you're ready to find out if your signal is strong enough to be heard by the repeater and activate its transmitter. Activating the repeater is called *hitting the repeater* (or machine). Start by adjusting your squelch control so that the noise is just cut off. Remember to listen first to be sure you won't be interrupting any conversations, then press the microphone's push-to-talk (PTT) switch and say your call sign following by "Testing." Release the PTT switch and watch the signal strength indicator on your radio. If you were successful, the indicator display will show you that the repeater's output signal is being received. The repeater's output signal will be present for a few seconds then you'll hear a "tssssschht" sound. That sound is a *squelch tail* — the noise output by the repeater's receiver with no input signal before its own squelch circuit activates and shuts off audio output. The short delay in cutting off the transmitter is to keep the repeater from turning on and off rapidly due to a weak signal at the input.

ID AND CONTROL TOPICS

There you are, listening to a repeater conversation and suddenly…nothing. The repeater just shuts off — no output signal or anything! Has the repeater transmitter failed? Most likely, the repeater has *timed out* and stopped retransmitting the input signals. If you listen a little longer, the repeater will again respond and may even make a synthesized voice announcement, "Time out." One of the long-winded operators will say, "We timed out the repeater…" and proceed.

Most repeaters start a timer when they begin transmitting. If the timer expires, typically in three minutes or so, without the transmitter turning off, the repeater turns off its transmitter. This prevents overheating of the transmitter and gives stations a chance to break in by keeping one signal from occupying the repeater for long periods without a break. The timer resets when the transmitter turns off because the receiver does not detect an input signal. To reset the timer during your conversation, let the repeater *drop*. That is, stop transmitting long enough to hear the repeater's squelch tail as the transmitter shuts down.

REPEATER SYSTEMS

To extend their range and to hear signals blocked by obstacles, repeaters often employ *remote receivers*. The signals from these receivers are then transmitted by an *auxiliary station* to the repeater's transmitter site for retransmission.

Repeaters can also be *linked* to other repeaters. That is, they share the audio signals each receives, retransmitting them over a wider area than any one repeater can cover. It is also common for repeaters to retransmit signals on other bands. For example, a 2 meter repeater linked to a 70 cm repeater allows stations on either band to contact each other.

If the repeaters are *co-located*, meaning located at the same site, the repeaters can be physically connected with cables. Otherwise, a *control link* is required.

Control links consist of a transmitter and receiver that only relay audio and control signals between repeaters. These *auxiliary stations* are not used for direct contacts and most are on the 1.25 meter and 70 cm bands. The signals carried by the links control various repeater features, usually enabling and disabling the retransmission of signals by the linked repeaters. Repeater *networks* or *systems* are made up of several linked repeaters that can be many miles apart. Control signals are used to configure the way in which the network relays signals between repeaters.

If you become a regular user of a linked repeater system, you may want to join the group operating the repeaters. You can be authorized to use the *control codes* (sequences of audio tones like those used by telephones) to configure the repeater network and even perform basic maintenance and test functions. This is a valuable service you can provide.

The repeater *controller* is a piece of equipment that regularly sends the repeater ID, operates the time-out timer, switches the transmitter on and off and so forth. Modern controllers have microprocessors and sophisticated electronics that provide advanced features, such as synthesized voice announcements, time and date, weather conditions and other interesting things. These are generally activated with control codes of their own. Some repeaters offer the ability to make phones calls via ham radio, a system called *autopatch*. These functions are generally activated and managed with control codes of their own.

To use control codes of any sort on the repeater requires that your transceiver be able to generate the appropriate tones or tone sequences. Check your radio's operating manual to see how that's done. It may be as simple as pressing the PTT switch and pressing the numeric keys on the radio. Be sure to identify your transmission by stating your call sign first. For example, "W1AW" followed by (tone) (tone) (tone). Each repeater system will have its own protocol for using the control codes.

OPEN, SPECIAL USE AND PRIVATE REPEATERS

If you look in a repeater directory, you may see a symbol in the listing that indicates a repeater is *closed*. That means the repeater is not available for public use — only authorized stations may use the repeater. Other repeaters are dedicated to a special purpose, such as emergency communications. In both cases, the repeater owners prefer to restrict the use of their repeater. This is perfectly legitimate and allowed under FCC rules.

Most repeaters are open and free for anyone to use. Closed repeaters usually require membership in a group that supports the expenses of operating and maintaining a repeater — it may have special features or capabilities. To find out how to join the group, enter the repeater's call sign into the "Call Sign Search" window on the ARRL's home page. Contact the repeater group via the mailing address shown.

DIGITAL REPEATER SYSTEMS

Ham radio and the Internet each have complementary advantages. Hams can roam freely, using repeaters from a vehicle, at home or on foot. The Internet provides a high-speed connection between two points — nearly anywhere on Earth. It's a natural to combine the two and several systems do just that:

- IRLP (Internet Radio Linking Project)
- Echolink
- WIRES II — a proprietary system of the Yaesu company
- D-STAR — a system based on the public D-STAR standard

The two most popular systems, IRLP and Echolink, use VoIP (Voice over Internet Protocol) technology to link repeaters as illustrated in **Figure 6.4**. [T8C13] Online

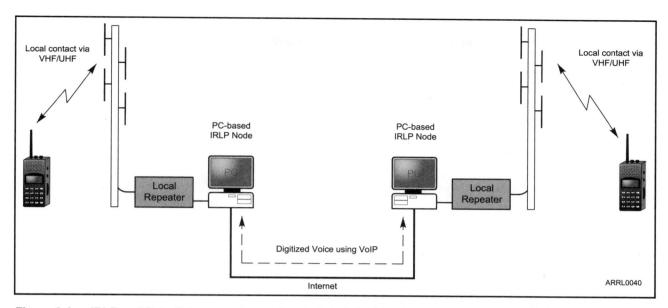

Figure 6.4 — IRLP and Echolink are systems of repeaters linked by the Internet protocol VoIP. Hams can use a local repeater and radio to make contacts worldwide by using control codes to connect to far-away repeaters.

telephone service vendors, such as Skype, use VoIP to deliver voice communications over the Internet by digital techniques. [T8C12] The main difference between IRLP and Echolink is that IRLP requires all audio to be transmitted into the system via a radio link. (IRLP does allow a PC user to *listen* to conversations.) That means you must be a licensed amateur to use repeaters linked by IRLP. Echolink allows audio to come from a PC and microphone, so a radio is not necessary but hams are required to send a copy of their license to the Echolink system administrators to be authorized to use the system.

If you are interested in using or listening to stations using VoIP, first explore the websites for each system. You'll find comprehensive information on how the system works, procedures for using it, and directories of access and control codes. The codes are entered by using your transceiver's keypad.

The D-STAR system provides voice communications and adds the ability to send data across the system, as well. In its high-speed form, D-STAR acts as an Ethernet network *bridge* and allowing a computer with a web browser to connect to the Internet or another D-STAR radio. The D-STAR standard was developed by the Japan Amateur Radio League (JARL) and is currently implemented in equipment manufactured by the Icom company.

Wires II uses a proprietary voice-only standard developed by radio manufacturer Yaesu. More protocols and systems are likely to appear in the coming years, so keep an eye on newsletters and magazines for information about them.

How does an IRLP or Echolink contact differ from a regular repeater contact? To initiate an IRLP or Echolink contact, the initiating station must know the repeater control code to request an IRLP connection — this is the ON code which is a sequence of DTMF (Dual-tone Multi-Frequency) tones, like dialing a phone number. [T8C06] The ON code varies from repeater to repeater and obtaining it may require membership in a club. Once the ON code is entered via your keypad, the four-digit code for the IRLP *node* — a destination repeater — is accessed. [T8C10] You will hear a confirmation tone or an error tone. If confirmed, announce your presence on the destination repeater as if you were operating locally to that repeater. IRLP nodes are listed in repeater directories and on the Internet. [T8C09]

At the destination repeater, operation is more like a regular repeater contact. You will hear a tone as the remote station connects via the IRLP network, then the connecting station will be heard. It can be disconcerting to be driving to work and hear a foreign call sign on your local repeater! Nevertheless, it's fun to meet hams from around the world. Much more information on Internet-linked systems is available in the ARRL book *VoIP: Internet Linking for Radio Amateurs*.

Before you go on, study these Technician exam questions from the question pool included at the back of this book or as a downloadable Study Guide version on the web:
T2A01 T2A03 T2B02 T2B04
T4B11
T8C06 T8C09 T8C10 T8C12 T8C13
If you have difficulty with any question, review the preceding section.

6.5 Nets

As you read about and experience more of ham radio, you will encounter frequent references to *nets*. What are these mysterious organizations and how can you participate? "Net" is just an abbreviation for *network*. Developed in the very early days of radio, these networks helped stations meet on the air to share news and exchange messages, called *traffic* in net lingo. The modern network, now associated with computers, is a direct descendent of the radio net and uses many of the same terms and concepts. There are lots of nets that are available to the Technician licensee.

TYPES OF NETS

Amateur Radio makes extensive use of nets and there are three major types: social, traffic and emergency or public service. Nets can serve a group as small as a club or have international coverage. Some are very formal and follow a strict procedure while others are more like a group conversation. Each net has a theme or purpose and has a regular schedule to convene at specific frequencies and times. Nets can be found on both HF and VHF/UHF bands, using the frequencies best suited to their coverage needs.

Social

The least formal and most common are the social nets, ham radio's on-the-air meetings. Themes of these nets vary widely, from hobbies (stamp collecting, model rocketry, chess and other interests) to award chasing (DX and county hunting) to "stay-in-contact" nets that support stations in RVs and aboard boats as they travel. These nets are easy to join and rarely require any training or special procedures. Just listen to how the net conducts its business and try to follow what you hear. Some nets maintain websites with useful information and operating procedures.

Traffic

This is the "original" net — an on-the-air method of exchanging and routing formal messages, known as *traffic*, across town or across the country. An extensive structure of traffic nets, the National Traffic System (NTS) is designed for efficient station-to-station *traffic handling*. The NTS is composed of local, regional and national nets through which the messages pass, station-to-station, until a local net member passes them to the recipient. The NTS is active every day, especially during emergency and disaster recoveries, such as from hurricanes. Traffic nets follow a set procedure that can be learned quickly. Local VHF traffic nets are easy to join and welcome newcomers.

Emergency and Public Service

Hams coordinate their emergency response activities with nets that spring into action whenever they're needed. In areas where severe weather is common, nets monitor weather conditions before a storm and assist with recovery in case damage results. Emergency nets are frequently dual-purpose: they pass emergency traffic and coordinate reporting and response activities. While waiting to be called to service, emergency nets hold regular information and training sessions. That is the time to join the group and learn the procedures, not after disaster strikes!

As a way to practice skills and provide service at the same time, hams also provide communications for public events, such as festivals, parades or sporting events. These public service activities are organized on the ground and on the air just as a disaster response would require. You can volunteer for one of these events and that is how many hams developed into skilled emergency response operators!

NET STRUCTURE AND PARTICIPATION

In order to participate effectively in a net, you need to know a little bit about how a net is organized. Knowing the common procedures and signals is also important. Knowing a few of the rules of the road keeps traffic flowing smoothly, both for vehicles and nets.

It is important to remember that no matter what the purpose or status of a net, a station with emergency traffic should break in at any time. If the net is operating on phone and you are reporting an emergency, break in by saying "Priority" or "Emergency," followed by your call sign. [T2C06] The NCS and members of the net should always immediately suspend any lower-priority operation and respond to the emergency.

Emergency Nets

There are some important differences in operations when a net is responding to a disaster or emergency. Efficiency and accuracy become the highest priorities so that the most important business is handled at all times. This requires all stations to maintain *net discipline*, following directions and net procedures at all times. Does this mean you must be an expert operator to participate in emergency nets? No, but you must be willing to listen and to follow directions!

Listening is a very important skill under these conditions. It is natural to want to contribute comments and suggestions under stressful circumstances, but doing so often causes delays and mistakes. Once you have checked into an emergency net, you should not transmit unless you are specifically requested or authorized to do so or a request is made for capabilities or information that you can provide. [T2C07]

Once an emergency net is established, the NCS must continue to make emergency communications the highest priority at all times. This includes emergency messages, such as radiograms with the Emergency status. Even if other messages have been waiting, emergency traffic of any sort has the highest priority. The order of priority for communications is Emergency, Priority, Health-and-Welfare and then Routine.

Table 6.6

Common Net Q-Signals and Procedural Signals

QNI	Net stations report in
QNU	The net has traffic for you
QRU	I have nothing for you
QRV	I am ready to copy
QSP	Please relay the following information
QTC	I have the following traffic
$\overline{AR}$	End of message
$\overline{AS}$	Stand by
K	Go ahead
CL	Closing station

Note: The overbar above characters, such as $\overline{AR}$, indicates a prosign. Prosigns are sent as Morse characters without any space between them.

The American Radio Relay League
RADIOGRAM
Via Amateur Radio

Number	Precedence	HX	Station of Origin	Check	Place of Origin	Time Filed	Date
207	P	E	W1FN	10	LEBANON NH	1200 EST	JAN 4

To:
MARK DOE
RED CROSS DISASTER OFFICE
123 MAIN ST
RUTLAND VT 05701

Telephone Number: 802-555-1212

This Radio Message was received at:

Amateur Station_____ Date_____
Name_____
Street Address_____
City, State, Zip_____

NEED   MORE   COTS   AND   SANITATION
KITS   AT   ALL   FIVE   SHELTERS
_____ _____ _____ _____ _____
_____ _____ _____ _____ _____
_____ _____ JOAN SMITH SHELTER MANAGER

	From	Date	Time		To	Date	Time
REC'D				SENT			

A licensed Amateur Radio Operator, whose address is shown above, handled this message free of charge. As such messages are handled solely for the pleasure of operating, a 'Ham' Operator can accept no compensation. A return message may be filed with the 'Ham' delivering this message to you. Further information on Amateur Radio may be obtained from ARRL Headquarters, 225, Main Street, Newington, CT 06111.

The American Radio Relay League, Inc. is the National Membership Society of licensed radio amateurs and the publisher of QST Magazine. One of its functions is promotion of public service communication among Amateur Operators. To that end, The League has organized the National Traffic System for daily nationwide message handling.

Figure 6.5 — The ARRL Radiogram form is the standard for originating and relaying messages. The preamble identifies the message and allows it to be tracked. The 25-word limit requires that the message be focused, balancing length against minimizing errors during transmission.

TRAFFIC HANDLING

The most important job during emergency and disaster net operation is the ability to accurately relay or "pass" messages exactly as written, spoken or received. [T2C08] It is common for messages to be formatted as *radiograms*. A typical radiogram is shown in **Figure 6.5**. A radiogram has three parts: the preamble (including the address), the body and the signature.

The preamble is made up of several bits of information about the message. These establish a unique identity for each message so that it can be handled and tracked appropriately as it moves through the Amateur Radio traffic handling system. [T2C10]

- Number — a unique number assigned by the station that creates the radiogram
- Precedence — a description of the nature of the radiogram: Routine, Priority, Emergency and Welfare
- Handling Instructions (HX) — for special instructions in how to the handle the radiogram.
- Station of Origin — the call sign of the radio station from which the radiogram was first sent by Amateur Radio. (This allows information about the message to be returned to the sending station.)
- Check — the number of words and word equivalents in the radiogram text. [T2C11]
- Place of Origin — the name of the town from which the radiogram started
- Time and Date — the time and date the radiogram is received at the station that first sent it
- Address — the complete name, street and number, city and state to whom the radiogram is going

Following the preamble is the text of the radiogram. You should take extra care to be sure the receiving station copies the message exactly. For example, proper names (such as "John Doe") and unusual words (such as material names or model identifiers) are spelled out using standard phonetics. [T2C03]

The ARRL radiogram limits the text to 25 total words. That doesn't sound like a lot, but the limit helps focus the message topic and eliminate unnecessary words. Longer messages also are more difficult to relay without errors so there must be a balance between length and accuracy. The signature follows the text and is usually the name of the person originating the message.

If you are asked to generate a radiogram, you should follow this format. If you cannot decide on what to use for some of the preamble contents, leave them blank. In an emergency, you may not always have all of the information. Another net station may be able to help you, or the message can be sent without the information. In an emergency, though, the one item that must be included is the name of the person originating the message.

If you check in to a traffic handling net, you will find that they make heavy use of the Q-signals you learned about earlier. Their clear meaning and short length greatly speed up

The radiogram has its origins in the telegraph message. Telegraph companies quickly learned that it was important to be able to track messages and minimize relaying errors. They developed the idea of adding a *preamble* to the message containing information describing the message and providing the means to trace it back to its origins.

The preamble concept proved so useful that it has been carried forward into the Internet and computer networks. The word "preamble" refers to the same information today — the information at the head of a transmission containing information about the message that follows. If you examine a transmission of data on a garden-variety Ethernet network, you will find that each packet of data has a preamble with a unique number and address, among other things.

Internet messages are also protected against error by the use of *checksums*, a word derived from the telegram's *check*. Both help the receiver detect errors. If a radiogram is received with a check different than the number of words in the text or if a network packet is received with a checksum that doesn't match what the receiver thinks it should be, then a request for retransmission is made.

Everywhere you look in modern network technology, you will find the echoes of the landline telegraphers and radio operators!

operations. They can be found in the *ARRL Operating Manual* which includes a chapter on traffic handling. The radiogram quick-reference card, ARRL FSD-218 can also be downloaded from the ARRL website.

FINDING NETS

You can find out what nets operate in your area by using the online *ARRL Net Directory*. Access the online net search in the ARRL website's "On the Air" section. To get an idea of the nets in your area, select "State Nets" and select your state from the list. You'll be surprised at how many there are to choose from! Click on the name of the net for detailed information about the net including an e-mail address for the net's manager. You can print out your search results for use in the field.

Before you go on, study these Technician exam questions from the question pool included at the back of this book or as a downloadable Study Guide version on the web:
  T2C03 T2C06 T2C07 T2C08 T2C10 T2C11
If you have difficulty with any question, review the preceding section.

6.6 Communications for Public Service

Thankfully, true emergencies and disasters are rare. Far more frequently, hams provide communications in support of public events such as parades or races. All of these are referred to as *public service* communications. Don't wait for an actual emergency or disaster to occur to participate in public service! It's a great exercise of your license privileges and provides excellent training to develop all kinds of communications skills.

When you are providing public service communications, remember that you are not allowed to receive payment for your services except for reimbursement of actual out-of-pocket expenses, including mileage. You may not charge an hourly fee or arrange a trade of your services for something else of value to you or your organization. Avoid providing communications when there is no benefit to the public, such as a private event.

PUBLIC SERVICE OPERATING GUIDELINES

Many hams are regularly involved with public service communications. It may be the reason you became interested in ham radio. When providing public service communications, you must also remember to operate efficiently and strive for the highest levels of performance. Here are some good public service operating practices:

- Don't become part of the event — you are there to assist, not to participate or act as an event manager.
- Maintain your safety — you are no help if you become injured.
- Maintain radio discipline — follow established protocols and refrain from idle conversation that might impede communications.
- Never speculate or guess — strive for 100% accuracy and don't be afraid to say "I don't know." Rumors are impossible to stop, once started.
- Protect personal information — never send confidential personal information via Amateur Radio without consent.
- Don't give out unauthorized information — reporters and members of the public are often hungry for information. Direct them to the appropriate spokesperson or information source.

TACTICAL COMMUNICATIONS

Tactical communications may be used to coordinate activities ("Go to the south parking lot"), report status ("The final float is leaving the staging area") or request resources ("First aid is needed at 2nd and the highway."). This type of message is rarely recorded and is not passed in radiogram format.

Tactical communication needs are usually satisfied by using VHF/UHF simplex or repeater channels. Mobile, portable and handheld radios are particularly useful when working with public safety and government agencies.

To increase efficiency and smooth coordination, stations engaged in tactical communications should use *tactical call signs*, such as "Command Post Three" or "School Kitchen" or "Judges Stand." These describe a function, location or organization. This allows operators to change without changing the call sign of the stations and frees non-amateur personnel from having to use amateur call signs. Tactical call signs do not, however, satisfy the FCC regulations for station identification, which must still be followed. Identify with your call sign every 10 minutes and at the end of the communication.

ARES AND RACES

The two largest Amateur Radio emergency response organizations are ARES® (Amateur Radio Emergency Service) sponsored by the ARRL and the Radio Amateur Civil Emergency Service (RACES). Both organizations provide emergency and disaster response communications. [T2C04] ARES consists of licensed amateurs who have voluntarily registered their qualifications and equipment for communications duty in the public service. [T2C12] Its members support local and regional government and non-governmental agencies such as the Red Cross, Salvation Army and National Weather Service. Any licensed amateur can participate in ARES.

RACES is a special part of the Amateur service created by the FCC to provide communications assistance to local, state, or federal government emergency management agencies during civil emergencies. [T2C05]. (See Part 97.407 of the FCC rules for more information on RACES.) Many amateurs are members of both ARES and RACES teams so that they can respond to either need.

EMERGENCIES AND DISASTER RELIEF

Providing communications assistance during an emergency or in response to a natural disaster is one of the Amateur Service's most important reasons for existing at all. In fact, in Part 97.1, the Basis and Purpose for the Amateur Service, emergency communications is the very first reason! The FCC places top priority on emergency communications which have priority over all other types of Amateur Radio communications on any frequency.

In a serious, widespread emergency, the FCC may declare a *temporary state of communications emergency*. The declaration will contain any special conditions or rules that are to be observed during the emergency. The declaration is in force until the FCC lifts it. Communications emergency declarations are distributed by the FCC through its website, via ARRL bulletins on headquarters station W1AW and the ARRL website, and through the National Traffic System and Official Relay Stations. Amateur websites and e-mail lists pick up the declarations and relay them throughout the amateur community.

The only exception to the "no one owns a frequency" rule is during a natural disaster or other communications emergency when you should avoid operating on or near frequencies used to provide disaster relief.

THREATS TO LIFE AND PROPERTY

The FCC also recognizes the need for flexibility in an emergency. For example, in defining emergency communications, Part 97.403 says:

"No provision of these rules prevents the use by an amateur station of any means of radiocommunication at its disposal to provide essential communication needs in connection with the immediate safety of human life and immediate protection of property when normal communication systems are not available."

If communications services are down and you're in the middle of a hurricane, tornado or blizzard, and you offer your communications services to the authorities, you are permitted to do whatever you need to do to help deal with the emergency. Public safety or medical personnel can use your radio. In an emergency situation where there is immediate risk to life or property and normal forms of communication are unavailable, you may use any means possible to address that risk, including operating outside the frequency privileges of your license. [T2C09] You are only prohibited from transmitting information on behalf of your employer or confidential personal information of a third party, such as a disaster victim, without their consent.

Similarly, in an emergency situation you may use whatever communications means is at hand to respond — any means on any frequency. If a fire department radio or marine SSB transceiver is all that's available, by all means use it to call any station you think might hear you!

This waiver of normal rules lasts as long as the threat to life and property remains imminent *and* there are no other means of communications than Amateur Radio. Once the threat has receded or normal communications become available, you must return to normal rules, even in support of public safety agencies. You are bound by FCC rules at all times, even if using your radio in support of a public safety agency. [T2C01] For example, while providing post-event communications at a fire department command post, you are not permitted to use a modified ham radio on fire department frequencies.

DISTRESS CALLS

If you are in immediate danger or require immediate emergency help, you may make a distress call on any frequency on which you have a chance of being heard. In these circumstances, here's what to do:

- On a voice mode, say "Mayday Mayday Mayday" or on CW send "$\overline{SOS}$ $\overline{SOS}$ $\overline{SOS}$" (Mayday should not be confused with the Pan-Pan urgency call) followed by "any station come in please"

- Identify the transmission with your call sign
- Give your location with enough detail to be located
- State the nature of the situation
- Describe the type of assistance required
- Give any other pertinent information

Then pause for any station to answer. Repeat the procedure as long as possible or until you get an answer. The reason for giving all of the information during each call is so that if you can't hear responding stations, they'll still learn where you are and what help you need. Under no circumstances make a false distress call or allow others to do so using your equipment. Your amateur license could be revoked and you could be subject to a substantial penalty or even imprisonment.

If you hear a distress call — on any frequency — you may respond. Outside the amateur bands, such as on the international marine distress calling frequency of 2182 kHz, be sure that no other station or vessel is responding before you call the station. Inside the amateur bands, suspend any other ongoing communications immediately. Record everything the station sends and then respond. If they hear you, let them know that you have copied their information, clarify any information as required and immediately contact the proper authorities. Stay on frequency with the station in distress until authorities are either on frequency or arrive at the scene.

EMERGENCY COMMUNICATIONS TRAINING

To be truly effective when responding to an emergency, you need some training and even better, some practice opportunities! Doesn't music sound a lot better when the musicians have learned the music and practiced it before the show? Don't let your license gather dust while you wait for the "Big One" — by getting on the air you can continuously improve your skills and have fun doing it.

Start by joining a local amateur emergency preparedness team. Your local radio club, ARES team, RACES team, county Search-and-Rescue or Salvation Army chapter (just to name a few) are all organizations to investigate. Choose the one that suits your interests.

Take advantage of any training the group might provide or recommend. For example, the ARRL offers basic and advanced emergency communications online training classes. The Federal Emergency Management Agency (FEMA) offers free emergency preparedness training courses on their website. The courses on the National Incident Management System (NIMS) are very helpful to learn how public safety agencies will be organized in a disaster.

When your group has a drill or exercise, try to participate, even if just in the planning and organization stage. The experience will serve you well when a real activation occurs. The ARRL sponsors an annual Simulated Emergency Test in October of every year. If your club participates in ARRL Field Day, be sure to attend. Almost every town has at least one public event with communications needs that ham radio could fill. Help plan to put an amateur team on the job!

Finally, test your own preparedness! Check your go-kit and emergency equipment every six months to be sure it's all together and working. Double check your power sources, especially batteries that might grow weak over time. You're only as effective as your equipment will let you be.

EMERGENCY COMMUNICATIONS AND YOUR EMPLOYER

Many emergency and public safety personnel have obtained an Amateur Radio license, attracted by our service's flexibility and adaptability. Many employers have also taken note of the service's capabilities. This has the potential to conflict with part 97.113(a)(3) that forbids amateurs from having a financial interest in their communications:

"No amateur shall transmit...Communications in which the station licensee or control operator has a pecuniary interest, including communications on behalf of an employer."

There are only two exceptions to this rule. They are part 97.113(c) and (d) — teachers who use ham radio as part of their instruction and operators employed to operate a club station that transmits bulletins and code practice at least 40 hours per week on at least six amateur bands.

Many radio clubs are made up of employees of a particular employer. This is permitted, as is communications while you are at work as long as it is not on behalf of your employer (and your employer permits it).

Participating in training exercises and drills organized by your employer is allowed but only according to the restrictions spelled out in 97.113(a)(i). The drills are limited in number and duration as a balance between readiness and keeping Amateur Radio free of commercial operating. If you have any questions about this type of drill, ask your ARES Emergency Coordinator.

It is important to preserve the "bright line" between the strictly-volunteer foundation of Amateur Radio and the many commercial and government uses of radio — worthy though they may be. Communications on a regular basis that could reasonably be furnished through other radio services are not permitted.

It is also important to note that news messages and reports are not considered emergency communications by the FCC. You are not allowed to relay such information on behalf of broadcasters via Amateur Radio. Inform reporters who ask you to relay reports that you can't do that under FCC rules.

If you are confused about what is and is not permitted, review the guidelines published in *The Commercialization of Amateur Radio: The Rules, The Risks, The Issues* and available on the ARRL website. If you are interested in getting involved with emergency communications in your area and have questions, your instructor can help you contact your local ARES emergency leadership, such as an ARES team Emergency Coordinator (EC).

Before you go on, study these Technician exam questions from the question pool included at the back of this book or as a downloadable Study Guide version on the web:

T2C01 T2C04 T2C05 T2C09 T2C12

If you have difficulty with any question, review the preceding section.

6.7 Special Activities, Modes and Techniques

We've just discussed a few of the many different ham operating styles and opportunities. The longer you stay with the hobby, the more different parts of it you discover! Here is a survey of some activities that you are likely to encounter as you enter the hobby.

DXING, AWARDS AND CONTESTING

Since the beginning of radio, even before amateurs appeared on the scene, operators strove to make contact over longer and longer distances. Marconi himself started by sending messages across a few hundred yards and gradually built up his capabilities to where he could span the Atlantic Ocean. Amateur Radio is no different. An enduring and popular pastime is to see if you can pull in far away signals from away over the horizon, exchanging QSL cards to confirm and remember the contact.

This is called DXing, where DX stands for "distant station." Distance is a relative thing.

DX means thousands of miles on HF and occasionally 6 meters. At VHF/UHF, any contact beyond the radio horizon is considered DX. Microwave operators scout out locations with unobstructed views to make contacts of many miles. Making DX contacts is best done on SSB or CW because of the efficiency of those modes.

VHF/UHF DX contacts tend to be short, since the time during which the band opens for long-distance propagation is usually brief. You'll need a multimode transceiver to use SSB, CW or digital modes for VHF/UHF DXing. [T7A09] When making a VHF/UHF DX contact, the most important piece of information to exchange is your grid square, discussed earlier in the section on locators. While a dipole or FM vertical will work at times, you'll get much better results by using a beam antenna that you can point in different directions. Your antennas should be horizontally polarized — the norm for VHF/UHF SSB, CW and digital operation. Again, take care to log your contacts properly with an accurate time.

Pursuing long-distance contacts really hones a ham's technical and operating skills. In the course of DXing, one learns many things about propagation, antennas and the natural environment! To recognize the achievements of DXers there are many awards offered by the ARRL and other organizations. For contacting the numerous countries the DXCC award is popular. VHF/UHF enthusiasts contact grid squares for the VUCC award. Contacting all of the US states (Worked All States — WAS) is popular around the world.

Aside from DXing, there are many awards for which a ham can qualify in the course of casual operating. Some radio clubs offer awards for contacting their members, for example. Organizations for low power enthusiasts offer awards. A popular one is the QRP ARCI certificate shown in **Figure 6.6**, awarded for making a contact that spans 1000 miles per watt of power. The list of awards is staggering. Ted, K1BV, compiles an online awards directory that lists over 3300 awards!

Focusing the competitive urge even more, radio *contests* are held in which the competitors try to make as many short contacts as possible in a fixed period of time. [T8C03] Some contests, called *sprints*, are very short. Others last an entire weekend. There are contests that use just one band or mode and others that span multiple bands and multiple modes.

If you encounter a contest on the air, jump in and make a few contacts. You'll be asked for some information called an *exchange*. It may consist of your location, a signal report and a *serial number* (that's the number of contacts you've made in the contest so far). Just ask, "What do you need?" The contest station will help you provide the right information. To keep things moving, send only the minimum information needed to identify your station and complete the exchange. [T8C04] You'll find that it's a lot of fun for even a casual operator!

As you might imagine, pursuing operating awards and participating in contests have the potential to create a very capable operator. The excellent stations and skills of DXers and contest operators are quite applicable to emergency operating and traffic

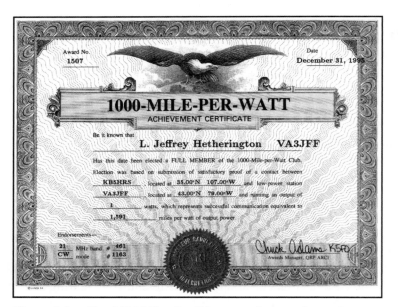

Figure 6.6 — The 1000 Miles Per Watt award recognizes a low-power station's ability to communicate over long distances. This is just one of thousands of operating awards available to hams for every imaginable type of operating achievement.

ARRL Field Day — The Biggest Amateur Event of All!

Every year on the fourth full weekend of June, North American hams head for the hills…and the fields and the parks and the backyards. It's Field Day! This is the annual emergency preparedness exercise in which more hams participate than any other. The basic idea — set up a portable station (or several) and try to make as many contacts with other ham groups as possible on as many amateur bands as possible. If you think the bands are busy on weekends, wait until you hear them during ARRL Field Day!

Some groups focus on the emergency preparedness aspect, others get into the competitive aspect trying for the most points, and some just treat it as the annual club picnic plus radio operating. Whatever your organization prefers, Field Day is a great way to see a lot of ham radio all in one spot and all at the same time. For more information, browse to ARRL Field Day web page, read the Field Day announcement in the May issue of *QST* magazine, or enjoy the Field Day summary and results that usually appear in the December issue. CQ, Field Day!

Over the River and Through the Woods

A different and more physical type of contest is known as *foxhunting*. Locating a hidden transmitter (the fox) has been a popular ham activity for many years. It has its practical side, too, training hams to find downed aircraft, lost hikers, and sources of interference or jamming. [T8C01] You don't need much in the way of equipment. You can get started with a portable radio with a signal strength indicator and a handheld or portable directional antenna, such as a small Yagi beam. [T8C02] One ham hides the transmitter (hams can be very devious and inventive when it comes to hiding places) and the rest drive, walk, or bike the area taking bearings and attempting to be the first to locate the transmitter.

In recent years a new type of outdoor radiosport has reached US shores from Europe and Asia — *radio direction finding*. Held as organized events, direction finding is a hybrid of the radio fox hunt using orienteering skills to navigate outdoors with map and compass. The US Amateur Radio Direction Finding organization is just one of a number of national groups in this worldwide sport, especially popular with teens and young adults. If you are a hiker or camper, then you might be interested in applying your outdoor skills to ARDF.

handling, as well. In fact, many of today's top contest operators got their start handling radiograms and participating in net operations! The ARRL and other organizations sponsor contests that run the gamut from international events attracting thousands to quiet, relaxed competitions to contact lighthouses or islands. You can find the rules for these events on the ARRL's Contest Calendar web page.

Special Events

In between the contesters and the DXers are the *special event* stations that operate for a short period to commemorate or publicize an activity of special significance. For example, a club might set up a station at a state fair or a sporting event.

Apart from the sheer novelty of it, the stations often offer a unique or colorful QSL card or certificate for contacting them. Many of them also obtain special call signs that can only be logged during their activity. You'll find these stations on the air from all around the world and collecting their QSL cards is a popular past time for many amateurs.

Before you go on, study these Technician exam questions from the question pool included at the back of this book or as a downloadable Study Guide version on the web:
T7A09
T8C01 through T8C04
If you have difficulty with any question, review the preceding section.

SATELLITES

Amateurs have built more than 50 satellites since 1961, launching them when extra space is available in a rocket payload. Amateur satellites are nicknamed *OSCAR* for *Orbiting Satellite Carrying Amateur Radio*. Their purpose is to enable amateurs to communicate with each other, not to provide navigation or other services such as autopatch calls. A satellite far enough above the Earth can even relay signals between countries. Some amateurs have even obtained the coveted DXCC award for contacting 100 different countries through satellites!

Amateur satellites relay signal between bands, usually on VHF and UHF, or act as FM repeaters that can be accessed with regular FM rigs. The ionosphere is usually transparent to signals at these frequencies, so the signals can pass between Earth and space easily. If two stations both have the satellite in view at the same time as shown in **Figure 6.7**, they can make contact via line-of-sight propagation to and from the satellite.

Communicating through an amateur satellite sounds like a very complicated, high-tech effort, but it can be quite simple. What you need is a radio that can transmit on one band and listen on another (most can, even handhelds). Satellite contacts, including contacts with the amateur station on the International Space Station can be made by any amateur licensed to transmit on the *uplink* frequency. [T8B01, T8B04] For example, a Technician licensee could communicate through a satellite that is listening for uplink signals on 2 meters and transmitting on a 10 meter *downlink* frequency even though a Technician is not permitted to transmit on 10 meters. Satellite uplink and downlink frequencies are restricted to the special satellite *sub-bands* listed in **Table 6.7**, segments of frequencies set aside for Earth-to-space communications.

Here are some terms that are commonly used regarding satellite communications:

- Apogee — The point of a satellite's orbit that is farthest from Earth
- Beacon — A signal from the satellite containing information about a satellite [T8B05]
- Doppler shift — A shift in a signal's frequency due to relative motion between the satellite and the Earth station [T8B07]
- Perigee — The point of a satellite's orbit that is nearest the Earth
- Keplerian elements — A set of numbers that describe the satellite's orbit so that it can be tracked
- LEO — Low Earth Orbit [T8B10]
- Elliptical orbit — An orbit with a large difference between apogee and perigee

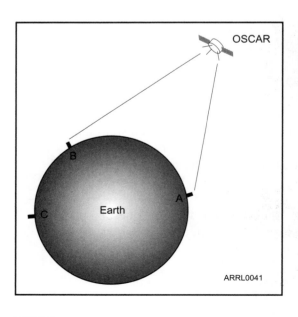

Figure 6.7 — Amateur satellites (OSCARs) can relay VHF and UHF signals between any two stations that both have the satellite in view at the same time. Contacts through the satellite are possible between station A and B, but not with station C.

Table 6.7

Selected Satellite Sub-bands

29.300-29.510 MHz

145.80-146.00 MHz

435.00-438.00 MHz

1260-1270 MHz

2400-2410 and 2430-2438 MHz

It is a pleasant surprise to learn that Amateur Radio has a place on the International Space Station (ISS)! Not only that but nearly all of the astronauts hold Amateur Radio licenses. Depending on their workload, astronauts can be active from their orbiting home on the ISS. Any amateur licensed to use the 2 meter band can join in the fun from Earth.

The astronauts frequently operate using FM voice. A packet bulletin board system (BBS) and a voice FM system are both on-board. One or the other is active at all times — you can check the mode of the station on the AMSAT website. The ISS also carries an APRS digipeater and a digital ATV station!

When the ISS is in view, you can connect to the packet BBS with a regular packet station using a TNC and your 2 meter radio. If you prefer, you can also just listen to 145.800 MHz in hopes of hearing one of the astronauts on voice. To call the space station, call sign NA1SS, set your radio to transmit on 145.990 MHz and listen on 145.800 MHz.

- Spin fading — Signal fading caused by rotation of the satellite and its antennas [T8B09]
- Pacsat — a satellite equipped with FM packet radio for digital communications [T8B11]

To find out when a satellite will make a *pass* above the horizon at your location and can be accessed, you'll need a *satellite tracking program*. The tracking program will need you to enter certain bits of data about the satellite's orbit called the *Keplerian elements*. [T8B06] "Keps" as they are often called are available online and the program may be able to download them automatically. Using those values, the software can provide real-time maps of the satellite's location, the trajectory the satellite will follow across the sky, and even the amount of *Doppler shift* the signals will experience. [T8B03]

It is a good practice technique to enter the elements for the International Space Station (ISS) and then find it visually in the sky, watching it pass overhead at sunset or dawn. The ISS also carries an amateur station that you can contact by packet, or if you're lucky, one of the astronauts by voice! You can also practice by listening for the beacons of satellites as they pass by.

Next, you'll need to determine the satellite's operational *mode* — the bands on which it is transmitting and receiving. Most satellites only have one mode, but some have several that can be controlled by ground stations. Mode is specified as two letters separated by a slash. The first letter indicates the uplink band and the second letter indicates the downlink band. For example, the uplink for a satellite in U/V mode is in the UHF band (70 cm) and a downlink is in the VHF band (2 meters). [T8B08]

When you are ready to try a satellite contact, known as *squirting the bird*, you'll get best results with a beam antenna that you can aim at the satellite as directed by your tracking program. A small beam is best for starting because it will not have to be pointed very precisely. Some satellites can be contacted with simple vertical antennas when they are directly overhead and the distance to them is low. As you get better at pointing the antenna, you can use a more powerful beam and contact the satellite closer and closer to the horizon, increasing the number of stations on Earth that are in view of the satellite at the same time as your station! Always use the minimum amount of transmitter power to contact satellites, since their relay transmitter power is limited by their solar panels and batteries. [T8B02]

For more information on amateur satellites, investigate the website of AMSAT. AMSAT is the organization that coordinates the building and launch of most amateur satellites. You'll find a lot of information about how satellites work and how to find them on the air and in the sky. There are also bulletins that you can receive to update you on satellite status and news about amateur satellites. Choose one of the active satellites listed on the website and start your quest for a satellite QSO!

Before you go on, study these Technician exam questions from the question pool included at the back of this book or as a downloadable Study Guide version on the web:
T8B01 through T8B11
If you have difficulty with any question, review the preceding section.

SPECIAL MODES

Hams try almost every possible way to communicate over the radio. After all, that's a big part of what Amateur Radio is about — experimentation and adaptation. Here are a few examples of the unusual methods hams have employed.

Video

Hams have two primary means of exchanging pictures or video in real-time, aside from exchanging data files of graphic images or video. *Slow-scan television* (SSTV) was invented by hams in the 1960s in order to send still images over conventional voice radios in about eight seconds. Modern SSTV signals are generated by computers and inexpensive digital cameras and modern SSTV "modes" transmit images in color. You can hear SSTV signals most often around 14.230 MHz. Learn more about SSTV in *The ARRL Operating Manual*, *The ARRL Handbook* or through the ARRL Technical Information Service (TIS) SSTV page.

You can also find *amateur television* (ATV) enthusiasts on the UHF bands at 430 MHz and higher. Because of the signal's wide bandwidth (6 MHz), the mode is restricted to the wide UHF bands. The ATV NTSC fast-scan color television signal is the same as an analog broadcast TV signal. [T8D04] These signals can even be received on a commercial analog TV receiver equipped with a suitable frequency converter to tune amateur frequencies. (Hams are beginning to use digital TV signals, as well, which can be received by digital TV receivers.) *The ARRL Operating Manual*, *The ARRL Handbook* and ATV web page provide more information on ATV.

Meteor Scatter and Moonbounce

You may recall from the previous material on propagation that radio signals are diffracted or reflected by conductive surfaces. Along with the conductive ionosphere layers, there are two more reflecting surfaces in the sky that hams use for communications.

Many thousands of meteoroids, ranging in size from dust particles to rocks, enter the Earth's atmosphere every day. As they burn up in the upper atmosphere, the heat creates a short-lived trail of ionized gases so hot that they can reflect radio signals. The trails last for less than a second to several seconds, but hams can bounce VHF and UHF signals off of them to other hams from 500 to more than 1000 miles away. This is called *meteor scatter*.

While the signals can be sent and received manually by a skilled operator, many hams use special software and the same data interface and sound card used for HF keyboard-to-keyboard modes. The best-known programs are *HSMS* and *WSJT*.

Another large rock in the sky is our lunar companion, the Moon, and yes, hams have bounced VHF and UHF signals off the Moon, too! Earth-Moon-Earth (EME) communications generally use CW or a digital mode optimized for recovery of extremely weak signals. You might be surprised to learn that a huge, steerable, NASA-sized dish is not required!

Radio Control

The final special mode is a hybrid of ham radio and modeling — *radio control* (RC). You are probably familiar with the remote control cars, trucks, boats and planes operated by RC hobbyists. You are probably not aware that special amateur frequencies on the 50 MHz band are set aside for radio control *telecommand* signals. If you are an RC modeler, getting a ham license enables you to avoid the congested non-licensed frequencies near 27, 72 and 75 MHz.

Telecommand signals are one-way transmissions intended to initiate, modify or terminate functions of the controlled device. [T1A13] Signals in the related category of *telemetry* are one-way transmissions as well, but these send back measurements or status information from a measuring instrument or system. [T1A07]. Amateur satellites usually send telemetry signals to ground controllers back on Earth.

Amateurs may transmit telecommand signals with an output power of up to 1 watt. [T8C07] Although the signals do not identify the licensee on the air, RC modelers are required to display their call sign and name and address on the RC transmitter. [T8C08] If you would like to find out more about Amateur Radio and RC modeling, there is more information in *The ARRL Handbook* and at RC hobby shops.

Before you go on, study these Technician exam questions from the question pool included at the back of this book or as a downloadable Study Guide version on the web:
T1A07 T1A13
T8C07 T8C08 T8D04
If you have difficulty with any question, review the preceding section.

Chapter 7

Licensing Regulations

In this chapter, you'll learn about:

- **How FCC rules are organized**
- **Amateur Radio's "mission"**
- **Types of licenses**
- **Licensing exams and Volunteer Examiners**
- **Responsibilities of licensees**
- **Frequency and emission privileges**
- **International radio rules**
- **Amateur call signs**

When you see the mouse, you'll find more information at www.arrl.org/ham-radio-license-manual

It's time for the rules of the road, ham radio style! In the preceding chapters, you've learned the technology and customs of Amateur Radio. Now you have the background to understand what the rules and regulations are intended to accomplish. In turn, that will make it a lot easier for you to learn (and remember!) the rules.

There are two chapters of the book that deal with the rules and regulations. This section deals with licensing regulations — bands and frequencies, call signs, international rules, how the licensing process works and so forth. These are administrative rules. The next chapter will deal with rules about operating.

7.1 Licensing Terms

In dealing with rules and regulations, it's very important everyone uses the same words to mean the same things, so we'll begin with a series of definitions. They are, after all, what the rules are built from! In case you want to look up a specific rule or definition, they're all online at the FCC Wireless Telecommunications Bureau (WTB) website. Better yet, get a copy of the ARRL's *FCC Rules and Regulations for the Amateur Radio Service*. It contains the latest rules and is a useful resource for your ham station.

PART 97

Each of the radio services administered by the FCC has its own section of rules and regulations. The amateur service is defined by and operates according to the rules in Part 97 of the FCC's rules. [T1A03] (The FCC rules are one part of Title 47 of the Code of Federal Regulations (CFR) which is the section on Telecommunications.) Each rule is defined separately and is assigned a number beginning with 97, such as Part 97.101, a set of General Standards. Within each rule, individual parts get additional designators, for example Part 97.101(a) that specifies good practices be used. This looks complicated at first, but the numbering system actually helps you find the exact rule quickly.

Basis and Purpose

Part 97.1 is the most important rule of all — it's the *Basis and Purpose* of Amateur Radio. This explains the "mission" of Amateur Radio, why we are allocated precious RF spectrum, and what Amateur Radio is intended to accomplish. Here's what Part 97.1 says, with a little explanation added:

"The rules and regulations in this Part are designed to provide an amateur radio service having a fundamental purpose as expressed in the following principles:

(a) Recognition and enhancement of the value of the amateur service to the public as a voluntary noncommercial communication service, particularly with respect to providing emergency communications."

An important word to remember is "noncommercial." Hams aren't allowed to be paid for their services (with a few exceptions) and must operate on a voluntary basis. That includes conducting or promoting one's business activities over the air. Hams are extremely valuable when they respond to emergencies and disasters to provide temporary communications, especially because they are volunteers. In fact, responding to emergencies may be the most important reason that the amateur service exists today — it is, after all, the very first reason given!

"(b) Continuation and extension of the amateur's proven ability to contribute to the advancement of the radio art."

Hams have a history of discovering and inventing that continues today. After World War I, hams were given all of the "worthless" shortwave bands, but soon discovered that they were perfect for long-distance communications. Even with all the communications research going on around the world, hams still invent useful systems and antennas. Ham radio's famous creativity pays back the public's investment of spectrum many times over.

"(c) Encouragement and improvement of the amateur service through rules which provide for advancing skills in both the communications and technical phases of the art."

Not only do hams tinker with radios, but they train to operate them in useful ways. Events such as Field Day and the myriad exercises held all around the world are ways in which amateurs keep their emergency response skills sharp. Competitive operating events, chasing awards, and station-building continually develop the ham's communications skills. [T1A01]

"(d) Expansion of the existing reservoir within the amateur radio service of trained operators, technicians, and electronics experts."

Having a bunch of folks around who are handy with radios has turned out to be a great idea over the years! There is a long list of ways in which hams have shown their communications skills to be a valuable resource to the public, to the military and to private industry.

"(e) Continuation and extension of the amateur's unique ability to enhance international goodwill."

It has been said that ham radio is an international "Passport to Friendship." There is nothing like a live connection with someone far away, whether by Morse code on the HF bands or an IRLP chat between two hams holding handheld radios. Hams are almost unique in their ability to "make contact" with people around the world every day, from all walks of life with little or no intervening systems or bureaucracy. [T1A05]

Through ham radio, you will encounter many different activities and events, from conducting radio experiments to simply communicating with other licensed hams around the world. As long as they satisfy one or more of these criteria, then, yes, they are "real" ham radio. We have by no means exhausted the possibilities! [T1A12]

Definitions

Let's start with the question, "Who makes and enforces the rules for the Amateur Radio service in the United States?" This is a pretty important thing to know! The answer, of course, is the *Federal Communications Commission* or *FCC*. [T1A02] No matter what you're doing on the air, the FCC rules must be followed, even if you are operating on behalf of another government agency! The FCC is also the agency that grants your Amateur Radio license. A license in the amateur service allows you to operate anywhere that the FCC regulates the amateur service — the 50 states and all possessions under US government control. (You can also operate in countries with which the US has *reciprocal operating authority* as described later in this chapter.)

With the FCC established as the body in charge of regulating ham radio, here are some fundamental definitions that the FCC uses to construct the rules:

Amateur service — 97.3(a)(4) *"A radiocommunication service for the purpose of self-training, intercommunication and technical investigations carried out by amateurs, that is, duly authorized persons interested in radio technique solely with a personal aim and without pecuniary interest."* A telecommunications *service* is governed by a set of rules and regulations that define and administer a specific type of communications activity, such as amateur, land mobile, or marine. The amateur service is one of many communications services. Pecuniary means "related to money or payment," including trade or barter, so amateurs are expected to use Amateur Radio only because they have a personal interest in radio and in radio communications.

Amateur operator — 97.3(a)(1) *"A person named in an amateur operator/primary license station grant on the ULS consolidated licensee database to be the control operator of an amateur station."* Until you have your amateur license, you are not authorized to operate a station according to the amateur service rules. If you're not authorized, then you're not an amateur operator.

Amateur station — 97.3(a)(5) *"A station in an amateur radio service consisting of the apparatus necessary for carrying on radiocommunications."* This may seem circular, but what it means is any radio station that complies with the rules of the amateur service is considered an amateur station. [T1A10]

These rules are where it all begins: the service, the operator and the station. There are many other definitions, of course. Some will be familiar words that may be used in unfamiliar ways. If you are in doubt about the meaning of any word in the rules, it's likely that a precise definition is already waiting for you in Part 97.3, including technical terms.

TYPE AND CLASSES OF LICENSES

Aside from passing the exam, are there any other qualifications a person must have to get an Amateur Radio license? Only one: they can't be a representative of a foreign government. (Citizens of other countries can and do get US amateur licenses.) There are no age, health, or fitness requirements — ham radio is truly equal opportunity!

An Amateur Radio license actually consists of two parts — an *operator license* and a *station license*. In most other services, they are granted separately, such as for broadcast stations where employees actually operate the equipment. The operator license gives you permission to operate an amateur station according to the rules of the amateur service. The station license authorizes you to have an amateur station. The combined license is an amateur *operator/primary station license*. Each person can have only one such license. **Figure 7.1** shows an actual Amateur Radio license of the type you will receive in the mail from the FCC. Don't lose the original — you are supposed to have it available for inspection at any time. Making a laminated copy of the business card-sized section allows you to have a copy of your license with you wherever you are.

There are three classes of Amateur Radio licenses being granted today: Technician, General and Amateur Extra. [T1C13] Each carries a different set of *frequency and operating privileges* that expand from Technician to General to Extra along with the comprehensiveness of the exams. As you pass harder exams, you get more privileges. There are other license classes — the Novice, Technician Plus, and Advanced — for which

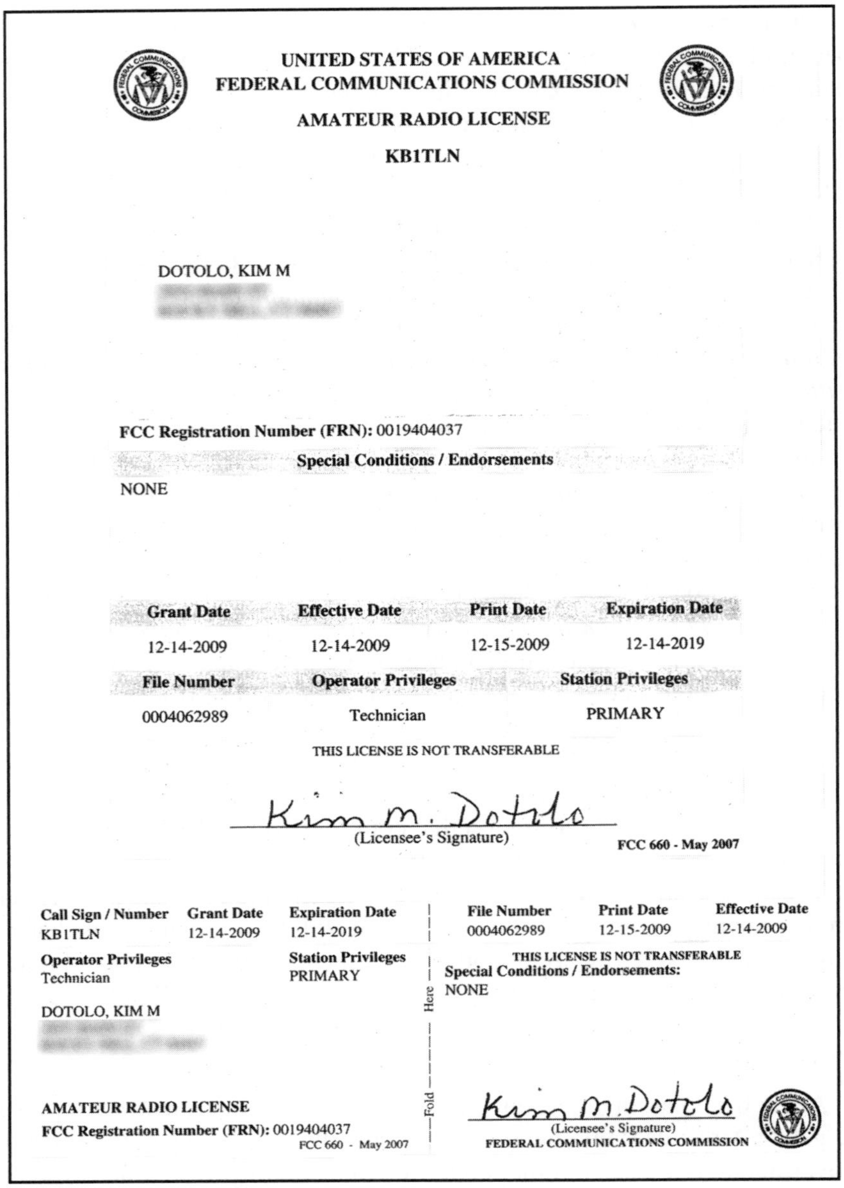

UNITED STATES OF AMERICA
FEDERAL COMMUNICATIONS COMMISSION

AMATEUR RADIO LICENSE

KB1TLN

DOTOLO, KIM M

FCC Registration Number (FRN): 0019404037

Special Conditions / Endorsements

NONE

Grant Date	Effective Date	Print Date	Expiration Date
12-14-2009	12-14-2009	12-15-2009	12-14-2019

File Number	Operator Privileges	Station Privileges
0004062989	Technician	PRIMARY

THIS LICENSE IS NOT TRANSFERABLE

Kim M. Dotolo
(Licensee's Signature)

FCC 660 - May 2007

Call Sign / Number	Grant Date	Expiration Date		File Number	Print Date	Effective Date
KB1TLN	12-14-2009	12-14-2019		0004062989	12-15-2009	12-14-2009

Operator Privileges	Station Privileges
Technician	PRIMARY

THIS LICENSE IS NOT TRANSFERABLE
Special Conditions / Endorsements:
NONE

DOTOLO, KIM M

AMATEUR RADIO LICENSE

FCC Registration Number (FRN): 0019404037

FCC 660 - May 2007

Kim M. Dotolo
(Licensee's Signature)
FEDERAL COMMUNICATIONS COMMISSION

Figure 7.1 — An FCC Amateur Radio license is both an operator and a station license. The printed license shown here has two sections: one for posting in your station and one to carry with you.

Table 7.1
Amateur License Class Examinations

License Class	Exam Element	Number of Questions
Technician	2 (Written)	35 (passing is 26 correct)
General	3 (Written)	35 (passing is 26 correct)
Amateur Extra	4 (Written)	50 (passing is 37 correct)

new licenses are no longer being granted. There are still people who hold these licenses — you'll meet them on the air.

The exams themselves are referred to as *elements*. A Technician license requires that you pass Element 2. Higher license classes require that you pass a specific element and all lower-class elements. For example, to obtain the General class license, you must pass Elements 2 and 3 in order. **Table 7.1** shows the elements that must be passed for each license class.

Before you go on, study these Technician exam questions from the question pool included at the back of this book or as a downloadable Study Guide version on the web:
T1A01 T1A02 T1A03 T1A05 T1A10 T1A12
T1C13
If you have difficulty with any question, review the preceding section

---

EXAMINATIONS

Unique among the various radio services, amateurs give their own exams, even making up the *question pool* used on the exams! This is part of ham radio's self-policing, self-regulating history. The organization of the amateur test administration is discussed in the sidebar, "Who's In Charge Here? — Amateur Exams." Amateur volunteers have run their own test programs for years and the results have been very good, with thousands and thousands of successful tests, as you will soon see.

Volunteer Examiners

An amateur who actually gives the exams and runs the *test sessions* is a *Volunteer Examiner* (VE). A Volunteer Examiner is accredited by one or more of the Volunteer Examiner Coordinators (VECs) to administer amateur license exams. Each test session requires at least three VEs be present that hold an amateur license with a class higher than those of the prospective licensees. For example, to give a Technician exam (Element 2),

Who's In Charge Here? — Amateur Exams

Amateur Radio exams are administered by volunteers organized by a *Volunteer Examiner Coordinator* (VEC) such as the ARRL/VEC. There are 14 VECs throughout the US. Some give tests in one region and others are nationwide. The list of VECs is available on the FCC's Wireless Telecommunications Bureau (WTB) website. (WTB is the branch of the FCC in charge of radio services.) Click "Amateur Radio" under "Wireless Services" and then "Volunteer Examiner Coordinators" for a list of VECs and their websites. Each VEC is responsible for training and registering Volunteer Examiners and registering and administering the test sessions.

The VECs also maintain the question pools that contain the questions for the various license exams. Once every four years, the question pool for a license class changes. The Technician exam questions changed in July of 2014 — the reason this new *License Manual* was written — and these questions are expected to remain in use through June of 2018. The General class exam questions will change in 2015 and for the Extra exam in 2016. To prepare a new question pool, the VECs get together as the *National Conference of Volunteer Examiner Coordinators* (NCVEC) and form a *question pool committee* (QPC) to create the new question pool. This allows license class teachers and authors and website designers to work with the questions and prepare teaching and exam aids.

VECs also process all of the paperwork associated with the test sessions. Records are kept of when and where the session was held, what VEs ran the session, who took exams and how they did, and all of the FCC applications for the successful candidates. The VEC makes sure that the paperwork is in order and then makes the necessary entries directly with the FCC's database of amateur licensees. This saves the FCC a lot of time and money and gets you your license quicker!

the VEs must hold a General class license or higher. The exception is that VEs holding Amateur Extra licenses may give Amateur Extra class exams.

Although Technicians can't be counted as part of the three required VEs, they can help process paperwork and assist with running the session, except for grading and handling General or Extra exams. This is a great way to help others get their license.

Taking the Exam

The first step is to find a test session by contacting the VECs in your area or checking their websites. You can find test sessions registered with the ARRL/VEC on the ARRL's licensing web page. Some test sessions are held on a regular basis and are open to the public. Others are held at events such as conventions and hamfests (an Amateur Radio flea market). Others are hosted by individuals or clubs at a private residence or facility.

Once you've selected a session, be sure to register or check if they accept *walk-ins*, or unregistered attendees. (They might not have extra exam materials.) Then show up on time and ready to take the exam. Take a check or cash for the exam fee — most test sessions cannot process credit cards.

The VEs will register you, check your ID (have at least two forms of identification, one being a photo ID) and set you up with the necessary test papers. The test is in multiple-question format. There will be 35 questions on the Technician exam and you have to answer 26 correctly to pass. Take your time and be sure you're happy with your answers before you turn in the test — there are no extra points given for speed!

CSCE and Form 605

Once you've turned in your test, the VEs will grade it while you relax a bit! Did you pass? Congratulations! If not, don't despair — check with the VEs running the test session about taking another version of the exam for that element. Lots of hams took the exam a couple of times before passing!

Those who pass will be given two forms to fill out: a *Certificate of Successful Completion of Examination* (CSCE) and a *NCVEC Quick Form 605* (**Figure 7.2**). The CSCE is your "receipt" of successfully passing the test for an element and what class of license you have earned. You should keep the CSCE until the FCC database has been updated with all of your new information. The CSCE is good for 365 days as proof that you have successfully passed one or more elements.

The NCVEC Quick Form 605 that you fill out as shown in Figure 7.2 will be filed with the FCC. It records your personal information and will be used to link your identity with a specific call sign. The test session VEs will help you fill out the form correctly. With both forms signed and your exam fee paid, are you a ham? Not quite — you have to wait until you are notified of your call sign as discussed in the sidebar. Once your information shows up in the FCC database, you are fully authorized to go on the air! [T1C10]

Obtaining Form 605

If you decide to "do the paperwork" on real paper instead of online, you'll need to get a blank FCC Form 605. This is not difficult! You can get FCC Form 605 with detailed instructions by contacting the FCC in any of these ways:

- FCC Forms Distribution Center, tel 800-418-3676
- FCC Forms "Fax on Demand" — tel 202-418-0177, ask for form number 000605
- FCC Forms On-Line

The ARRL also offers a FCC Form 605 package geared to amateur needs Write to: ARRL/VEC, FCC Form 605, 225 Main St, Newington CT 06111-1494. Include a large business-sized stamped, self-addressed envelope with your request.

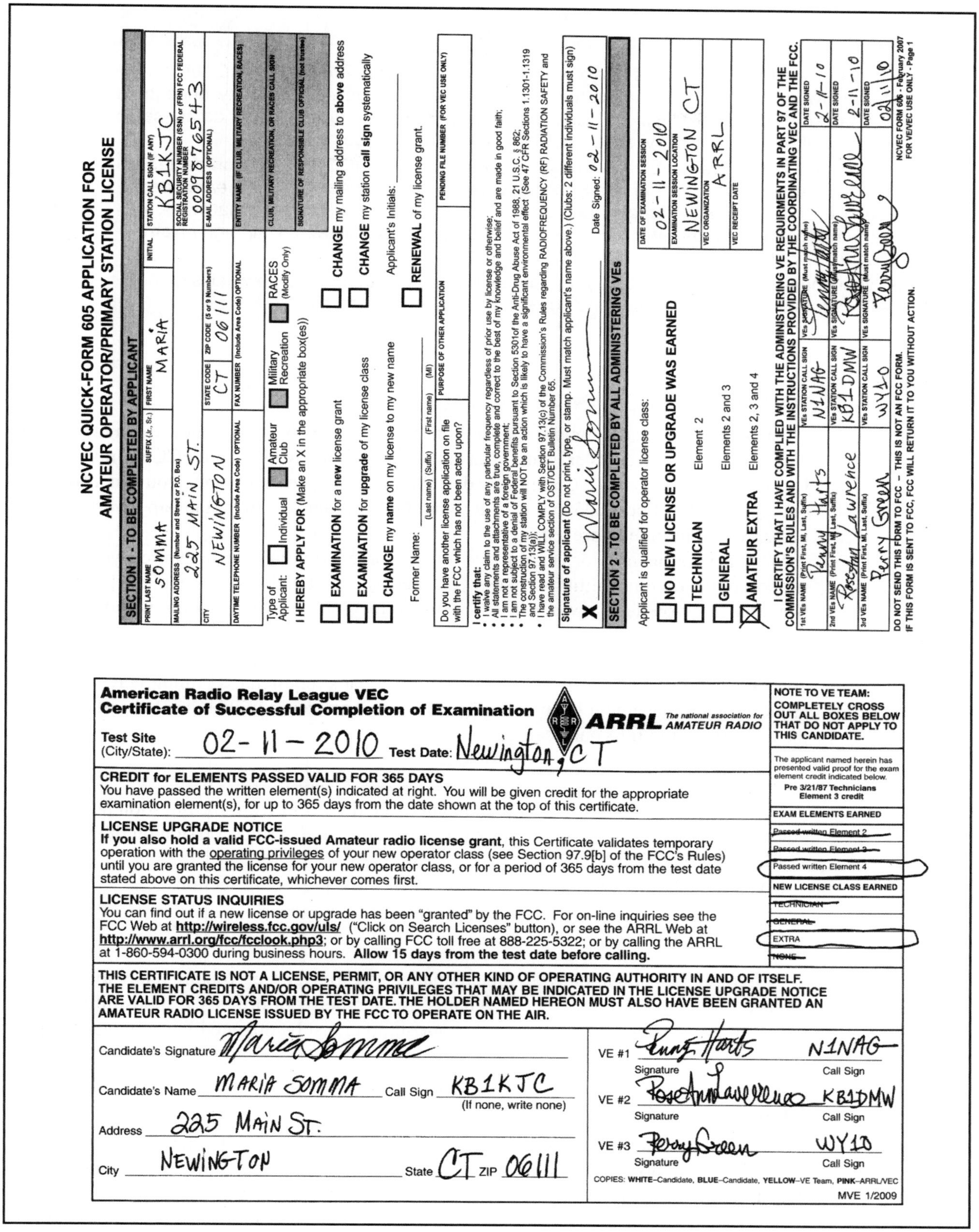

NCVEC QUICK-FORM 605 APPLICATION FOR AMATEUR OPERATOR/PRIMARY STATION LICENSE

SECTION 1 - TO BE COMPLETED BY APPLICANT

PRINT LAST NAME: SOMMA FIRST NAME: MARIA SUFFIX (Jr., Sr.): INITIAL:

STATION CALL SIGN (IF ANY): KB1KJC

MAILING ADDRESS (Number and Street or P.O. Box): 225 MAIN ST.

SOCIAL SECURITY NUMBER (SSN) or (FRN) FCC FEDERAL REGISTRATION NUMBER: 0001876543

CITY: NEWINGTON STATE CODE: CT ZIP CODE (5 or 9 Numbers): 06111

E-MAIL ADDRESS (OPTIONAL):

DAYTIME TELEPHONE NUMBER (Include Area Code) OPTIONAL: FAX NUMBER (Include Area Code) OPTIONAL:

ENTITY NAME (IF CLUB, MILITARY RECREATION, RACES):

Type of Applicant: ☐ Individual ☐ Amateur Club ☐ Military Recreation ☐ RACES (Modify Only)

CLUB, MILITARY RECREATION, OR RACES CALL SIGN:

I HEREBY APPLY FOR (Make an X in the appropriate box(es))

SIGNATURE OF RESPONSIBLE CLUB OFFICIAL (but not trustee)

☐ EXAMINATION for a new license grant

☐ EXAMINATION for upgrade of my license class

☐ CHANGE my name on my license to my new name

Former Name: _____ (Last name) (Suffix) (First name) (MI)

☐ CHANGE my mailing address to above address

☐ CHANGE my station call sign systematically

Applicant's Initials: _____

☐ RENEWAL of my license grant.

PENDING FILE NUMBER (FOR VEC USE ONLY)

Do you have another license application on file with the FCC which has not been acted upon? PURPOSE OF OTHER APPLICATION

I certify that:
- I waive any claim to the use of any particular frequency regardless of prior use by license or otherwise;
- All statements and attachments are true, and correct to the best of my knowledge and belief and are made in good faith;
- I am not a representative of a foreign government;
- The construction of my station will NOT be an action which is likely to have a significant environmental effect (See 47 CFR Sections 1.1301-1.1319 and Section 97.13(a));
- I am not subject to a denial of Federal benefits pursuant to Section 5301 of the Anti-Drug Abuse Act of 1988, 21 U.S.C. § 862;
- I have read and WILL COMPLY with Section 97.13(c) of the Commission's Rules regarding RADIOFREQUENCY (RF) RADIATION SAFETY and the amateur service section of OST/OET Bulletin Number 65.

Signature of applicant (Do not print, type, or stamp. Must match applicant's name above.) (Clubs: 2 different individuals must sign)

X _Maria Somma_ Date Signed: 02-11-2010

SECTION 2 - TO BE COMPLETED BY ALL ADMINISTERING VEs

DATE OF EXAMINATION SESSION: 02-11-2010

EXAMINATION SESSION LOCATION: NEWINGTON CT

VEC ORGANIZATION: ARRL

VEC RECEIPT DATE:

Applicant is qualified for operator license class:

☐ NO NEW LICENSE OR UPGRADE WAS EARNED

☐ TECHNICIAN — Element 2

☐ GENERAL — Elements 2 and 3

☒ AMATEUR EXTRA — Elements 2, 3 and 4

I CERTIFY THAT I HAVE COMPLIED WITH THE ADMINISTERING VE REQUIREMENTS IN PART 97 OF THE COMMISSION'S RULES AND WITH THE INSTRUCTIONS PROVIDED BY THE COORDINATING VEC AND THE FCC.

1st VEs NAME (Must match name): Penny Harts VEs STATION CALL SIGN: N1NAG VEs SIGNATURE (Must match name): _____ DATE SIGNED: 2-11-10

2nd VEs NAME (Print First, MI, Last, Suffix): RoseAnn Lawrence VEs STATION CALL SIGN: KB1DMW VEs SIGNATURE (Must match name): _____ DATE SIGNED: 2-11-10

3rd VEs NAME (Print First, MI, Last, Suffix): Perry Green VEs STATION CALL SIGN: WY1O VEs SIGNATURE (Must match name): _____ DATE SIGNED: 02-11-10

DO NOT SEND THIS FORM TO FCC – THIS IS NOT AN FCC FORM.
IF THIS FORM IS SENT TO FCC, FCC WILL RETURN IT TO YOU WITHOUT ACTION.

NCVEC FORM 605 - February 2007
FOR VE/VEC USE ONLY - Page 1

American Radio Relay League VEC
Certificate of Successful Completion of Examination

ARRL — The national association for AMATEUR RADIO

Test Site (City/State): 02-11-2010 Test Date: Newington, CT

NOTE TO VE TEAM:
COMPLETELY CROSS OUT ALL BOXES BELOW THAT DO NOT APPLY TO THIS CANDIDATE.

The applicant named herein has presented valid proof for the exam element credit indicated below.

Pre 3/21/87 Technicians Element 3 credit

CREDIT for ELEMENTS PASSED VALID FOR 365 DAYS
You have passed the written element(s) indicated at right. You will be given credit for the appropriate examination element(s), for up to 365 days from the date shown at the top of this certificate.

EXAM ELEMENTS EARNED
~~Passed written Element 2~~
~~Passed written Element 3~~
Passed written Element 4

LICENSE UPGRADE NOTICE
If you also hold a valid FCC-issued Amateur radio license grant, this Certificate validates temporary operation with the operating privileges of your new operator class (see Section 97.9[b] of the FCC's Rules) until you are granted the license for your new operator class, or for a period of 365 days from the test date stated above on this certificate, whichever comes first.

NEW LICENSE CLASS EARNED
~~TECHNICIAN~~
~~GENERAL~~
EXTRA
~~NONE~~

LICENSE STATUS INQUIRIES
You can find out if a new license or upgrade has been "granted" by the FCC. For on-line inquiries see the FCC Web at http://wireless.fcc.gov/uls/ ("Click on Search Licenses" button), or see the ARRL Web at http://www.arrl.org/fcc/fcclook.php3; or by calling FCC toll free at 888-225-5322; or by calling the ARRL at 1-860-594-0300 during business hours. **Allow 15 days from the test date before calling.**

THIS CERTIFICATE IS NOT A LICENSE, PERMIT, OR ANY OTHER KIND OF OPERATING AUTHORITY IN AND OF ITSELF. THE ELEMENT CREDITS AND/OR OPERATING PRIVILEGES THAT MAY BE INDICATED IN THE LICENSE UPGRADE NOTICE ARE VALID FOR 365 DAYS FROM THE TEST DATE. THE HOLDER NAMED HEREON MUST ALSO HAVE BEEN GRANTED AN AMATEUR RADIO LICENSE ISSUED BY THE FCC TO OPERATE ON THE AIR.

Candidate's Signature: _Maria Somma_

Candidate's Name: MARIA SOMMA Call Sign: KB1KJC (If none, write none)

Address: 225 MAIN ST.

City: NEWINGTON State: CT ZIP: 06111

VE #1 Signature: _Penny Harts_ Call Sign: N1NAG

VE #2 Signature: _RoseAnn Lawrence_ Call Sign: KB1DMW

VE #3 Signature: _Perry Green_ Call Sign: WY1O

COPIES: WHITE–Candidate, BLUE–Candidate, YELLOW–VE Team, PINK–ARRL/VEC

MVE 1/2009

Figure 7.2 — The CSCE (Certificate of Successful Completion of Examination) is your test session receipt that serves as proof that you have completed one or more exam elements. It can be used at other test sessions for 365 days. The NCVEC Form 605 that you fill out will be filed with the FCC.

TERM OF LICENSE AND RENEWAL

Amateur licenses are good for a 10-year term. [T1C08] You can renew them indefinitely without ever taking another exam. You can renew online by using the FCC's Universal Licensing System (ULS). Up until 90 days before your license expires, you can also fill out a paper FCC Form 605 and mail it to the FCC. (After that time, the paperwork may not be processed in time to prevent expiration.)

What if your license does expire? People do forget! If your license expires, you are supposed to stop transmitting because your license is not valid after it expires. [T1C11] Nevertheless, you have a two-year grace period to apply for a new license without taking the exam again. [T1C09]

If your license is lost or destroyed, you can request a replacement from the FCC. You don't have to fill out a Form 605 — a letter is all that's required explaining why you are requesting a replacement license. A new copy of your license will be printed and mailed to the address on file for your license.

Before you go on, study these Technician exam questions from the question pool included at the back of this book or as a downloadable Study Guide version on the web:
T1C08 through T1C11
If you have difficulty with any question, review the preceding section.

RESPONSIBILITIES

Congratulations on a job well done, Amateur Radio licensee! You are ready to join the ranks of more than 700,000 other US hams! Remember your primary responsibility as the holder of an Amateur Radio license — your station must be operated in accordance with the FCC rules.

Unauthorized Operation

This includes preventing improper use of your station equipment when you're not present. For example, unlicensed family members aren't allowed to operate in your absence because you're not there to ensure proper operation — even if they operate with you on a regular basis! There are several ways to secure your station from unauthorized operation (such as locked doors), but the most common and recommended method is to simply disconnect the microphone and power cables when you're not around. That removes the temptation and prevents operation by unlicensed persons with access to your station.

Personal Information

The FCC requires you to maintain a valid current mailing address in their database at all times. This is so you can be contacted by mail, if needed. If you move or even change PO boxes, be sure to update your information using the FCC ULS online system. If you do not maintain a current address and mail to you is returned to the FCC as undeliverable, your license can be suspended or revoked and removed from the database. [T1C07]

The other piece of information that might be unfamiliar is the FRN (Federal Registration Number). This is an identification number assigned to you as a licensee. You can use your Social Security Number as the Taxpayer ID Number. Registering with the FCC is covered in the next section.

Station Inspection

As a federal licensee, you are obligated to make your station available for inspection upon request by an FCC representative. By accepting the FCC rules and regulations for the amateur service, you agree that your station could be inspected any time. These visits are very rare and only occur when there is reason to believe that your station has been operated improperly. Remember to keep your original license available for inspection, too! [T1F13]

Before you go on, study these Technician exam questions from the question pool included at the back of this book or as a downloadable Study Guide version on the web:
T1C07   T1F13
If you have difficulty with any question, review the preceding section.

7.2 Working with the FCC

While you can still use paper forms and letters to interact with the FCC, working online is much quicker and often simpler. As you become accustomed to working online, you can also tap a wealth of information stored in the FCC database.

THE FCC ULS WEB SITE

The FCC has developed a comprehensive website for all of its licensees to use: It's called the *Universal Licensing System* or ULS. The ULS home page is shown in **Figure 7.3**. You can use the ULS in several ways:
- Register for online access to your license information
- Make simple changes to your address and other information
- Renew your license
- Search for licensees by name, call sign or location

Here's a sample of how to use the ULS Search function. Browse to the site and click the LICENSES button next to "Search." When the License Search page loads, click on "Amateur." When the search form is presented, enter your ZIP code and click "Search" at the bottom of the page. You might be surprised at the results!

REGISTERING

To use the ULS site for managing your license, you'll need to register with the FCC. You can register whether or not you have a license. Registering is done through *CORES* — the Commission Registration System. Click the REGISTER button next to "New Users" on the ULS home page to begin the registration process. You will be assigned a unique FCC identification number, your *Federal Registration Number* or FRN. The ARRL has a guide to the registration process online.

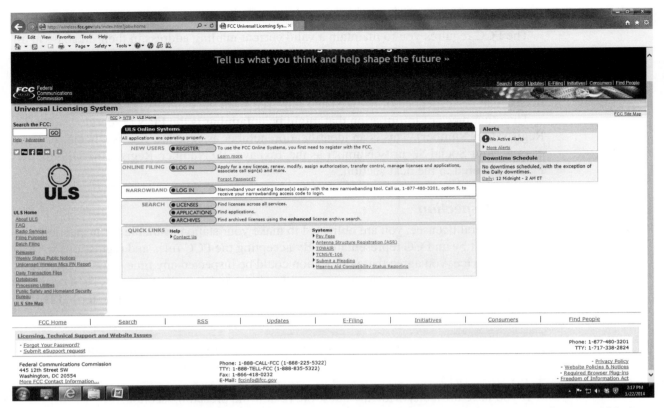

Figure 7.3 — The FCC's Universal Licensing System (ULS) is the preferred method for hams to interact with the FCC. After registering, you can use the ULS to renew your license, change your address or even change your call sign!

USING YOUR FRN

The FRN is your key to unlocking all of the license management services available via the ULS system. You must use it on paper forms, such as FCC Form 605. Having an FRN means you don't have to use your Social Security Number when filing forms. When you register, you will receive a letter that contains your FRN and a starting password. Use the same FRN for all of your personal licensing applications. If you become the FCC's contact for a club or repeater call sign, you must set up a unique FRN for the club or repeater. Do not use your individual FRN number.

Log on to the ULS site and click the LOG IN button next to "Online Filing." Enter the FRN and password in the letter. Once logged in you can then change your password to something you prefer. Don't lose either your FRN or your password!

7.3 Bands and Privileges

The frequencies and modes and methods that hams are allowed to use are all known to the rules and regulations as *privileges*. What gives the FCC authority to grant privileges is the Communications Act of 1934. What then grants these privileges to you is your license. By signing Form 605 and applying for a license, you agree to be bound by the FCC rules and that means staying within the privileges of your license.

FREQUENCY PRIVILEGES

The most important privileges are *frequency privileges*. Once upon a time, at the dawn of radio, hams and broadcasters and commercial and military stations were all mixed in to-

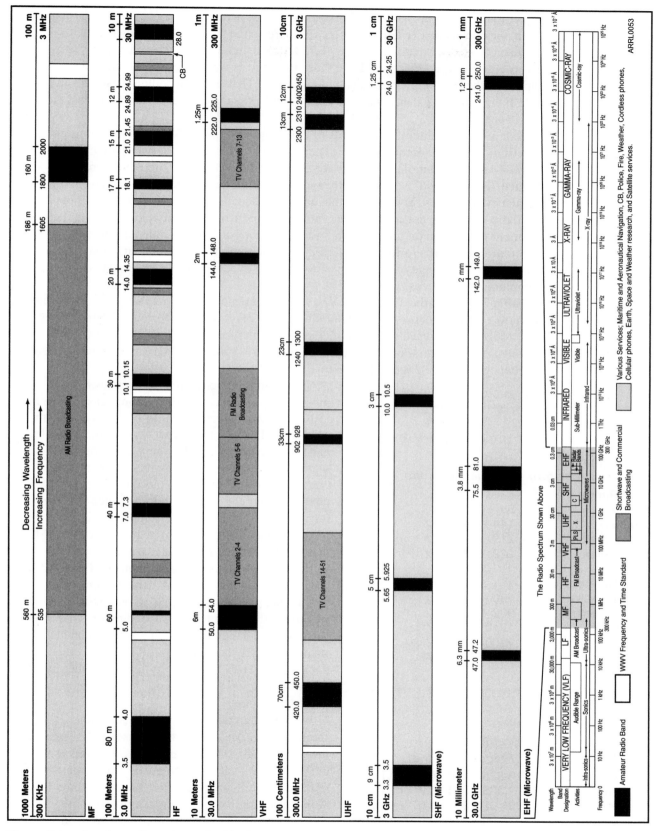

Figure 7.4 — Amateur allocations, the black rectangles in the chart, are distributed relatively evenly throughout the radio spectrum. The variations in propagation at these different frequencies give hams a lot of opportunities to experiment with different types of communications systems and methods.

gether in one big band. That didn't work very well when the one big band got crowded. Soon, the powers that be decided to restrict hams to the "worthless" frequencies with wavelengths shorter than 200 meters (1.5 MHz and higher). That worked fine for hams because they quickly discovered that those frequencies were precisely the ones that supported the best long-distance communications! That situation couldn't last long and it didn't; the shortwave bands were carved up among the different users and hams had their first "bands."

Today, there are literally hundreds of bands and dozens of different types of radio spectrum users. The frequency privileges granted to the various services are called *allocations*. For example, amateurs are allocated 144-148 MHz, the 2 meter band. **Figure 7.4** is a grand overview of the radio spectrum and where amateurs have frequency privileges. You can see that amateur allocations are sprinkled throughout the radio spectrum, not concentrated in or excluded from any one region. As you recall

from our discussion on propagation, radio signals of different frequencies propagate differently. Thus, it's fortunate that spectrum planning has resulted in amateurs having access to a wide range of frequencies in which to experiment and to apply to different communications needs.

Table 7.2 shows the Technician VHF/UHF frequency privileges that you are expected to know for your license exam. Remember that a band can be referred to by frequency ("50 MHz") or by wavelength ("6 meters"). Use the formula f (in MHz) = 300 / wavelength (in meters) or wavelength (in meters) = 300 / f (in MHz) to convert between frequency and wavelength. [T1B03 to T1B07]

You should memorize the frequencies for the most common bands used by Technicians: 6 meters (50-54 MHz), 2 meters (144-148 MHz) and 70 cm (420-450 MHz). Hams keep a chart of their privileges handy, since not many of us have every one of them memorized. To help you remember your privileges, copy the information onto a piece of paper and tape it in your car or near your computer or on the refrigerator. Take every opportunity to recite the information, reinforcing it time after time. You'll find that it's not nearly as hard as it seems at first!

The HF privileges for Technicians in **Table 7.3** are useful and interesting. (A full-page chart of Technician privileges is provided on this book's website and in Chapter 1.) Depending on solar activity, the 10 meter band can provide contacts worldwide and you can try out RTTY and other HF data modes. The CW-only privileges on other bands will acquaint you with "classic" ham radio on the shortwave bands.

Within the amateur HF bands, access to frequencies is determined by license class. From the Technician class, as higher class licenses are obtained, more and more frequency privileges are granted until all amateur privileges are granted to Amateur Extra licensees. For example, on the 80 meter band, Technicians may use CW from 3.525-3.600 MHz.

Table 7.2
VHF and UHF Technician Amateur Bands

ITU Region 2

Band (Wavelength)	Frequency Limits
VHF Range	
6 meters	50 – 54 MHz
2 meters	144 – 148 MHz
1.25 meters	219 – 220 MHz
1.25 meters	222 – 225 MHz
UHF Range	
70 centimeters	420 – 450 MHz
33 centimeters	902 – 928 MHz
23 centimeters	1240 – 1300 MHz
13 centimeters	2300 – 2310 MHz
13 centimeters	2390 – 2450 MHz

Table 7.3
Technician HF Privileges
200 watts PEP maximum output

Band (Wavelength)	Frequency (MHz)
80 meters	3.525-3.600 (CW only)
40 meters	7.025-7.125 (CW only)
15 meters	21.025-21.200 (CW only)
10 meters	28.000-28.300 (CW, RTTY and data)
	28.300-28.500 (CW and SSB)

As you read about amateur rules and regulations, you will occasionally encounter *emission mode designators*, such as A1A for amplitude-modulated CW for aural reception or J3E for single-sideband, suppressed-carrier telephony. It is not necessary to memorize these codes, but it is a good idea to know where to look them up if you need to. A table of designators for the most common amateur emission types can be found on this book's website and a complete discussion of emission types is contained in *The ARRL Handbook*.

EMISSION PRIVILEGES

Within most of the ham bands, additional restrictions are made by mode or *emission type*. (Emission is the formal name for any radio signal from a transmitter.) Just as a frequency privilege is permission to use a specific frequency, an *emission privilege* is permission to communicate using a particular mode, such as phone, CW, data or image. **Table 7.4** lists the modes that can be used by amateurs — as a Technician class licensee, you can use all of them.

The combination of frequency, license class and emission privileges makes for a fairly complicated division of the amateur bands into *sub-bands*. Parts of the ham bands in which only certain modes can be used are called *mode-restricted*. As a Technician licensee, though, your situation is very simple: There is a small CW-only sub-band occupying the bottom 100 kHz of the 6 and 2 meter bands. The segment of the 1.25 meter band from 219-220 MHz is restricted to digital message forwarding only. [T1B13] For all amateur allocations above 222 MHz, there are no other sub-bands! **Table 7.5** shows all of the subdivisions of amateur bands through 23 cm. [T1B10, T1B11] (A full-page chart of Technician privileges is provided on this book's website and in Chapter 1.)

Why have mode-restricted sub-bands? Because the methods of operating for the different modes are sometimes not compatible. CW and phone operation, for example, are conducted quite differently and the signals interfere with each other. In the past few years, with the increasing number of digital modes, incompatibilities between digital and CW signals have led to interference between these two groups of operators. Hams have worked around this problem by using narrow-bandwidth modes, such as CW, at the low-frequency end of the bands and wider-bandwidth signals, from data through voice, higher in the band. That is the price of flexibility to experiment and use all the different modes!

Table 7.4

Amateur Emission Types

Emission	Description
CW	Morse code telegraphy
Data	Computer-to-computer communication modes, usually called digital modes
Image	Television (fast-scan and slow-scan) and facsimile or fax
MCW	Tone-modulated CW, Morse code generated by keying an audio tone
Phone	Speech or voice communications
Pulse	Communications using a sequence of pulses whose characteristics are modulated in order to carry information
RTTY	Narrow-band, direct-printing telegraphy received by automatic equipment, such as a computer or teleprinter
SS	Spread-spectrum communications in which the signal is spread out over a wide band of frequencies
Test	Transmissions containing no information

Table 7.5

VHF/UHF (1500 W PEP maximum)

6 meters (50-54 MHz)

50	50.1		54
CW only	Phone, Image, RTTY, Data, CW		

2 meters (144-148 MHz)

144	144.1		148
CW only	Phone, Image, RTTY, Data, CW		

1.25 meters (219-220 and 222-225 MHz)

219	220	222	225
Fixed digital message forwarding systems only		Phone, Image, RTTY, Data, CW	

70 cm (420-450 MHz)

420	430	435	438	450
Not available north of Line A		Satellite		
Phone, Image, RTTY, Data, CW				

33 cm (902-928 MHz)

902	928
Phone, Image, RTTY, Data, CW	

23 cm (1240-1300 MHz)

1240	1300
Phone, Image, RTTY, Data, CW	

HF (200 W PEP maximum)

80 meters (3.5-4.0 MHz)

3.5	3.525	3.600	4.000
	CW only		

40 meters (7.0-7.3 MHz)

7.0	7.025	7.125	7.300
	CW only		

15 meters (21.0-21.45 MHz)

21.0	21.025	21.200	21.450
	CW only		

10 meters (28.0-29.7 MHz)

28.000	28.300	28.500	29.700
CW, RTTY, Data	SSB phone and CW		

ARRL0544

POWER LIMITS

The maximum power an amateur is allowed to generate at the output of the transmitter or amplifier is 1500 watts of *peak envelope power* (PEP). PEP is the average power during one RF cycle of the radio signal at the very peak of a modulating waveform, such as for speech. For a CW signal, PEP is measured during the *key-down period* in which the transmitter is ON. FM is a constant-power mode, so it does not matter whether you are speaking or not.

Amateurs are expected to use the minimum power required to carry out the desired communication as long as it does not exceed the maximum power allowed. [T2A11] That doesn't mean you have to turn down the output power until you can just barely be heard — that's probably not the desired communication! What it means is that if you can carry out your intended communications with less power, you should do so.

Transmitter output power is measured at the output of the last amplifier, whether internal to the transmitter or an external piece of equipment, at the input to the antenna feed line — not at the antenna or anywhere along the feed line. The limit of 1500 watts is allowed nearly everywhere on the ham bands except on the following frequencies:

- Novice and Technician licensees are limited to 200 watts PEP on HF bands.
- All amateurs are limited to 200 watts PEP on the 30 meter band.
- All amateurs are limited to 50 watts PEP in the 219-220 MHz segment of the 1.25 meter band
- Stations being operated as beacons are limited to 100 watts PEP
- Stations operating in the 70 cm band near certain military installations may be limited to 50 watts PEP or less.
- There are other restrictions for Novice class licensees and for stations operating on the 60 meter band.

Most amateurs rarely use or *run* more than a few hundred watts on the VHF and UHF bands. Exceptions would be while pursuing very weak-signal methods, such as Earth-Moon-Earth (EME) or tropospheric propagation where high power is required to establish and maintain contact. High power levels at these frequencies can create safety hazards. We discuss RF safety in the **Safety** chapter of this book.

Before you go on, study these Technician exam questions from the question pool included at the back of this book or as a downloadable Study Guide version on the web:
T1B03 through T1B07   T1B10   T1B11   T1B13
T2A11
If you have difficulty with any question, review the preceding section.

PRIMARY AND SECONDARY ALLOCATIONS

It would be nice if every type of radio user had exclusive access to their allocations. Many amateur bands are exclusively allocated to hams, worldwide. Because spectrum is scarce and many services have valid needs for radio communications, occasionally two services receive *shared allocations,* including some of the amateur bands. When this happens, one group is generally given priority and these are called *primary allocations.* Groups that have access to spectrum on a lower priority receive *secondary allocations.* The groups that receive the allocations are *primary and secondary services.*

The primary service is *protected* from harmful interference by signals from secondary services. The secondary service gains access to the frequencies in the allocation with the understanding that it must not cause harmful interference to primary service users and it must accept interference from primary users. For example, amateurs have a secondary allocation in the 70 cm band and must avoid interfering with radiolocation stations that have primary status. [T1B08]

By sharing the bands in this way, more frequencies are available for more users than if every frequency was exclusively allocated to one service alone. Hams share several of our bands and enjoy wider access to frequencies than would otherwise be the case.

All of the UHF and higher-frequency bands have some kind of sharing arrangements. They may apply to certain geographic areas or around certain military installations. A good example is the restriction on 70 cm band operations for amateurs north of "Line A" — approximately 50 miles south of the Canadian border. Because Canada has allocated 420-430 MHz to other services, to prevent international interference the FCC has ruled that US amateurs may not use that segment of the band near the Canadian border. Frequency sharing of amateur HF bands is less common, but the 60 meter band is shared with US government services and 40 meters is shared with shortwave broadcast stations above 7.2 MHz.

Not all US amateur allocations are allocated to amateurs worldwide. Where there are competing allocations, the amateur service is considered to be secondary use. For example, if you learn that 23 cm transmissions from your station are interfering with a radiolocation service outside the US, you should stop operating or figure out how to stop the interference. [T1A14]

Part 97.303 lists all of the frequency-sharing requirements for US hams and is available on the ARRL website. It is worth familiarizing yourself with sharing requirements to avoid interference either to or from your station!

BAND PLANS

Finding each different type of activity will become much easier if you follow the *band plans* that organize the different types of activity by frequency. This helps hams find specific types of operation. By grouping similar activities together, the spectrum is used more effectively, as well. Band plans are developed over time by amateurs themselves, not by the FCC. The ARRL maintains the current band plan online.

Band plans are voluntary arrangements by amateurs for using different modes or for different activities. [T2A10] They are not rules from the FCC, although the FCC considers them to be "good practice" for amateurs. Band plans apply during normal conditions and do not guarantee the use of any frequency at any time. Except for repeater frequencies, amateurs are expected to be flexible on a day-to-day basis.

REPEATER COORDINATION

It's a natural question, "Who decides what repeater can use a specific pair of frequencies?" You may be surprised to learn that hams themselves decide, and the FCC has nothing to do with it. This is part of the Amateur Radio tradition of self-policing and self-administration. Hams developed a system of regional *frequency coordination* to insure that repeaters use the amateur bands wisely and avoid interference to the greatest degree possible.

Repeaters and their auxiliary stations are grouped together into one or two segments of a band. Their input and output *frequency pairs* are fixed and have a common offset in each region. This enables the maximum number of repeaters to use the limited amount of spectrum. To keep order, a committee of volunteers known as a *frequency coordinator* recommends transmit and receive frequencies. [T1A08] Where regions overlap, the coordinators work together to minimize interference and keep the coordination process orderly. A list of frequency coordinators is available on the website of the National Frequency Coordinators' Council.

Because repeaters cover wide areas, it is also necessary for the frequency coordinator to consider other operating parameters such as transmit power, the height of antennas, and whether a repeater should employ access tones. The frequency coordinator representatives are selected by the local or regional amateurs whose stations are eligible to be auxiliary or repeater stations. [T1A09]

A *coordinated* repeater uses frequencies approved by a regional coordinator. *Uncoordinated* repeaters are strongly discouraged because they often cause interference. Before putting their repeater on the air, repeater owners apply to their region's coordinator for an available pair of frequencies. The coordinators determine what frequencies are best suited for the repeater's location. Once the frequency pair is assigned, the repeater owner can then turn on the repeater's transmitter.

Before you go on, study these Technician exam questions from the question pool included at the back of this book or as a downloadable Study Guide version on the web:

T1A08 T1A09 T1A14

T1B08

T2A10

If you have difficulty with any question, review the preceding section.

7.4 International Rules

In the early days of radio, every country made up its own radio rules. As equipment got better and long-distance communications became common, it became clear that coordination was required. After World War I, governments got together and started making international treaties that specified how the countries were to regulate radio communications. This worked reasonably well until World War II, during which communications technology made major advances. After the war ended, a new method of managing radio was needed.

ITU (INTERNATIONAL TELECOMMUNICATION UNION)

FCC doesn't have authority in other countries — every country has its equivalent agency. How do they coordinate? It would be chaos if every country made up their own allocations, since radio waves don't stop at international borders! Realizing the need for international coordination, the *International Telecommunication Union* (ITU) was formed as an agency of the United Nations (UN) in 1949. The ITU is an administrative forum for working out international telecommunications treaties and laws, including frequency allocations. The ITU also maintains international radio laws that all UN countries agree to abide by. [T1B01]

Regions

The ITU divides the world into the three regions shown in **Figure 7.5** and organizes frequency allocations accordingly. North and South America, including Alaska and Hawaii, form Region 2. Some US possessions and territories in the Pacific are located in Region 3.

Because VHF and UHF signals frequently do not travel far beyond the radio horizon, it was decided that allocations in these frequency ranges did not have to align precisely. As a result, the amateur allocations vary between regions. Table 7.2 shows allocations for Amateur Radio above 50 MHz in Region 2. Allocations on those bands are different for amateurs in Regions 1 and 3.

HF allocations are more consistent between regions, although there are some variations that affect US amateurs. For example, allocations for US hams operating in Region 3 such as on Guam (KH2) or Wake Island (KH9) are different than for mainland US stations. (You can see these differences in the tables of Part 97.301 Authorized frequency bands.)

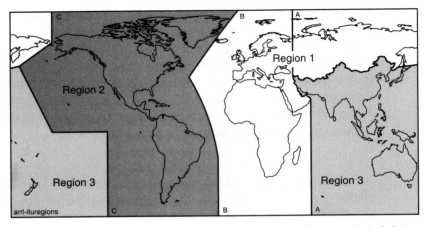

Figure 7.5 — The map shows the world divided into the three administrative regions of the International Telecommunication Union. This helps the ITU and member nations manage frequency allocations around the world.

[T1B02] Privilege allocation rules are fairly consistent within a single region, although there are some exceptions such as those in Part 97.307 that allow Hawaiian hams to communicate with stations in Region 3 using phone and image transmissions between 7.075 and 7.100 MHz.

Radio rules change at region boundaries, regardless of your citizenship or of the ownership of a vessel or aircraft. If you operate maritime mobile, for example, when you cross from Region 2 to Region 1, you will then have to operate according to the Region 1 rules. That's true even if you are a citizen of a Region 2 country and the vessel is "flagged" in a country from Region 3. [T1B12]

INTERNATIONAL OPERATING

Operating in a foreign country can be a lot of fun! You can meet local hams, and if you are licensed for HF operation you can become "DX" and attract a crowd on the air. To operate at all, the foreign country must permit amateur operation. [T1C04] In addition, you must have permission and when you are inside a country's national boundaries, including territorial waters, you are required to operate according to their rules. You may also operate from any vessel or craft that is documented or registered in the United States. [T1C06] If the vessel is in territorial waters, regulations of the host country and those of the vessel's registry both apply.

To operate using your US amateur license, there must be a reciprocal operating agreement between the countries. There are three ways of getting operating permission: reciprocal operating authority, an International Amateur Radio Permit (IARP), and the European Conference of Postal and Telecommunications Administrations (CEPT) license agreement. For more information on these agreements, information is available on the ARRL's website.

Regardless of which avenue is available, don't forget the final part of the amateur service's basis and purpose — to foster goodwill. You are a ham radio ambassador while on the air!

RECIPROCAL OPERATING AUTHORITY

Many countries have entered into *reciprocal operating authority* agreements with the United States, recognizing each other's amateur licenses. This is a government-to-

International Amateur Radio Union (IARU)

Just as the countries of the world support the ITU, so do the amateur "countries" support the IARU. Each country with a national society, such as the ARRL in the US, is part of the IARU, which is organized in three regions, just like the ITU. Formed in 1925 as national governments began forming radio law, the IARU acts as the worldwide amateur voice to government and international rules making bodies, such as the ITU.

government agreement recognizing each other's amateur licenses. No additional information is required; just take a copy of your US license with you. Check with local licensing authorities about filing an application and any fees before operating. You must follow your host country's rules and regulations.

IARP — INTERNATIONAL AMATEUR RADIO PERMIT

In some Central and South American countries, an IARP allows US amateurs to operate without seeking a special license or permit to enter and operate from that country. The IARP is issued by a member-society of the *International Amateur Radio Union* (IARU) — for the US, the IARU member society is the American Radio Relay League (ARRL). The IARP can either be Class 1 (equivalent to the US Amateur Extra) or Class 2 (equivalent to the US Technician).

CEPT

A CEPT license allows US General and Extra class licensees to travel to and operate from most European countries or their possessions without obtaining an additional licensee or permit. When traveling to a CEPT country, you'll need to have your original US license, proof of US citizenship such as a passport, and a copy of the FCC's Public Notice about CEPT licenses. The two CEPT license classes are the same as described for the IARP.

PERMITTED CONTACTS AND COMMUNICATIONS

Unless specifically prohibited by the government of either country, any ham can talk to any other ham. [T1D01] International communications must be limited to the purposes of the amateur service or remarks of a personal nature. [T1C03] Some countries do not recognize Amateur Radio, although the number is very small. Other countries prohibit contacts between their citizens and those of specific other countries. Again, this is quite uncommon.

Inside the US, the FCC may impose restrictions on a US ham as part of a judgment or administrative ruling. This is unusual and most often the result of some kind of bad behavior by the individual amateur. Remember that in a communications emergency, you can still talk to any ham anywhere if needed to prevent loss of life or property.

Before you go on, study these Technician exam questions from the question pool included at the back of this book or as a downloadable Study Guide version on the web:

T1B01 T1B02 T1B12
T1C03 T1C04 T1C06
T1D01

If you have difficulty with any question, review the preceding section.

7.5 Call Signs

Call signs are our "radio names" and each amateur's is unique. Call signs all have a common structure and once you learn it, figuring out the nationality of call sign (or *call*) is easy.

PREFIX AND SUFFIX

Every amateur call sign has a *prefix* and a *suffix*. In the US, an amateur call sign prefix consists of one or two letters and one numeral. The suffix consists of one to three letters.

For example, W3ABC is a valid US amateur call sign, while KDKA and KMA3505 are call signs from other US radio services. [T1C02]

Every country is assigned at least one unique *block* of prefixes. For example, US amateur call signs begin with K, N, W or the two-letter combination AA through AL. No matter what, if you hear a call sign beginning with those letters, you know it's a US call sign. Most Canadian hams use VA through VG, French hams use F, Japanese amateur call signs begin with a J, and hams from Singapore have calls beginning with 9V.

The suffix of a call sign is the unique part that identifies the particular station and consists of only letters. In the call W1AW, "AW" is unique among all other calls beginning with "W1" (known as "W1 calls"). Suffixes are one, two or three letters. The combination of prefix and suffix uniquely identify a station anywhere on Earth. Within a country, the call signs can be assigned to indicate license class or location or other special characteristics.

US Call Districts and Call Signs

In the US, the number in the call sign's prefix indicate in which one of the 10 districts shown in **Figure 7.6** the call was originally assigned. The call sign is permanently assigned to the individual operator in the US and remains the same, no matter where the operator moves in the country. W3IZ, for example, originally got his call in the third district but now lives in Connecticut.

A US call sign is further classified by the number of letters in the prefix and suffix. A

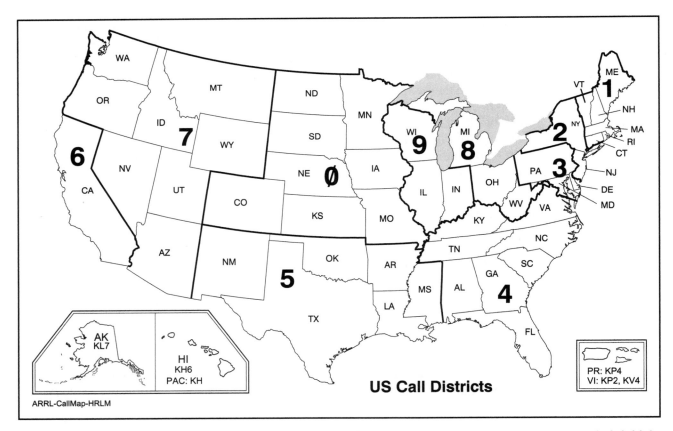

Figure 7.6 — There are 10 US call districts in the continental United States. When an amateur passes their initial license exam, the numeral of the district in which he or she resides becomes part of their assigned call sign. Alaska and the US possessions in the Pacific and Caribbean form three more districts and have calls with unique prefixes.

Table 7.6

US Amateur Call Signs in Districts 0-9

Group	License Class	Format
Group A	Amateur Extra	Prefix K, N or W with two-letter suffix (1×2), or two-letter prefix beginning with A, N, K or W and one-letter suffix (2×1), or two-letter prefix beginning with A and a two-letter suffix (2×2)
Group B	Advanced	Two-letter prefix beginning with K, N or W and a two-letter suffix (2×2)
Group C	General	One-letter prefix beginning with K, N or W and a three-letter suffix (1×3)
Group D	Technician, Novice, and Club	Two-letter prefix beginning with K or W and a three-letter suffix (2×3)

call such as AA5BT with a two-letter prefix, "AA", and a two-letter suffix, "BT", is called a *2-by-2 or 2×2*. WA1ZMS is a 2×3. W1A is a 1×1. Regular US call signs have only one numeral in the prefix. The FCC grants these calls by license class as shown in **Table 7.6**. When you receive your Technician license, you will receive a Group D call sign. Calls are always assigned in sequential order — you get the next one on the stack, just like vehicle license plates.

PORTABLE AND UPGRADE

Assuming your call was granted in the continental US, you can operate anywhere within the US with the same call. You are properly identified as a US ham on the air and there is no requirement for you to indicate where in the country you happen to be. However, hams often add a *portable designator* following the call sign if they are operating somewhere outside the district that would otherwise be indicated by the numeral in the prefix. For example, if NØAX operates while driving through California, he might give his call as NØAX/6, spoken as "N Zero A X portable 6" or "NØAX/6" if using CW. The word "portable" used here may be replaced by "slant" or "slash" or some other word that separates the call sign and the indicator.

If you are operating outside the US, you must add the prefix of the country to your call because you are no longer in your home country. If you travel to Canada, for instance, you are then required to sign your call and follow it with "portable VE#" where the # indicates the Canadian province in their call sign system (for example, NØAX/VE3). If you go to Hawaii — a US state, but not part of the continental US — you would

---

The Slashed Zero — Ø

You will notice right away that hams from the "zero" or tenth district write their calls with a forward slash through the zero — Ø. (The tenth district was created in the late 1930s when the ninth district's ham population got too big!) The slash was needed because the typewriters of those days made it hard to distinguish between 0 (zero) and O (capital letter O). It's still hard today in many fonts. Commercial and military operators adopted the custom of backspacing one space and typing a slash over the zero. Not in the usual set of characters used by word processing programs, the slashed zero is assigned the sequence of keyboard keys ALT-0216 in many character sets.

sign your call followed by "portable KH6." Similar requirements exist in other countries. Check their rules for how you are supposed to identify on the air.

Hams often append a designator to indicate that they are engaging in a particular activity, such as adding "/MOB" or saying "Mobile" after your call when operating from a moving vehicle. These *self-assigned* designators are allowed, as long as they are not the same as a designator that would conflict with the prefix of another country. For example, adding "/VE" to your call because you are a Volunteer Examiner is not OK because it would make listeners think you were in Canada!

When you upgrade your license, as soon as you receive your CSCE from the VEs administering the exam, you can append a portable designator to indicate your new license class. When your new license class appears in the FCC database, you can drop the upgrade designator. Here are the two designators:

• General — say "portable AG" or send "/AG" on CW
• Amateur Extra — say "portable AE" or send "/AE" on CW

The designator lets other stations know that you have upgraded your license although the database does not yet show the change.

CHOOSING A CALL SIGN

You can also choose your own call sign! You can have almost as much fun choosing a call from the *vanity call* program as in choosing a vanity license plate for your car. Many hams pick a call with their initials in the suffix or one that forms a short word. Licensed hams can pick any available call authorized for their license class as shown in Table 7.6. There are lots of available calls for Technician licensees to choose from in Group C (1-by-3) and Group D (2-by-3). [T1C05, T1C12]

CLUB AND SPECIAL EVENT CALL SIGNS

Clubs can also have their own call signs. There are some rules about the club, however. The club must have at least four members and the FCC can ask for documentation showing that the club exists and has meetings. [T1F12] Club licenses are granted to the person the club designates as the *trustee* of the club station. Your club's trustee can apply for a call sign by contacting a Club Station Call Sign Administrator. [T1C14] Once the club call has been assigned from Group D, the club can also use the vanity call program to change the call sign.

Any FCC-licensed amateur or club can also obtain a special 1×1 (1-by-1) call sign such as K3X or W6P for a short-duration special event of significance to the amateur community. These call signs are unique because they only have one letter in both the prefix and suffix. [T1C01] Application for the call sign must be made to a Special Event Call Sign Coordinator (the ARRL is one of five coordinators) and a call sign will be granted for a short time, usually 15 days. The usual call of the amateur or club requesting the special event call must be given once per hour and the special event call used for regular identification. Special event call signs are very popular and contacts with them are sought out by hams, worldwide.

Before you go on, study these Technician exam questions from the question pool included at the back of this book or as a downloadable Study Guide version on the web:
T1C01   T1C02   T1C05   T1C12   T1C14
T1F12
If you have difficulty with any question, review the preceding section.

Chapter 8

Operating Regulations

In this chapter, you'll learn about:

- **Control operators**
- **Guest operating and privileges**
- **Identification on the air**
- **Tactical call signs**
- **Rules about interference**
- **Third-party communications**
- **Remote and automatic control**
- **Prohibited communications**
- **Broadcasting**

When you see the mouse, you'll find more information at www.arrl.org/ham-radio-license-manual

You might be surprised to learn that there really aren't very many regulations about operating! The FCC relies on hams to come up with procedures and methods that work. As long as the ham's signals meet the technical requirements and those few operating rules are followed, the FCC trusts Amateur Radio to be self-regulating. Contrary to what you might think, the FCC steps in only rarely to resolve disputes or otherwise intractable problems. As a result, hams have a great deal of freedom to innovate and adapt within the rules with just a handful of simple do's and don'ts.

8.1 Control Operators

All transmissions must be made under the control of a *control operator*, a properly-licensed operator who is responsible for making sure all FCC rules are followed. [T1E01] There can be only one control operator for a station at a time. That's the person responsible for station operation, no matter who is actually speaking into the microphone.

DEFINITIONS

Reading the rules themselves can be a bit confusing, so let's be sure to clearly define the terms. There are two basic ideas on which amateur rules are based: a control operator responsible for creating a signal and a *control point* at which control is asserted by the control operator.

- A control operator is the amateur designated as responsible for making sure that transmissions comply with FCC rules. The control operator doesn't have to be the station licensee and doesn't even have to be physically present at the transmitter, as you will see below, but all amateur transmissions are the responsibility of a control operator. The station licensee is responsible for designating the control operator. [T1E03]
- A control operator must be named in the FCC amateur license database or be an alien with reciprocal operating authorization. (An alien is a citizen of another country.) This is a simple requirement — the FCC has to know who you are, that you are licensed and

Rebecca, KS4RX (left, an Extra class licensee), acts as control operator for Amanda, KG4NBF (a Technician class licensee), during Field Day. Rebecca is responsible for making sure all FCC rules are followed when the station is operating on bands and modes not available to Technicians.

where you can be contacted. Any licensed amateur can be a control operator. [T1E02]

- The control point is where the station's control function is performed. Usually, the control point is at the transmitter and the control operator physically manipulates the controls of the transmitter. The control point can be remotely located and connected by phone lines, the Internet or a radio link. [T1E05]

PRIVILEGES AND GUEST OPERATING

As the control operator, you may operate the station in any way permitted by the privileges of your license class. [T1E04] It doesn't matter what the station owner's privileges are, only the privileges of the control operator. When the station owner and the control operator are the same person, responsibilities are easy to understand.

Being a guest operator is very common — you may allow another amateur to use your station or you may be the guest. Either way, it's important to understand what sets the control operator's privileges. A guest operator hosted by a higher-class licensee can operate using the host's privileges only if the host is the control operator. If the host is not the control operator, the guest is restricted to the privileges of their license.

Here's an example — you, a Technician class licensee, are invited to spend the afternoon at the station of your Elmer, who holds an Extra class license. While your Elmer is supervising and acting as control operator, you can operate the station on any amateur band and mode. This is very common and is a good way to learn about the HF bands and styles of operating not used on VHF/UHF. However, if your Elmer decides to step out of the shack or run an errand, you are restricted to your Technician privileges. [T1E12]

What if you are the guest and have a higher-class license than the host? A guest operator hosted by a lower-class licensee can use their higher-class privileges as the control operator whether the host is present or not. In this case, there are special identification rules described in the section on Guest Operators.

Regardless of license class, though, both the guest operator and station owner are responsible for proper operation of the station. [T1E07] The control operator is responsible for the station's transmissions. The station owner is responsible for limiting access to the station only to responsible licensees who will follow the FCC rules. Note that the FCC will presume the station licensee to be the control operator unless there is a written record to the contrary. [T1E11]

UNLICENSED OPERATORS

Here's a situation that you may already have experienced. Have you visited a club or personal station and made a contact or two? Perhaps you took part in Field Day or a radio contest with licensed hams. Maybe you just took the microphone to say hello to a friend over the air. If so, you have acted as an unlicensed operator under the supervision of a control operator.

There's nothing at all wrong with an unlicensed operator using an amateur station — as long as a control operator is present when transmissions are made. Unlicensed operators, including family members of licensed hams, may not act as control operators. This is sometimes difficult for family to understand — even though they can probably turn on the rig and use the microphone, they are not allowed to do so because they are not licensed. No license equals no control operator equals no unsupervised transmissions.

It's not okay for family members to operate while the licensed control operator is elsewhere on the property, doing a chore, or grabbing a snack. The control operator must be present and ensuring that all FCC rules are met. One thing they *can* do — unlicensed operators can use the equipment to receive at any time!

Before you go on, study these Technician exam questions from the question pool included at the back of this book or as a downloadable Study Guide version on the web:
  T1E01 through T1E05   T1E07   T1E11   T1E12
  If you have difficulty with any question, review the preceding section.

8.2 Identification

In most contacts the other party can't see you — they have no other way of identifying your signals other than your call sign. It's also important that other stations be able to determine who is transmitting. Your call sign is your identity on the air and "to identify" or "identification" means to send or speak your call sign over the air.

Identification rules apply whether you're operating at home, on the road or floating in a hot-air balloon. Proper identification is important not only so the FCC knows who's transmitting, but for any other station that wants to contact you or to know where your signal is coming from. This is why identification is one of the few areas in which the FCC tells you how to operate!

NORMAL IDENTIFICATION

The first rule of identification is that unidentified transmissions are not allowed, no matter how brief. [T1D11] Unidentified means that no call sign was associated with a transmission. If you need to make a short transmission to test an antenna or make adjustments to your radio, just stating or sending your call sign will suffice. For example, to see if you are in reach of a distant repeater, don't just key the microphone and listen for the repeater's signal — that's called *kerchunking* because of the sound that everyone monitoring will have to listen to. Just state your call sign once as you transmit. No problem!

The identification rules are simple — give your call sign at least once every 10 minutes during a contact and when the communication is finished. [T1F03] What? Not at the beginning of the contact? You generally need to give your call to establish contact, so that's a moot point. What if the contact is too short for the 10-minute rule? Just give your call sign as you end the contact.

You don't have to give the other station's call sign, either. The purpose of identification is to identify the source of *your* signal, not those of other stations. Giving the other station's call sign is for convenience and is considered good practice. Here's a tip: It's not necessary to say, "For ID," since whenever you give your call sign, you ID!

Your call can be given in Morse code, by voice, in an image, or as part of a digital transmission. Video and digital call signs will need to be transmitted in a standard protocol or format so that anyone can receive it. If you are using phone, you are required to identify in English, even if you are communicating in a language other than English. [T1F04] The FCC recommends the use of phonetics when you identify by voice — that avoids confusing letters that sound alike. [T2B09] You may also identify by CW even if using phone. [T1F05]

Tactical Calls

Tactical call signs (or tactical IDs) are used to help identify where a station is and what it is doing. Examples of tactical calls include "Waypoint 5," "First Aid Station," "Hollywood and Vine" and "Fire Watch on Coldwater Ridge." [T1F01] Tactical calls can be used at any time, but are usually used in emergency and public service operation when providing communications.

Tactical calls don't replace regular call signs and the regular identification rules apply — give your FCC-assigned call sign every 10 minutes and at the end of the communication. [T1F02] Tactical calls allow consistent identification that streamlines communication based on function. It would be really confusing if everyone had to remember which individual call sign was performing which function. It's common, for example, for a station to have different operators at different times. When a new operator takes over, he or she simply gives their FCC-assigned call sign along with a tactical call, such as "This is N1OJS at Race Headquarters." They use "Race Headquarters" as the tactical call from then on, giving their regular call sign (N1OJS) once every 10 minutes.

Self-Assigned Indicators

When operating away from your home station, you should add information to your call sign so that other stations are aware of your location. For example, an Alaskan station would add some extra information when operating in the lower 48 contiguous states. Otherwise, the special Alaskan prefixes (AL7, KL7, NL7, WL7) would cause confusion about the location of the station. For example, if KL7CC is operating from a location in the 3rd district, he could give his call sign as "KL7CC/W3." The added "/W3" is called a *self-assigned indicator*.

FCC Part 97.119(c) says, "*One or more indicators may be included with the call sign. Each indicator must be separated from the call sign by the slant mark (/) or by any suitable word that denotes the slant mark. If an indicator is self-assigned, it must be included before, after, or both before and after, the call sign.*" For example, on phone, KL7CC could identify as W3/KL7CC or KL7CC/W3, using "stroke," "slash" or "portable" between the indicator and the call sign. [T1F06]

Note that the indicator is not allowed to conflict with some other indicator specified in the FCC rules, such as those that indicate an upgraded license class as in the next section. It may not conflict with a prefix assigned to some other country. For example, the indicator /M to indicate mobile operation is not allowed because M is an English prefix, but /MOB is permitted because MOB is not a prefix allocated to another country.

Upgrade Indicators

After you receive your Technician call sign, it's a pretty good bet that you will start

Licensee or Trustee?

There seems to be a lot of confusion as to whether the person whose name is on the license for a repeater or a remote base is the "licensee" or the "trustee" of the station. The operator could be either one! It depends upon the type of license the station is operating under.

If the repeater or remote base is operating under the auspices, and using the call sign of, an individual amateur's personal station license, then the operator is the "licensee" of the station, not the "trustee."

If it is operating under the auspices of an FCC-issued club station license, and using the FCC-issued club call sign, then the person whose name appears on the license is the "trustee," not the "licensee."

thinking about advancing to the next class of license. When you do pass that next test and *upgrade*, you can begin using your new privileges right away by adding the appropriate indicator shown below to your call sign on the air. The FCC requires you to add the two-letter indicator after your call sign to let someone receiving your transmissions that you have upgraded your license class, even though the information may not appear in the FCC's database yet. [T1F08]

- Novice to Technician: add "/KT"
- Technician to General: add "/AG"
- General or Advanced to Extra: add "/AE"

Before you go on, study these Technician exam questions from the question pool included at the back of this book or as a downloadable Study Guide version on the web:

T1D11

T1F01 through T1F06   T1F08

T2B09

If you have difficulty with any question, review the preceding section.

GUEST OPERATORS

When you are visiting another station, unless your host "lends" you the station, you must identify using the call sign of the host. It doesn't matter who has the highest-class license. If the guest operator has a higher class license than the host, the guest identifies with the call sign of the host followed by their own call sign.

For example, if KD7FYX (a General class licensee) uses the station of KD7PFA (a Technician class licensee) on the 20 meter voice band where Technicians do not have privileges, he must sign "KD7PFA/KD7FYX." On a VHF band, signing "KD7PFA" would be sufficient since both licensees have the same privileges on VHF.

This is a little easier to understand if you think about it from the standpoint of a listening station. If they heard KD7PFA on 20 meter phone and looked up the call in the FCC database, they would think KD7PFA was using a frequency not allowed for Technician-class licensees. Adding KD7FYX makes it clear whose privileges are being used. Needless to say, identifying is more efficient if KD7PFA temporarily lends the station to KD7FYX.

Bob Raymond, WA1Z, is an active contester. He knows and practices all the FCC rules that govern Amateur Radio, including the need to identify your station. The purpose of identification is to identify the source of your signal, not those of other stations. Giving the other station's call sign is for convenience and is considered good practice.

MISCELLANEOUS IDENTIFICATION RULES

There are two exceptions to the identification rules: remote control signals and signals retransmitted through *space stations*. If you are controlling a model aircraft, for example, you don't have to send your call sign. Remote control signals are weak and don't travel long distances, so a call sign is not of much use.

Space stations are amateur stations located more than 50 km above the Earth's surface, such as amateur satellites and stations on the International Space Station (ISS). They do not have to identify themselves although the station on the ISS uses the call signs NA1SS and RUØSS. Amateur satellite *transponders* that receive an entire range of received frequencies and retransmit them on a different band are not required transmit their call sign either.

Test Transmissions

Identification rules apply to on-the-air test transmissions, as well, no matter how brief. The call sign must be given once every 10 minutes and at the end of transmissions. [T2A06, T2A07] Test transmissions must be kept brief to avoid causing interference or to keep from occupying an otherwise useful frequency. The usual method of identifying during a test transmission is to say "W1AW testing" or send "W1AW VVV", where "V" is usually used as a Morse code test signal.

Automatic Identification

Stations under automatic control (see section 8.5 later in this chapter) must also identify themselves. This is part of the requirement that automatically-controlled stations be controlled by procedures or devices that make sure FCC rules are followed. Repeaters are the most common type of station that operates under automatic control. Repeaters identify themselves by transmitting the station call sign by voice, by Morse code (20 WPM or slower), or as an image in a standard video signal format if retransmitting video signals.

Special Event Stations

If an amateur club with an FCC license arranges to use a special event call sign, such as the popular 1-by-1 calls (see the chapter on **Licensing Regulations**), both the regular call and the special event call must be given on the air. The usual call sign of the amateur or club requesting the special event call sign must be given once per hour. This allows listeners to determine the identity of the station, since special event call signs are short-lived and can be reused by different groups. The ARRL's Special Event Station calendar lists upcoming special call sign use.

Before you go on, study this Technician exam question from the question pool included at the back of this book or as a downloadable Study Guide version on the web:
T2A06   T2A07
If you have difficulty with any question, review the preceding section.

8.3 Interference

Interference is caused by "noise" and by "signals." Noise interference is caused by natural sources, such as thunderstorms (atmospheric static is referred to as *QRN*), or by signals unintentionally radiated by appliances, industrial equipment and computing equipment. The type of interference discussed in this section is caused by signals from other transmitters.

Interference from nearby signals, or *QRM*, is part of the price of frequency flexibility. If hams operated on assigned and evenly-spaced channels, there would be much less interference. The channels would also be frequently overloaded! Interference is not necessarily illegal, just inconvenient. Most interference is manageable! Hams have learned various ways of dealing with QRM starting with the following:

• Common sense and courtesy help avoid many problems
• Be sure to equip your radio with good filters to reject interference
• Remember that no one owns a frequency — be flexible and plan ahead
• Be aware of other activities, such as special events, DXpeditions and contests

HARMFUL INTERFERENCE

If a transmission seriously degrades, obstructs or repeatedly interrupts the communications of a regulated service, that's considered harmful interference. [T1A04]

Interference, otherwise known as *QRM*, causes severe distortion to this image sent by "slow scan" TV. While harmful interference can be vexing, accidental interference is common, as is the case here.

Every ham should make sure to both transmit and receive in a way that minimizes the possibility of causing harmful interference. Reports of interference such as transmitting off-frequency or generating spurious signals (splatter and buckshot) should be checked out. When testing or tuning a transmitter, use a dummy load and always keep your test transmissions short.

While harmful interference can be vexing, accidental interference is common. For example, propagation on a band can change due to ionospheric or atmospheric conditions. A signal that wasn't there a few minutes ago might suddenly become strong enough to disrupt your contact. Changing an antenna direction can allow a previously rejected signal to be heard, or the new heading might transmit a signal toward other stations. Sometimes, an operator will begin listening during a pause in activity and start transmitting thinking the frequency is available. These things happen — you shouldn't expect a perfectly clear frequency!

What should you do if harmful interference occurs to your contact? Assuming the interfering station isn't intentionally trying to cause interference, can you change frequency a little bit or change antenna direction? Common courtesy should prevail but remember that no one has an absolute right to any frequency. [T2B08] Be flexible — it's one of ham radio's greatest strengths! What should you do when you cause harmful interference? If it's your fault, apologize, identify and take the necessary steps to reduce interference — change frequency, reduce power or move your antenna.

Harmful interference is prohibited in some circumstances described in the FCC's rules on Frequency Sharing Requirements (Part 97.303). For example, amateurs are not allowed to create interference that endangers the functions of radionavigation services, even accidentally. [T1A06]

Don't Be Too Sensitive!

Harmful interference is not necessarily intentional; it may simply be due to an overloaded receiver! Modern receivers are tremendously sensitive, but it's expecting too much of them to run at full sensitivity while rejecting strong signals nearby. Signal processing features such as noise blankers and preamplifiers can create problems where there are none. What is perceived as harmful interference can often be reduced or eliminated with good receiver operating technique

A frequent source of problems is the noise blanker. Most noise blankers operate by sensing short, sharp noise pulses. They look at an entire band, not just what is coming through the narrow signal filters. A strong nearby signal can confuse a noise blanker to the point of nearly shutting down a receiver or causing what sounds like severe splatter over many kilohertz. Unless you have really strong local line or ignition noise, turn off your noise blanker. If the band is full of strong signals, noise blankers are useless or worse.

The RF attenuator can be your biggest friend when dealing with strong nearby signals. It's surprisingly easy for a strong signal to overload a receiver. Overload causes the receiver to create spurious signals and noise up and down the band.

Switching on the attenuator cures a surprising number of ailments because your receiver is no longer being overloaded. Remember that the goal is to maximize understandability by increasing the ratio of signal to noise. Try out your attenuator and you may be surprised at how much it cleans up a band!

The RF gain control can make your receiver very sensitive but also susceptible to overloading. Experiment with reducing RF gain to see if it improves your receiver's performance on a busy band. Even during casual operating, turning down the RF gain can dramatically reduce background noise.

Does your receiver have passband tuning, IF shift, variable bandwidth or similar controls? All those new digital signal processing (DSP) features you paid for can also clean up noise and attenuate low-frequency or high-frequency audio. Read the receiver's manual and learn what these controls do. By effectively using the capabilities of a modern receiver, you will surely find that the band is quieter and nearby signals less disruptive. In fact, you will find yourself making better use of your receiver's controls every day!

WILLFUL INTERFERENCE

Intentionally creating harmful interference is called *willful interference* and is never allowed. [T1A11] The interference doesn't have to be aimed at one specific contact or group. Any time communications are deliberately disrupted, that's willful interference. For example, intentionally transmitting spurious signals by overmodulating is willful interference. Luckily, willful interference is pretty rare on the ham bands since most people have the good sense and maturity to not do it.

Before you go on, study these Technician exam questions from the question pool included at the back of this book or as a downloadable Study Guide version on the web:
T1A04 T1A06 T1A11
T2B08
If you have difficulty with any question, review the preceding section.

8.4 Third-Party Communications

Ham radio is frequently used to send messages, written or not, on behalf of unlicensed persons or organizations. One of the oldest activities in ham radio is the sending of messages, relaying them from station to station until delivered by a ham near the addressee. This is *third-party communication*. Because third-party communications bypass the normal telephone and postal systems, many foreign governments have an interest in controlling it. Looking at third-party communications from the ham radio side, the FCC does not want the amateur service to become a non-commercial messaging system. So, we have some rules specifically governing third-party communications.

DEFINITIONS AND RULES

Let's start by defining the important aspects of third-party communications:

- The entity on whose behalf the message is sent is the "third party" and the control operators that make the radio contact are the first and second parties. A third party can also be the recipient of a message generated by a ham.
- A licensed amateur capable of being a control operator at either station is not considered a third party.
- The third party need not be present in either station. A message can be taken to a ham station or a ham can transmit speech from a third-party's telephone call over the ham radio — this is called a *phone patch*.
- The communications transmitted on behalf of the third party need not be written. Spoken words, data or images can all be third-party communications.
- The third party may participate in transmitting or receiving the message at either station. An unlicensed person in your station sends third-party communications when they speak into the microphone, send Morse code or type on a keyboard.
- An organization, such as a church or school, can also be a third party.

Simplifying the definition, any time that you send or receive information via ham radio on behalf of any unlicensed person or an organization, even if the person is right there in the station with you — that's third-party communications.

Third party communication can be fun! In this photo, Neil Foster, N4FN, is assisting an unlicensed student during the School Club Roundup contest.

The FCC recognizes that third-party communications is a vital part of ham radio and its mission, specifically to train operators and to provide an effective emergency communications resource. Handling messages, or *traffic*, phone patches and live conversations are all part of both normal and emergency communications. As a result, third-party communications may be exchanged between any amateur stations operating under FCC rules with the constraint that the communications must be noncommercial and of a personal nature.

When signals cross borders, the rules change. International third-party communications are restricted to those countries that specifically allow third-party communications with US hams. [T1F11] **Table 8.1** shows which countries have *third-party agreements* with the United States. If the other country isn't on this list, third-party communication with

Table 8.1
Third-Party Agreements

The United States has third-party agreements with the following nations.

V2	Antigua/Barbuda
LO-LW	Argentina
VK	Australia
V3	Belize
CP	Bolivia
E7	Bosnia-Herzegovina
PP-PY	Brazil
VE, VO, VY	Canada
CA-CE	Chile
HJ-HK	Colombia
D6	Comoros (Federal Islamic Republic of)
TI, TE	Costa Rica
CM, CO	Cuba
HI	Dominican Republic
J7	Dominica
HC-HD	Ecuador
YS	El Salvador
C5	Gambia, The
9G	Ghana
J3	Grenada
TG	Guatemala
8R	Guyana
HH	Haiti
HQ-HR	Honduras
4X, 4Z	Israel
6Y	Jamaica
JY	Jordan
EL	Liberia
V7	Marshall Islands
XA-XI	Mexico
V6	Micronesia, Federated States of
YN	Nicaragua
HO-HP	Panama
ZP	Paraguay
OA-OC	Peru
DU-DZ	Philippines
VR6	Pitcairn Island
V4	St. Kitts/Nevis
J6	St. Lucia
J8	St. Vincent and the Grenadines
9L	Sierra Leone
ZR-ZU	South Africa
3DA	Swaziland
9Y-9Z	Trinidad/Tobago
TA-TC	Turkey
GB	United Kingdom
CV-CX	Uruguay
YV-YY	Venezuela
4U1ITU	ITU - Geneva
4U1VIC	VIC - Vienna

that country is not permitted. [T1F07] The ARRL maintains a current Third-Party Agreement List on their website.

This is all much clearer if illustrated by some examples:

- A message from one ham to another ham is not third-party communications, whether directly transmitted or relayed by other stations.
- Letting an unlicensed neighbor make a contact under your supervision is third-party communications, even if the contact is short and for demonstration or training purposes.
- If you contact a DX station that asks you to pass a message to a family member in your state, doing so would be third-party communications. Check to be sure the DX station's country has a third-party agreement with the US before accepting the message.
- Making a contact to allow a visiting student to talk to his family in South America is third-party communications even if both the student and the family are present at the stations involved. Be sure there is a third-party agreement in place.

Before you go on, study this Technician exam question from the question pool included at the back of this book or as a downloadable Study Guide version on the web:

T1F07   T1F11

If you have difficulty with any question, review the preceding section.

8.5 Remote and Automatic Operation

Many stations, such as repeaters and beacons, operate without a human control operator present to perform control functions. It is also becoming common to operate a station via a link over the Internet or phone lines. These two types of operation are specially defined in the rules, but the requirement remains the same — the station must be operated in compliance with FCC rules, no matter where the control point is located.

DEFINITIONS

Local control — a control operator is physically present at the control point. This is the situation for nearly all amateur stations, including mobile operation. Any type of station can be locally controlled. [T1E09]

Remote operation — the control point is located away from the transmitter, but a control operator is present at the control point. The control point and the transmitter are connected by some kind of control link, usually via Internet, phone line or radio. Many stations operate under remote control over an Internet link. [T1E10] Any station can be remotely controlled.

Automatic operation — the station operates completely under the control of devices and procedures that ensure compliance with FCC rules. A control operator is still required, but need not be at the control point when the station is transmitting. Repeaters, beacons and space stations are allowed to be automatically controlled. Digipeaters that relay messages, such as for the APRS network, are also automatically controlled. [T1E06, T1E08]

RESPONSIBILITIES

No matter what type of control is asserted — local, remote or automatic — the station must operate in compliance with FCC rules at all times. No excuses! Automatic stations might not have a control operator controlling the station at all times, but a control operator must be responsible for the station's operation.

Repeater owners must install the necessary equipment and procedures for automatic control that ensures the repeater operates in compliance with FCC rules. If automatic control results in rule violations, the FCC can require a repeater to be placed on remote control (meaning that a control operator must be present when the repeater is operating). Repeater users are responsible for proper operation via the repeater, however. [T1F10]

Because digital protocols are designed to operate automatically, there are special rules for automatic control when using them. Stations using a data mode (including RTTY) may operate under automatic control in certain segments of the HF bands and above 50 MHz as listed in Part 97.221(b). Data stations are the only type of automatically-controlled station allowed to forward third-party communications. (Note that it is okay to pass third-party messages using a repeater.)

Before you go on, study these Technician exam questions from the question pool included at the back of this book or as a downloadable Study Guide version on the web:
T1E06   T1E08   T1E09   T1E10
T1F10
If you have difficulty with any question, review the preceding section.

8.6 Prohibited Transmissions

Not many types of transmissions are specifically prohibited because amateurs are given wide latitude to communicate within the technical and procedural rules. While other services are very tightly regulated, hams are encouraged to experiment and be flexible. There are limits — here are four types of communications prohibited for reasons that are hopefully obvious:

- Unidentified transmissions — any transmission without an identifying call sign during the required time period.
- False or deceptive signals — transmissions intended to deceive the listener, such as using someone else's call sign.
- False distress or emergency signals — because of the legal requirement to respond, these are taken very seriously by the FCC and other authorities.
- Obscene or indecent speech — avoid controversial topics and expletives. [T1D06]

Generally speaking, regular communications that could reasonably be performed through some other radio service are also prohibited. For example, regularly directing boat traffic on a lake should be done on marine VHF channels, not ham radio. Communications in return for some kind of compensation is also prohibited as we discuss below. None of these prohibitions are unreasonable and help keep Amateur Radio a useful and rewarding activity free of commercial intrusion.

BUSINESS COMMUNICATIONS

No transmissions related to conducting your business or employer's activities are permitted. This is, after all, *amateur* radio, and there are plenty of radio services available for commercial activities. However, one's own personal activities don't count as "business" communications. For example, it's perfectly okay for you to use ham radio to talk to your spouse about doing some shopping or to confer about what to pick up at the store. You can order things over the air, as long as you don't do it regularly or as part of your normal income-making activities. It is also okay to advertise equipment for sale as long as it pertains to Amateur Radio and it's not your regular business. [T1D05] Here are some examples of acceptable and non-acceptable activities:

OK

- Using a repeater's autopatch to make or change a doctor's appointment
- Advertising a radio on a swap-and-shop net
- Describing your business as part of a casual conversation

NOT OK

- Using a repeater's autopatch to call a business client and change an appointment
- Selling household or sporting goods on a swap-and-shop net
- Regularly selling radio equipment at a profit over the air
- Advertising your professional services over the air

Another broad prohibition is being paid for operating an amateur station. Your employer can set up an emergency amateur station, and even pay you to build it, but you can't be paid for time you spend operating it. This is also true for employees of public safety and medical organizations as you learned when reading about emergency communications.

One exception to the profession or business prohibition is that teachers may use ham radio as part of their classroom instruction. In that case, they can be a control operator of a ham station, but it must be incidental to their job and can't be the majority of their duties. [T1D08]

Before you go on, study these Technician exam questions from the question pool included at the back of this book or as a downloadable Study Guide version on the web:
T1D05 T1D06 T1D08
If you have difficulty with any question, review the preceding section.

ENCRYPTED TRANSMISSIONS

Amateur Radio is a public form of communication. That is part of the agreement we make in trade for our operating privileges. As a consequence, it is not okay to transmit secret codes or to obscure the content of the transmission if the intent is to prevent others from receiving the information.

Translating information into data for transmission is called *encoding*. Recovering the encoded information is called *decoding*. Most forms of encoding are okay because they are done according to a published digital protocol. Any ham can look up the protocol and develop the appropriate capabilities to receive and decode data sent with that protocol.

Encoding that uses codes or ciphers to hide the meaning of the transmitted message is called *encryption*. Recovering the encrypted information is called *decryption*. Amateurs may not use encryption techniques except for radio control and control transmissions to space stations where interception or unauthorized transmissions could have serious consequences. [T1D03]

The difference between encoding and encryption are sometimes not always clear-cut. If a transmission is encoded according to an obscure and little-used protocol, for most hams it might as well be encrypted. But as long as the protocol is published and available to the public, that transmission is acceptable. The general rule to remember is that no ham should be prevented from receiving the communications of another ham because the necessary information has been withheld.

BROADCASTING AND RETRANSMISSION

Non-hams often refer to ham transmissions as "broadcasting" but that is inaccurate. Broadcasting consists of *one-way transmissions* intended for reception by the general public. [T1D10] Hams are not permitted to make this type of transmission except for the purposes of transmitting code practice, information bulletins for other amateurs, or when necessary for emergency communications. [T1D12]

The prohibition on broadcasting includes repeating and relaying transmissions from other communications services. Hams are also specifically prohibited from assisting and participating in news gathering by broadcasting organizations. [T1D09]

The prohibition against transmission of music (and other entertainment-type material in video and image transmissions) extends to incidental retransmission of music from a nearby radio. This means that you should turn down the car radio or music player when you're using the ham radio! Music can only be rebroadcast as part of an authorized rebroadcast of space station transmissions — a rather unusual circumstance. [T1D04]

Retransmitting the signals of another station is also generally prohibited, except when you are relaying messages or digital data from another station. Some types of stations (repeaters, auxiliary stations and space stations) are allowed to automatically retransmit signals on different frequencies or channels. [T1D07]

SPECIAL CIRCUMSTANCES

Ham communications must be intended for reception by hams. This leads to some exceptions from the normal broadcasting rules — hams may retransmit weather and propagation information from government stations, but not on a regular basis.

Hams like to operate from unusual places and while in motion, so what about operating from a plane or boat? These circumstances are covered in Part 97.11. You may operate, but only with the approval of the captain. (All commercial passenger flights have strict FAA bans on transmitting from inside the aircraft while in flight.) In the case where you do get permission, such as on a private plane, you have to use your own radio equipment and can't use any of the aircraft's or boat's radio equipment. Amateur transmissions may not interfere with any of the other radio systems on-board, including navigation systems or aircraft used in a boat's operation (such as a helicopter).

In general, hams can't communicate with non-amateur services, but the FCC may allow hams to talk to non-ham services at certain times or during a declared communications emergency. RACES operators may also communicate with government stations during emergencies. Once a year, the FCC permits ham-to-military communication on Armed Forces Day during May — see *QST* magazine or the ARRL's Special Events web calendar. [T1D02]

Before you go on, study these Technician exam questions from the question pool included at the back of this book or as a downloadable Study Guide version on the web:
T1D02 T1D03 T1D04 T1D07 T1D09 T1D10 T1D12
If you have difficulty with any question, review the preceding section.

Chapter 9

Safety

In this chapter, you'll learn about:

- **Working safely with electricity**
- **Safety grounding**
- **Lightning protection**
- **RF exposure rules**
- **Evaluating your station**
- **Reducing exposure to RF**
- **Safely installing antennas**
- **Working on and around towers**

When you see the mouse, you'll find more information at www.arrl.org/ham-radio-license-manual

There is nothing particularly risky about working with electricity, even though you can't see it. Compared to many activities, radio is one of the safest hobbies for people of all ages. Most hams go through an entire lifetime of ham radio without having a serious safety incident. This is because they educate themselves about safety and follow simple rules.

Safety is just as important for radio as it is for house wiring or working on an engine or using power tools. The key to safety is understanding the potential hazards, taking steps to *mitigate* (avoid or eliminate) them, and being able to respond to an injury in the unlikely event that one occurs. By being informed and prepared, your exposure to electrical hazards is greatly reduced.

Depending on your background, some of this material will be a review — nothing wrong with learning safety twice! Other topics specifically about radio will likely be new. In either case, safety is important enough that you'll encounter questions about it on your exam. We'll begin with basic electrical safety information before moving on to radio. Safety coverage concludes by reviewing some of the mechanical aspects of your radio activities.

Remember, this information is not here to frighten you. Radio and electricity are not always unsafe. You are probably learning about radio as an unfamiliar technology. Wouldn't it be a good idea to learn the appropriate safety techniques at the same time? Of course it would!

9.1 Electrical Safety

Working safely with electricity mostly means avoiding contact with it! Most modern ham radio equipment is solid-state and uses low voltage dc power, but the ac line voltage that powers most equipment is dangerous. You may also encounter vacuum tubes and the high voltages they use. Treat electricity with respect.

ELECTRICAL INJURIES

Electrical hazards can result in two types of injury: shocks and burns. Whenever electricity can flow through any part of your body, both can occur. Shocks and burns can

Table 9.1
Effects of Electric Current in the Human Body

Current	Reaction
Below 1 milliampere	Generally not perceptible
1 milliampere	Faint tingle
5 milliamperes	Slight shock felt; not painful but disturbing. Average individual can let go. Strong involuntary reactions can lead to other injuries.
6-25 milliamperes (women) 9-30 milliamperes (men)	Painful shock, loss of muscular control*; the freezing current or "can't let-go" range.
50-150 milliamperes	Extreme pain, respiratory arrest, severe muscular contractions. Death is possible.
1000-4300 milliamperes	Rhythmic pumping action of the heart ceases. Muscular contraction and nerve damage occur; death likely.
10,000 milliamperes	Cardiac arrest, severe burns; death probable

* If the extensor muscles are excited by the shock, the person may be thrown away from the power source.
Source: W.B. Kouwenhoven, "Human Safety and Electric Shock," Electrical Safety Practices, Monograph, 112, Instrument Society of America, p 93. November 1968.

be caused by ac or dc current flowing through the body.

Depending on the voltages present, shocks and burns can range from insignificant to deadly. Voltage is what causes the current to flow but doesn't shock all by itself. Just like in a regular resistor, as the voltage applied across your body varies so does current. While parts of the body such as hair and fingernails are not good conductors, the interior of the body conducts electricity quite well, being mostly salty water.

Electrical current through the body can disrupt the electrical function of cells. Currents of more than a few milliamps can also cause involuntary muscle contractions which leads to the jerking and jumping image on TV and in the movies. No joking matter, muscle spasms can cause falls and sudden large movements. The sudden pulling back of an outstretched hand or finger that comes in contact with an energized conductor is a result of arm muscles contracting. **Table 9.1** lists some of the effects of current in the body. [T0A02]

While any shock can be painful, the most dangerous currents are those that travel through the heart, such as from hand-to-hand or hand-to-foot. Electrical currents of 100 mA or more can disrupt normal heart rhythm. Depending on the resistance of the path taken by the current, voltages as low as 30 volts can cause enough current flow to be dangerous.

Burns caused by dc current or low-frequency ac current are a result of resistance to current in the skin, either through it to the body's interior or along it from point to point. The current creates heat and that's what results in the burn.

Avoiding Electrical Hazards

Shocks and burns are completely preventable if there is no way for you to come in contact with an energized conductor — simply prevent current flow through the body! Start by never working on "live" equipment unless it is necessary for troubleshooting or testing. Remove, insulate, or otherwise secure loose wires and cables before testing or repairing equipment. Never assume equipment is off or de-energized before beginning your work. Check with a meter or tester first.

If you do need to work on equipment with the power on — sometimes there's no way around it — follow these simple safety steps:

- Keep one hand in your pocket while probing or testing energized equipment and wear insulating shoes. This gives current nowhere to flow in or along your body.
- It's also easy to fall into bad habits after working with low-voltage or battery-powered

equipment that poses few hazards. Be extra careful when changing to work around higher voltages.

- Never bypass a safety interlock during testing unless specifically instructed to do so. Safety interlocks remove power when access panels, covers, or doors are opened to hazardous areas in equipment. They are intended to prevent unintentionally opening a cabinet or enclosure where dangerous voltages or intense RF may be present.

- Capacitors in a power supply can store charge after a charging circuit is turned off, presenting a hazardous voltage for a long time. This includes small-value capacitors charged to high-voltage! Make sure all high-voltage capacitors are discharged by testing them with a meter or use a *grounding stick* to shunt their charge to ground. [T0A11]

- Storage batteries release a lot of energy if shorted, leading to burns, fire, or an explosion. Keep metal objects such as tools and sheet metal clear of battery terminals and avoid working on equipment with the battery connected. [T0A01]

- Remove unnecessary jewelry from your hands because metal is an excellent conductor. Rings can also absorb RF energy and get quite hot in a strong RF field, such as inside an amplifier, filter, or antenna tuner.

- Avoid working alone around energized equipment.

- Remember that electricity moves a *lot* faster than you! Even your quickest touch is plenty long enough for electricity to flow.

 The *ARRL Handbook* has an excellent discussion of workbench and radio shack safety. You can also follow the links on the *Ham Radio License Manual* web page and the ARRL's Technical Information Service web page to more articles on all kinds of radio and electrical safety, including first aid for electrical injuries.

Response to Electrical Injury

If you were shocked in your radio shack, would others be able to help? The first step in any first aid response to electrical injury is to remove power. Install a clearly-labeled master ON/OFF switch for ac power to your station and workbench. It should be located away from the electrically-powered equipment where you are likely to be. Show your family and friends how to turn off power at the master switch and at your home's circuit-breaker box.

 It's also a good idea for all sorts of reasons for you and your family to get CPR training and to learn how to administer first aid for electrical injuries. To learn more about responding to electrical injuries browse the online WebMD first aid section on "electric shock treatment."

AC SAFETY GROUNDING

A large part of electrical safety is to not create hazards in the first place! This is why the National Electrical Code and your local building codes were created — to prevent common electrical hazards. A home wired "to code" has properly sized outlets and wiring and a *safety ground* to help prevent shocks. The safety ground is a connection to a ground rod at your home near the main electrical service entry. It provides a path to earth ground for current in case of an accidental short-circuit between either the hot or neutral wires and an appliance's metal enclosure or *chassis*.

Most ham stations don't require new wiring and can operate with complete safety when powered from your home's ac wiring. That is, as long as you follow simple guidelines: [T0A06]

- Use three-wire power cords that connect the chassis of your equipment to the ac safety ground.

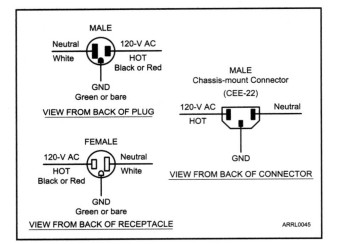

Figure 9.1 — The correct wiring technique for 120 V ac power cords and receptacles. The white wire is neutral and the green wire is the safety ground. The hot wire can be either black or red. These receptacles are shown from the back, or wiring side.

- Use ground fault circuit interrupter (GFCI) circuit breakers or circuit breaker outlets.
- Verify ac wiring is done properly by using an ac circuit tester.
- Never replace a fuse or circuit breaker with one of a larger size.
- Don't overload single outlets.

If you do decide to run new wiring for your station as it grows, either have a licensed electrician do the wiring or inspect it. Be sure to follow the standard hot-black (occasionally red)/neutral-white/ground-green/bare wiring shown in **Figure 9.1**. [T0A03] Use cable and wire sufficiently rated for the expected current load as shown in **Table 9.2**. Use the proper size fuses and circuit breakers. If you build your own equipment and power it from the ac lines, be sure to always install a fuse or circuit breaker in series with the ac hot conductor. [T0A08]

Table 9.2
Current-Carrying Capability of Some Common Wire Sizes

Copper Wire Size (AWG)	Allowable Current (A)	Max Fuse or Circuit Breaker (A)
6	55	50
8	40	40
10	30	30
12	25 (20)[1]	20
14	20 (15)[1]	15

[1]The National Electrical Code limits the fuse or circuit breaker size (and as such, the maximum allowable circuit load current) to 15 A for #14 AWG copper wire and to 20 A for #12 AWG copper wire conductors.

RF Burns

A home's safety ground is adequate to control shock hazards for 60 Hz ac power systems. At radio frequencies, though, the safety ground wiring usually acts more like an antenna than a ground as we discussed in the **Amateur Radio Equipment** chapter. The only safety hazard that results from poor RF grounding is an "RF burn."

RF burns result from contact with a "hot spot" — a location where high RF voltage is present on the outside of a connector, cable or equipment enclosure. The RF voltage creates currents in the skin at the point of contact. While they can be painful, RF burns generally don't do much damage. *Bonding* (connecting) your equipment together with wire or strap keeps all of the radio equipment at the *same* RF voltage — even if it's not exactly at ground potential — minimizing hot spots and the possibility of an RF burn.

LIGHTNING

Even though amateur antennas and towers are generally struck by lightning no more frequently than tall trees or other nearby structures, it is wise to take some precautionary steps. This is especially true for stations in areas with frequent severe weather and lightning. Lightning protection is intended to provide fire protection for your home since most of the damage to home resulting from a lightning strike is from fire.

Starting at your antennas, all towers, masts, and antenna mounts should be grounded according to your local building codes. [T0B11] This is done at the base, or in the case of roof mounts, though a large-diameter wire to a ground rod. Ground connections should be as short and direct as possible — avoid sharp bends. [T0B10, T0B12] Where cables and feed lines enter the house, use lightning arrestors grounded to a common plate that is in turn connected to a nearby external ground such as a ground rod. [T0A07] *The ARRL Ham Radio License Manual's* web page lists links to resources on lightning protection.

When lightning is anticipated, the best protection is to disconnect all cables outside the house and unplug equipment power cords inside the house. This interrupts the lightning's path to get to ground. In fact, it's not a bad idea to disconnect both power and phone cords to household appliances and long network and signal cables to your computing equipment. If you think you will unplug your equipment frequently, it might be a good idea to use or make power strips so that you can unplug many pieces of equipment with a single cable. Don't just turn them off — lightning jumps across switches quite easily. Physically unplug the power cable.

Determine whether your renter's or homeowner's insurance will cover you for damage from a lightning strike and whether the presence of antennas modifies that coverage. Your insurance agent will be able to help you determine the exact coverage and whether any special riders or amendments are needed. You may want to investigate the equipment insurance available to ARRL members, as well.

Regardless of how much protection you install on your antenna system, operating during a thunderstorm is a bad idea. Even a nearby strike can create a voltage surge of thousands of volts in a power or phone line, causing equipment damage or setting the house on fire!

Before you go on, study these Technician exam questions from the question pool included at the back of this book or as a downloadable Study Guide version on the web:
T0A01   T0A02   T0A03   T0A06   T0A07   T0A08   T0A11
T0B10   T0B11   T0B12
If you have difficulty with any question, review the preceding section.

9.2 RF Exposure

In recent years, there has been a lot of discussion about whether there are health and safety hazards from exposure to electromagnetic radiation (EMR). Many studies have been done at both power line frequencies (50 and 60 Hz) and RF (both shortwave and mobile phone frequencies). No link has been established between exposure to low-level EMR and health risks, including those frequencies used by amateurs.

RF radiation is not the same as *ionizing radiation* from radioactivity because the energy in signals at radio frequencies is far too low to cause an electron to leave an atom (ionize) and therefore cannot cause genetic damage. [T0C12] At these relatively low frequencies, RF energy is *non-ionizing radiation*. [T0C01]

Nevertheless, even in the absence of evidence that RF fields pose a health risk, it is prudent to avoid unnecessary exposure to high levels of RF. The FCC regulations set limits on the *Maximum Permissible Exposure* (MPE) from radio transmitters of any sort. To abide by these rules without requiring expensive testing, hams are expected to evaluate their stations to see if their operation has the potential to exceed MPE levels. The evaluation process is covered later in this section. The ARRL's *RF Exposure and You* contains a detailed treatment of RF exposure rules and safety techniques.

RF energy can only cause injury to the human body if the combination of frequency and power causes excessive energy to be absorbed. The only demonstrated hazards from exposure to RF energy are *thermal effects* (heating). *Biological* (athermal) effects have not been demonstrated at amateur frequencies and power levels. Measurable heating occurs only for very strong fields or in fields that originate very close to the body.

RF safety techniques involve making sure that persons are not exposed to high-strength fields in one of two ways:

- Preventing access to locations where strong fields are present.
- Making sure strong fields are not created in or directed to areas where people might be present.

RF burns caused by touching or coming close to conducting surfaces with a high RF voltage present are also an effect of heating. While these are sometimes painful, they are rarely hazardous. As discussed above, RF burns can be eliminated by proper grounding techniques or by preventing access to an antenna. [T0C07]

Heating as a result of exposure to RF fields is caused by the body absorbing RF energy. Absorption occurs because the RF energy causes the molecules to vibrate at the same frequency. The energy of the vibrations is dissipated within the body as heat. The stronger the field, the more the molecules vibrate and the more heating of the body's tissues results. Absorption also varies with frequency because the body absorbs more RF energy at some frequencies than others. [T0C05] The total amount of heating then depends on both the RF field's intensity and frequency and is called the *specific absorption rate* or SAR.

Power Density

The intensity of an RF field is called *power density*. Power density is the amount of energy per unit of area. The most common way of stating power density is in milliwatts per square centimeter (mW/cm^2). The power density of an RF field is highest near antennas and in the directions where antennas have the most gain. Power density can also be very high inside transmitting equipment.

Increasing transmitter power increases power density everywhere around an antenna to the same degree that transmitter power increased. Increasing distance from an antenna lowers power density in proportion to the square of the distance from the antenna. For example, at twice the distance from an antenna, power density is divided by four. Controlling these two factors, power and distance, forms the basis for amateur RF safety.

EXPOSURE LIMITS

Safe levels of exposure based on demonstrated hazards have been established by the FCC. These are the Maximum Permissible Exposure (MPE) levels. Because the specific absorption rate (SAR) varies with frequency, so does the MPE as shown in **Figure 9.2** and **Table 9.3**. Where SAR is high, MPE is low.

SAR depends on the size of the body or body part affected and is highest where the body and body parts are naturally resonant. An adult-size body is resonant at about 35 MHz if the person is grounded and 70 MHz if they are not grounded. Body parts are resonant at higher frequencies (smaller wavelengths). For example, an adult's head is resonant at the much higher frequency of 400 MHz.

Above and below the ranges of highest absorption, the body responds less and less to the RF energy, just like an antenna responds poorly to signals away from its natural resonant frequency. Frequencies at

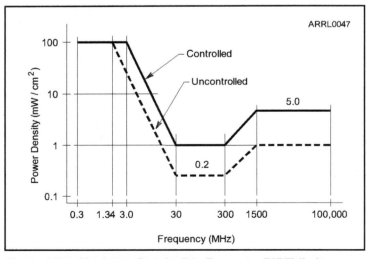

Figure 9.2 — Maximum Permissible Exposure (MPE) limits vary with frequency because the body responds differently to energy at different frequencies. The controlled and uncontrolled limits refer to the environment in which people are exposed to the RF energy.

Table 9.3
Maximum Permissible Exposure (MPE) Limits

Controlled Exposure (6-Minute Average)

Frequency Range (MHz)	Power Density (mW/cm²)
0.3-3.0	(100)*
3.0-30	$(900/f^2)$*
30-300	1.0
300-1500	f/300
1500-100,000	5

Uncontrolled Exposure (30-Minute Average)

Frequency Range (MHz)	Magnetic Field Power Density (mW/cm²)
0.3-1.34	(100)*
1.34-30	$(180/f^2)$*
30-300	0.2
300-1500	f/1500
1500-100,000	1.0

f = frequency in MHz
* = Plane-wave equivalent power density

which the body has the highest SAR are from 30 to 1500 MHz. These are the regions on the MPE graph where the limits for exposure are the lowest. For example, when comparing MPE for amateur bands at 3.5, 50, 440 and 1296 MHz, you can see that MPE is lowest at 50 MHz and highest at 3.5 MHz. [T0C02]

Controlled and Uncontrolled Environments

You'll notice in Figure 9.2 that there are two sets of lines, one called *controlled* and the other *uncontrolled*. These refer to the type of environments in which the exposure to RF fields takes place.

• People in controlled environments are aware of their exposure and can take the necessary steps to minimize it.

• People in uncontrolled environments are not aware of their exposure, such as in areas open to the general public or your neighbor's property.

The FCC has determined that the higher controlled-environment limits generally apply to amateur operators and members and guests in their immediate households, provided that they are aware of RF fields being used. If this is the case, the controlled-environment limits apply to your home and property — wherever you control physical access.

AVERAGING AND DUTY CYCLE

Since the effects from RF exposure are related to heating and take place over many seconds, the MPE limits are based on *averages*, not *peak* exposure. This allows exposure to be averaged over fixed time intervals.

• The averaging period is 6 minutes for controlled environments.

• The averaging period is 30 minutes for uncontrolled environments.

The difference in averaging periods reflects the difference in how long people are expected to be present and exposed. People are assumed less likely to stay in an uncontrolled environment receiving continuous exposure, so the averaging period is much longer.

During the averaging period, a transmitter may only be generating RF for a fraction of the time. For most amateur contacts, the transmitter output is no more than 50% of the time and usually much less. This pattern lowers the *duty cycle* of the emissions. Duty cycle is the ratio of the transmitted signal's on-the-air time to the total operating time during the measurement period and has a maximum of 100%. Stated simply, duty cycle is the percentage of time a transmitter is transmitting. [T0C11] (*Duty factor* is the same as duty cycle expressed as a fraction, instead of percent, for example 0.25 instead of 25%.)

Since duty cycle affects the average power level of transmissions, it must be considered when evaluating exposure. [T0C10] The lower the duty cycle (less transmitting), the higher the transmitter output can be and still have an average value within the exposure limits. For example, what is the result if a transmitted signal in a controlled environment is present for 3 minutes and then absent for the remaining 3 minutes of the averaging period? Because the signal is only present for ½ of the time (50% duty cycle), the signal power can be twice as high and still have the same average power as it would if transmitted continuously with a duty cycle of 100%. [T0C13]

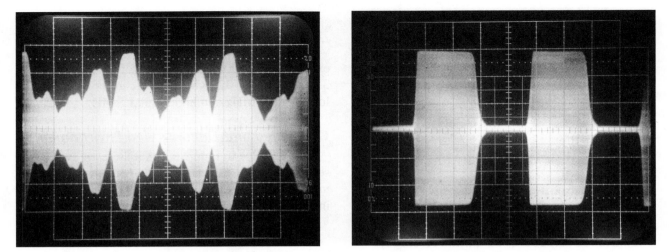

Figure 9.3 — The SSB signal on the left and the Morse code signal on the right both have the same peak power, but the average power of the SSB signal is lower.

Table 9.4
Operating Duty Cycle of Modes Commonly Used by Amateurs

Mode	Duty Cycle	Notes
Conversational SSB	20%	1
Conversational SSB	40%	2
SSB AFSK	100%	
SSB SSTV	100%	
Voice AM, 50% modulation	50%	3
Voice AM, 100% modulation	25%	
Voice AM, no modulation	100%	
Voice FM	100%	
Digital FM	100%	
ATV, video portion, image	60%	
ATV, video portion, black screen	80%	
Conversational CW	40%	
Carrier	100%	4
Digital (PSK31, RTTY)	100%	

Note 1: Includes voice characteristics and syllabic duty cycle. No speech processing.

Note 2: Includes voice characteristics and syllabic duty cycle. Heavy speech processor employed.

Note 3: Full-carrier, double-sideband modulation, referenced to PEP. Typical for voice speech. Can range from 25% to 100%, depending on modulation.

Note 4: A full carrier is commonly used for tune-up purposes.

Some modes have lower average power than others as illustrated in **Figure 9.3**. For example, while sending Morse code (CW), the transmitter is off between the individual dots and dashes. SSB signals only reach peak power for short periods at voice peaks and so have the lowest duty cycle. FM, however, is a constant-power mode and so the signal is continuously at full power when the transmitter is on. The *operating duty cycle* for typical uses of each mode (also called the *emission duty cycle*) is shown in **Table 9.4**.

Because most amateur operation is intermittent, the time spent transmitting on the air is low. For example, during a roundtable contact among three stations, each is likely to be transmitting only one-third or 33% of the time. This further reduces average exposure, regardless of the mode being used.

For a given peak envelope power (PEP), an emission with a lower operating duty cycle produces less RF exposure. PEP multiplied by the mode's operating duty cycle and the fraction of the time spent transmitting gives the resulting overall average power during the exposure period.

Average power = PEP × operating duty cycle × (time transmitting / averaging period)

For example, let's say that your 100-watt transmitter is generating conversational SSB without speech processing. Table 9.4 shows an operating duty cycle of 20% for that mode.

During your operating period you transmit for 1 minute out of every 3. Your average power during the evaluation period is:

100 watts × 20% for conversational SSB × (1 min / 3 min) =
   100 × 0.2 × 0.33 = 6.6 watts

During a 2 meter net as net control, using your 50-watt VHF FM transmitter, you transmit and listen for equal periods. Your average power is:

50 watts × 100% for FM × 50% on/off = 25 watts

Effect of Antenna Gain

There is one additional effect that has to be taken into account. As you've learned, beam antennas focus radiated power toward one direction, creating gain. Gain has the effect of increasing your average power in the preferred direction. (It also decreases your average power in other directions.) This means that there are four factors that affect RF exposure: transmitter power and frequency, distance to the antenna, and the radiation pattern of the antenna.

If you use an antenna with gain, you will need to include the effect of gain in your exposure evaluations. For example, if your antenna has 6 dBi of gain, corresponding to a four-fold increase in power radiated in the preferred direction, you would multiply your average power by four when calculating RF exposure in the antenna's forward direction.

EVALUATING EXPOSURE

According to FCC rules, all fixed stations must perform an exposure evaluation. (Mobile and handheld transceivers are exempt.) There are three ways of making this evaluation. By far the most common evaluation uses the techniques outlined in the FCC's OET Bulletin 65 (OET stands for *Office of Engineering Technology*). [T0C06] This method uses tables and simple formulas to evaluate whether your station has the potential of causing an exposure hazard.

You could also obtain RF power density instrumentation and actually measure the power density of your transmissions. It is also acceptable to make computer models of your station and use those results. Both of these methods are rarely used due to the expense or effort required.

Once you've done an evaluation, you don't need to re-evaluate unless you change equipment in your station that affects average output power, such as increasing transmitter power or antenna gain. You'll also need to re-evaluate if you add a new frequency band. [T0C09]

Before you start, check to see if your station is exempt from the evaluation requirement. If the transmitter power (PEP) to the antenna is less than the levels shown in **Table 9.5** on the frequencies at which you operate, then no evaluation is required! [T0C03] The FCC has determined that the risk of exposure from these power levels is too small to create an exposure risk. So, if you have a 25-watt VHF/UHF mobile rig or a 5-watt handheld transceiver, there's no need for an evaluation of any sort.

If you do need to do an evaluation, don't despair. It's not as complicated as it seems. The *Ham Radio License Manual* web page lists resources that make the job a lot easier, such as on-line exposure calculators and pre-calculated tables you can use for common antennas. You'll need information on the RF signal's frequency and power level, distance from the antenna and the antenna's radiation pattern. [T0C04]

Table 9.5

Power Thresholds for RF Exposure Evaluation

Band	Power (W)
160 meters	500
80	500
40	500
30	425
20	225
17	125
15	100
12	75
10	50
6	50
2	50
1.25	50
70 cm	70
33	150
23	200
13	250
SHF (all bands)	250
EHF (all bands)	250

The general procedure consists of several steps:

- Start with the average power from each transmitter on each band. Use the same process discussed above, starting with full PEP and then applying the various corrections for mode and patterns of use.
- If you have long coaxial feed lines, you may want to subtract feed line losses, particularly in the 30 to 1500 MHz frequency range.
- Then use the ARRL tables to include the effects of antenna gain and height.
- Finally, use the tables to determine the distance required from the antenna to comply with MPE limits.

The process is the same whether you do it manually using tables or online with a web page calculator. Both require the same information from you about your station and use the same tables.

Don't forget to do the evaluation for each frequency band and antenna used on that band. You can save yourself some work by performing the evaluation for the highest average power, mode, and usage on each antenna. That will show the minimum or worst-case separation requirements under all circumstances for that frequency and antenna.

Once the evaluation is complete, compare the minimum separation with your actual installation. Chances are, you'll find no hazard exists — most stations are simply not capable of causing a health hazard.

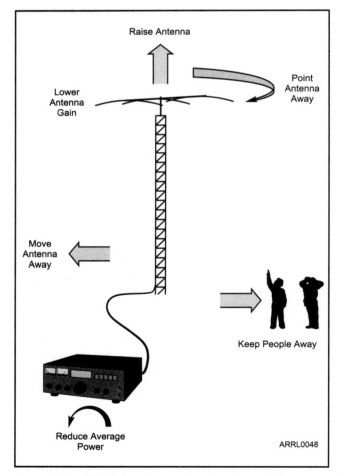

Figure 9.4 — There are many ways to reduce RF exposure to nearby people. Whatever lowers the power density in areas where people are will work. Raising the antenna will even benefit your signal strength to other stations as it lowers power density on the ground!

EXPOSURE SAFETY MEASURES

What if you do find a potential hazard? What if you are just beginning to build a station and want to avoid creating a hazard? You have plenty of options as shown in **Figure 9.4**:

- Locate antennas away from where people can get close to them and away from property lines. This is always a good idea since touching an antenna energized with even low-power signals can result in an RF burn. [T0C08]
- Raise the antenna. This is another good idea because it usually improves your signal in distant locations, as well.
- If you have a beam antenna, avoid pointing the antenna where people are likely to be.
- Use a lower gain antenna to reduce radiated power density or reduce transmitter power. You may find that you're able to make contacts just as well with less power or gain.
- Limit the average power of your transmissions by transmitting for shorter periods or even using a mode with a lower duty cycle.

Any of these techniques will reduce RF exposure to you and your neighbors. You'll likely be able to find a combination that has a minimum effect on your operations yet still makes sure you are within the MPE limits.

Even though emissions from mobile and handheld transmitters are exempt from evaluation, there are some good ways to minimize unnecessary RF exposure:

- Place mobile antennas on the roof or trunk of the car to maximize shielding of the passengers.
- Use a remote microphone to hold a handheld transceiver away from your head while transmitting.

RF exposure safety measures are easy to apply and part of good amateur practices. By understanding the reason for exposure limits and how to mitigate RF exposure hazards, you will be able to make more informed choices about designing, building and operating your station.

Before you go on, study these Technician exam questions from the question pool included at the back of this book or as a downloadable Study Guide version on the web:
T0C01 through T0C13
If you have difficulty with any question, review the preceding section.

9.3 Mechanical Safety

Just as workshop safety is important, there are plenty of mechanical aspects to Amateur Radio that generate their own safety concerns. Amateurs have been building and installing radios and antennas for more than 100 years, developing a large body of knowledge about the safe way to do things. The following sections provide some guidelines as you build and maintain your radio equipment.

Most importantly, follow the manufacturer's directions and recommendations. For example, how tight should guy wires be? The tower manufacturer will tell you — do what they say to do. This holds true for all types of installation, antenna and tower work.

MOBILE INSTALLATIONS

Putting a radio in a vehicle sounds pretty straightforward and it can be. However, there are a few common safety hazards that are often overlooked by a first-time installer. The most important consideration, even beyond good RF performance, is to preserve the safety of you and your passengers.

Anything loose in a passenger compartment can become a deadly projectile in an accident — imagine being knocked on the head by a loose radio traveling at 30 miles per hour or more! Secure *all* equipment in a vehicle, including accessories such as diplexers, switches and microphones. If possible, use *control heads* (detachable front panels) that connect to the radio with a long cable. Mount the heavier radio under a seat or in the trunk where it can't move.

Don't install the radio where it diverts your attention from the road. Don't block your vision by placing equipment on dashboards or in your field of view. (It's also a good idea to keep radios out of direct sunlight.) Place the radio or control head where the controls can be easily seen without taking your attention away from the road for prolonged periods.

Even with your radio properly mounted you still need to be a safe driver! Follow safe operating practices by adhering to these simple rules:
- Don't operate in heavy traffic. Hang up the microphone and resume the contact later.
- Pull over to make complicated adjustments to the radio. Fumbling through a radio's menu or trying to press two buttons at once is a sure way to risk an accident.

Know the traffic laws in your state that concern operation of two-way radios while driving. State legislatures have been making laws in response to careless mobile phone use. While Amateur Radio operations are usually exempt, certain types of operating may be restricted. It may be illegal to use headphones while driving or have a speaker on too loud to hear emergency vehicles. Scanners and radio equipment that can receive public safety transmissions may be illegal in vehicles, even though amateur transceivers are often exempted. It is a good idea to carry your amateur license and a copy of the state regulations exempting Amateur Radio with you whenever you are operating from your car.

PUTTING UP ANTENNAS AND SUPPORTS

The plan for installing your antenna that looked pretty good on the ground often has to be modified once you're dealing with the actual installation. Gravity has a way of making things more difficult than you expect! Above all, follow the manufacturer's directions — they want you to have a successful experience with their product and often provide useful information in product manuals and on their websites. If you haven't put up an antenna before, enlist the help of a more experienced ham.

Before you can start, you should be sure your plans satisfy any local zoning codes or covenants or restrictions in your deed or lease. If you are putting up a very tall tower (greater than 200 feet) or an antenna near an airport, check the rules about maximum height of structures near an airport. The Federal Aviation Administration (FAA) and FCC have specific regulations about towers in these locations.

When you're ready to put up the antenna, look carefully at the area around your antenna and any supporting structures it requires. Of course, people should not be able to come in contact with the antenna accidentally. If an antenna is to be mounted at ground level, consider surrounding it with a wooden or plastic fence. If you're installing a wire antenna, make sure the feed line does not sag below head height to snag an unwary passerby. If you're in a rural area, be sure that deer or other antlered animals won't catch the feed line with their headgear.

Power lines are the enemy of antenna installers. Place all antennas and feed lines well clear of power lines, including the utility service drop to your home. [T0B04] **Figure 9.5** illustrates the idea. Be sure that if the any part of the antenna or support structure falls, it cannot fall onto power lines. A good guideline is to separate the antenna from the nearest power line by 150% of total height of tower or mast plus antenna — a minimum of 10 feet of clearance during a fall is a must. [T0B06] Never attach an antenna or guy wire to a utility pole, since a mechanical failure could result in contact with high-voltage wires. [T0B09]

Trees are often used as wire supports. If you decide to throw or shoot a supporting line through or over trees, be sure the projected flight path is completely safe and clear of people and power lines. A line that breaks or snags can whip or rebound, often with a lot of energy, so wear protective gloves and goggles.

Once your antenna is in place, secure the feed line with tape or plastic wire ties. Keep all

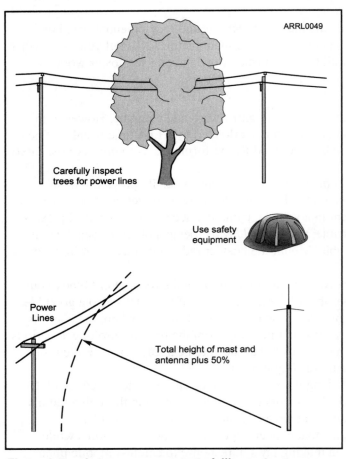

ARRL0049

Carefully inspect trees for power lines

Use safety equipment

Power Lines

Total height of mast and antenna plus 50%

Figure 9.5 — Antennas or supports falling onto a power line can result in electrocution. Take extra care in locating and erecting your antenna to avoid a deadly accident!

supporting guy lines above head height, if possible. Where someone can walk into guy wires, surround the guy anchor point with a fence or flag the wires with colored warning or survey tape.

Grounding rules for antennas and supports must be followed according to your local electrical code. Towers should be grounded with separate 8-foot long ground rods for each tower leg, bonded to the tower and each other. [T0B08] A smaller antenna mast should be grounded with a heavy wire and ground rod. Guy wires must be installed according to the tower manufacturer's instructions.

Tower and Climbing Safety

It's a fact of life that antennas tend to be in high places. At VHF and UHF, the higher an antenna, the greater the distance to the radio horizon. Noise from sources at ground level is reduced, as well. At HF, the vertical angle at which the most power is radiated generally decreases with height above ground, so most antennas are mounted as high as is practical. Once beyond the capability of a mast or pipe support — about 40 feet or so — a steel or aluminum tower is the usual solution.

While you may not immediately decide to put up a tower yourself, it is common for amateurs to help each other with antenna projects that involve tower work. Whether you decide to work on the tower itself or as part of the ground crew, safety is absolutely critical. The following safety guidelines will help you safely contribute to tower projects of your own or of others. For more information about working safely on and around towers, *The ARRL Handbook* and *The ARRL Antenna Book* cover the subject in some detail.

Starting with personal preparation, both climbers and ground crew should wear appropriate protective gear any time work is under way on the tower. Each member of the crew should wear a hard hat, goggles or safety glasses and heavy duty gloves suitable for working with ropes. [T0B01] If you are the climber, use an approved climbing harness and work boots to protect the arches of your feet. [T0B02] Don't use a "lineman's belt" as they are unsafe and no longer approved for tower work. Many climbers prefer footwear with a steel shank that supports the foot while standing on a narrow rung. **Figure 9.6** shows a properly equipped climber ready for tower work.

Before climbing or starting work, perform a thorough inspection of all equipment and installed hardware. Surprises during tower work are rarely a good thing!

- Inspect all tower guying and support hardware. Repair or tighten as necessary before anyone goes up!
- Crank-up towers must be fully nested and blocked, if necessary.

Figure 9.6 — The well-dressed tower climber. Note the waist D-rings for positioning lanyard attachment as well as the suspenders and leg loops. The lower photo shows an adjustable positioning lanyard. Be sure to wear protective gear — hard hat, goggles and gloves — whether you are working on the tower or as part of the ground crew. [Steve Morris, K7LXC, photos]

Never climb a crank-up tower supported only by the cable that supports the sections. [T0B07]

- Double-check all climbing belts and lanyards before climbing. Make sure clips and carabineers work smoothly without sticking open or closed. Replace or discard frayed straps and slings.
- Make sure all ropes and load-bearing hardware are in good condition before placing them in service.
- Use a gin pole (a temporary mast used to lift materials such as antennas or tower sections) so that you do not have to hoist things directly. [T0B05]
- Double-check the latest weather report, since you don't want to get caught on the tower in a storm.
- It's a good idea to visit the bathroom before starting your climb and don't forget the sunblock lotion!

Having a ground crew is important; avoid climbing alone whenever possible because it's never safe. [T0B03] If you do climb alone, take along a handheld radio. A ground crew should have enough members to do the job safely, including rendering aid if necessary. While everyone is on the ground, review the job in detail and agree on who gives instructions. If hand signals will be used, make sure everyone understands them! It also helps to rehearse the steps so that everyone knows the sequence. Does everyone know the proper knots and rope-handling technique? If not, make sure those who do are the ones who will be responsible for handling the lines.

During the job, keep distracting chatter to a minimum. One member of the crew should always be watching the climber or climbers. Stay clear of the tower base unless you need to be there because that's where dropped objects are likely to land. *Never* remove your hard hat while work is proceeding on the tower — an object dropped from 60 feet will be traveling a bit over 40 mph when it lands. Ouch!

By participating as part of a team, you'll learn how to perform tower work safely. Even if you never set foot on a tower yourself, knowing how to help can make a contribution to these significant projects.

Before you go on, study these Technician exam questions from the question pool included at the back of this book or as a downloadable Study Guide version on the web:
T0B01 through T0B09
If you have difficulty with any question, review the preceding section.

Chapter 10

Glossary

Words in definitions that are *italicized* have a separate glossary entry.
A word in **bold** is defined in that entry.

The ARRL website also provides a glossary with broader coverage of ham radio terms.

Active filter — See *filter.*

Adapters — *Connectors* that convert one type to another.

Allocations — Frequencies authorized for a particular FCC telecommunications *service.*

Alternating current or voltage (ac) — Electrical *current* or *voltage* with a direction or polarity, respectively, that reverses at regular intervals.

Amateur operator — A person named in an amateur operator/primary license station grant on the ULS consolidated licensee database to be the control operator of an amateur station.

Amateur Radio Emergency Service (ARES®) — An organization of amateur volunteers that is sponsored by the ARRL and provides emergency communication services to groups such as the American Red Cross and local Emergency Operations Centers (EOC).

Amateur service — A radio communication service for the purpose of self-training, intercommunication and technical investigations carried out by amateurs, that is, duly authorized persons interested in radio technique solely with a personal aim and without *pecuniary* interest.

Amateur station — A station licensed in the amateur service, including necessary equipment, used for amateur communication.

Amateur television (ATV) — Analog fast-scan television using commercial transmission standards (NTSC in North America).

American Radio Relay League (ARRL) — The national association for Amateur Radio.

Ammeter — A test instrument that measures *current.*

Ampere (A) — The basic unit of electrical *current*, also abbreviated amps. One ampere is the flow of one *coulomb* of charge per second.

Amplifier — A device or piece of equipment used to *amplify* a signal.

Amplify — Increasing the strength or *amplitude* of a signal.

Amplitude — The strength or magnitude of a signal.

Amplitude modulation (AM) — The process of adding information to a signal or *carrier* by varying its amplitude. Transmissions referred to as **AM phone** are usually composed of two sidebands and a carrier. Shortwave broadcast stations use this type of AM, as do stations in the Standard Broadcast Band (535-1710 kHz). AM in which only one sideband is transmitted is called *single-sideband* or *SSB* and is the most popular voice mode on the *high frequency (HF)* bands

AMSAT (Radio Amateur Satellite Corporation) — Organization that manages many of the amateur satellite programs.

Analog (linear) signal — A signal (usually electrical) that can have any amplitude (voltage or current) value, and whose amplitude can vary smoothly over time. Also see *digital signal*.

Antenna — A device that radiates or receives radio frequency energy.

Antenna matching network — see *impedance matching network*.

Antenna switch — A switch used to connect one transmitter, receiver or transceiver to several different antennas.

Antenna tuner — See *impedance matching network*.

Apogee — The point in a satellite's orbit at which it is farthest from the Earth. See *perigee*.

AGC — See *automatic gain control*.

ALC — See *automatic level control*.

Automatic Packet Reporting System (APRS) — A system by which amateurs report their position automatically by radio to central servers from which their locations can be observed.

Amateur Radio Direction Finding (ARDF) — Competitions in which amateurs combine orienteering with *radio direction finding*.

Anode — The more positively charged *electrode* of a *diode* or *vacuum tube*.

Antenna analyzer — A portable instrument that combines a low-power signal source, a frequency counter and an *SWR meter*. Also known as an **SWR analyzer**.

Array — An antenna with more than one *element*. In a **driven array** all elements are *driven elements*. In a **parasitic array** some elements are *parasitic elements*.

Attenuate — To reduce the strength of a signal. An **attenuator** is a device that attenuates a signal.

Audio frequency (AF) signal — An ac electrical signal in the range of 20 hertz to 20 kilohertz (20,000 hertz). This is called an audio signal because human hearing responds to sound waves in the same frequency range.

Automatic control — A station operating under the control of devices or procedures that ensure compliance with FCC rules.

Automatic gain control (AGC) — A circuit that automatically adjusts RF Gain in a receiver to maintain a relatively constant output volume.

Automatic level control (ALC) — A circuit that automatically controls transmitter power to reduce distortion of the output signal that can cause interference to other stations.

Automatic Repeat Request (ARQ) — The method of requesting a retransmission of data if the data is received with errors. Also known as **Automatic Repeat Query**.

Autopatch — A device that allows users to make telephone calls through a *repeater*.

Auxiliary station — A station that operates in support of another station, such as a *repeater*, by transmitting control information or relaying audio.

Balanced line — *Transmission line* in which none of the conductors is connected directly to *ground*. See *open-wire line*.

Balun — Contraction of "balanced to unbalanced" and pronounced "BAHL-un." A device to transfer power between a balanced load and an unbalanced feed line or device, or vice versa.

Band — A range of frequencies. An **amateur band** is a range of frequencies on which amateurs are allowed to transmit.

Band-pass filter (BPF) —A *filter* designed to pass signals within a range of frequencies called the **pass-band**, while attenuating signals outside the pass-band.

Band plan — Voluntary organization of activity on an amateur band under normal circumstances.

Band-stop filter — See *notch filter.*

Bandwidth — (1) Bandwidth is the range of frequencies that a radio signal occupies. (2) FCC Part 97 defines bandwidth for regulatory purposes as "The width of a frequency band outside of which the mean power is attenuated at least 26 dB below the mean power of the transmitted signal within the band." [Part 97.3 (8)]

Base — (1) A station at a fixed location. (2) See *transistor.*

Battery — A package of one or more *cells.*

Battery pack — A package of several individual *cells* connected together (usually in series to provide higher voltages) and treated as a single battery.

Baud — The rate at which symbols are transmitted in a *digital mode.*

Baudot — The code used for *radioteletype* (RTTY) characters.

Beacon station — An amateur station transmitting communications for the purposes of observation of propagation and reception or other related experimental activities.

Beam antenna — See *directional antenna.*

Bit error rate (BER) — The rate at which bit-level errors occur in a stream of digital data.

BJT — See *transistor.*

Block diagram — A drawing using boxes to represent sections of a complicated device or process. The block diagram shows the connections between sections. A block diagram shows the internal functions of a complex piece of equipment without the detail of a *schematic diagram.*

BNC — A type of *RF connector.*

Bonding — Connecting equipment or circuits together to keep them at the same voltage.

Break-in — Switching between transmit and receive during CW operation so that you can listen to the operating frequency between Morse *elements* (full break-in) or during short pauses in your transmissions (semi-break-in).

Breaking in — The term for joining an ongoing contact by transmitting your *call sign* during a pause in the contact.

Broadcasting — One-way transmissions intended to be received by the general public, either direct or relayed.

Bug — A mechanical Morse *key* that uses a spring to send dots automatically.

Bus — An electrical conductor for distributing power or to provide a common connection.

Call — (1) Abbreviated form of *call sign*. (2) Attempt to make contact.

Call district — The ten administrative areas established by the FCC.

Call sign — The letters and numbers that identify a specific amateur and the country in which the license was granted.

Calling frequency — A frequency on which amateurs establish contact before moving to a different frequency. Usually used by hams with a common interest or activity.

Capacitance — A measure of the ability to store energy in an *electric field*. Capacitance is measured in *farads*.

Capacitor — An electrical *component* that stores energy in an *electric field*. Capacitors are made from a pair of conductive surfaces called *electrodes* that are separated by an *insulator* called the *dielectric*.

Carrier — The unmodulated RF signal to which information is added during *modulation*. Also see *modulate*.

Cathode — The more negatively charged *electrode* of a *diode* or *vacuum tube*.

CB — Citizen's Band. An unlicensed radio service operating near 27 MHz intended for use by individuals and businesses over ranges of a few miles.

Cell (electrochemical cell) — A combination of chemicals and *electrodes* that converts chemical energy into electrical energy. See *battery*.

Centi (c) — The metric prefix for 10^{-2} or division by 100.

Certificate of Successful Completion of Examination (CSCE) — A document that verifies that an individual has passes one or more exam *elements*. A CSCE is good for 365 days and may be used as evidence of having passed an element at any other amateur license exam session.

Channel — (1) A range of *frequencies* used for one radio or communications signal. (2) The structure connecting the *source* and *drain* of an *FET* and through which *current* flows.

Channel spacing — The difference in frequency between *channels*.

Characteristic impedance — The ratio of RF *voltage* to *current* in a *transmission line* that is *matched*.

Charge — Store energy in a *battery* by reversing the chemical reaction in its *cells*.

Chassis ground — The common connection for all parts of a *circuit* that connect to the metal enclosure or chassis (pronounced "*CHAA-see*") of the circuit.

Check in — Register your station's presence on a *net* with the *net control station*.

Checksum — A method of detecting errors in digital data by including a calculated value with the data.

Choke filter — A type of *low-pass filter* that blocks RF current.

Circuit — A conductive path through which *current* can flow.

Circuit breaker — A protective component that "breaks" or opens a *circuit* or *trips* when an excessive *current* flow occurs.

Closed repeater — A *repeater* that restricts access to members of a certain group of amateurs. See *open repeater*.

Closed circuit — An electrical circuit with an uninterrupted path for the current to follow. Turning a switch on, for example, closes or completes the circuit, allowing current to flow. Also called a **complete circuit**.

Coaxial cable — Coax (pronounced *KOH-aks*). A type of *transmission line* with a single center conductor inside an outer shield made from braid or solid metal and both sharing a concentric central axis. The outer conductor is covered by a plastic **jacket**.

Color code — A system in which numerical values are assigned to various colors. Colored stripes or dots are painted on the body of *resistors* and other *components* to represent their value.

Collector — See *transistor*.

Common — Term for the shared reference for all voltages in a circuit. Also referred to as *circuit* common. See *ground* and *bonding*.

Common mode — Currents that flow equally on all conductors of a multiconductor cable, such as speaker wires or telephone cables, or on the outer surface of shielded cables.

Communications emergency — A situation in which communications is required for immediate safety of human life or protection of property.

Component — (1) A device having a specific quantity of an electrical property (such as *resistance*) or that has a specific electrical function. (2) One signal of a group that makes up a *composite* signal.

Composite signal — A signal with information encoded by a group of *component* signals. For example, an AM signal is a composite signal that consists of three components: the carrier and the upper and lower sidebands.

Compression — See *speech compression*.

Conductor — A material in which *electrons* move freely in response to an applied *voltage*.

Connector — A *component* used to connect and disconnect electrical circuits and equipment.

Continuous wave (CW) — Radio communications transmitted by on/off *keying* of a continuous radio-frequency signal. Another name for international *Morse code*.

Control code — Information in the form of data or tones used to adjust a station under *remote control*.

Control link — The means by which a control operator can make adjustments to a station operating under *remote control*.

Control operator — The person designated by the licensee of a station to be responsible for the transmissions of an amateur station.

Control point — The location at which a station's *control operator* function is performed.

Controlled environment — Any area in which an RF signal may cause radiation exposure to people who are aware of the radiated electric and magnetic fields and who can exercise some control over their exposure to these fields. The FCC generally considers amateur operators and their families to be in a controlled RF exposure environment to determine the *maximum permissible exposure* levels. See *uncontrolled environment.*

Conventional current — See *current.*

Core — In an *inductor*, the core is the material or space the wire is wound around or passed through.

CORES — Commission Registration System of the *FCC.*

Coulomb (C) — The basic unit of electrical charge. One coulomb is 6.25×10^{18} *electrons.* 1 *ampere* equals the flow of 1 coulomb of electrons per second.

Courtesy tone (beep) — A short burst of audio transmitted by a *repeater* to indicate that the previous station has stopped transmitting. It can also be used to indicate that the *time-out timer* has been reset.

CQ — "Calling any station," the general method of requesting a contact with any station.

Crossband — Able to receive and transmit on different amateur frequency bands. For example, a *repeater* might receive a signal on 70 cm and retransmit it at 2 meters.

CTCSS — Continuous Tone Coded Squelch System. A low frequency tone (also called subaudible tone) required to access many *repeaters.* See *PL.*

Current (electrical) — The movement of electrons in response to an electromotive force, also called *electronic current. Conventional current* is the flow of positive charge that moves in the opposite direction of electronic current.

Cutoff frequency — The frequency at which a *filter's* output *power* is reduced to one-half the input power.

Cycle — One complete repetition of a repeating waveform, such as a *sine wave*

CW (Morse code) — See *continuous wave.*

D region — The lowest region of the *ionosphere.* The D region (or layer) acts mainly to absorb energy from radio waves as they pass through it.

Data (digital) mode — Computer-to-computer communication, such as by *packet radio* or *radioteletype (RTTY)*, in which information is exchanged as data characters or digital information.

DC voltage — A voltage with a constant *polarity.* See *direct current.*

Deceptive (or false) signals — Transmissions that are intended to mislead or confuse those who may receive the transmissions. For example, distress calls transmitted when there is no actual emergency are false or deceptive signals.

Decibel (dB) — In electronics decibels are used to express ratios of *power, voltage,* or *current.* 1 dB = 10 $\log_{10}$ (power ratio) or 20 $\log_{10}$ (voltage or current ratio).

Deci (or lower case d) — The metric prefix for 10^{-1} or division by 10.

Delta loop — A loop antenna in the shape of a triangle.

Degree — A measure of angle or phase. There are 360 degrees in a circle or *cycle*.

Demodulate — To recover the information from a modulated signal by reversing the process of *modulation*. See *modulate*.

Designator — Letters and numbers used to identify a specific electronic *component*.

Detect — (1) To determine the presence of a signal. (2) To recover the information directly from a modulated signal.

Detector — The stage in a receiver in which the *modulation* (voice or other information) is recovered from a *modulated* RF signal.

Deviation — The change in *frequency* of an FM *carrier* due to a *modulating* signal. Also called **carrier deviation.**

Dielectric — The *insulating* material in which a *capacitor* stores electrical energy.

Diffraction — To alter the direction of a radio wave as it passes by edges of or through openings in obstructions such as buildings or hills. **Knife-edge diffraction** results if the dimensions of the edge are small in terms of the wave's wavelength.

Digipeater — A type of *repeater* station that retransmits or forwards digital messages.

Digital mode — See *data mode*.

Digital signal — (1) A signal (usually electrical) that can only have certain specific amplitude values, or steps — usually two; 0 and 1 or ON and OFF. (2) See *data mode*.

Digital signal processing (DSP) — The process of converting an *analog signal* to *digital* form and using a microprocessor to process the signal in some way such as filtering or reducing noise.

Diode — An electronic *component* that allows electric *current* to flow in only one direction.

Diplexer — A device that allows radios on two different bands to share a single antenna. Diplexers are used to allow a dual-band radio to use a single dual-band antenna. See *duplexer*.

Dipole— As used in Amateur Radio, the term usually refers to a *half-wave dipole* antenna.

Direct conversion — A type of receiver that recovers the modulating signal directly from the modulated RF signal.

Direct current (dc) — Electrical *current* that flows in only one direction.

Direct detection — A device acting as an unintentional receiver by converting a strong RF signal directly to voltages and currents internally, usually resulting in *radio frequency interference* to the receiving device.

Directional antenna — An antenna with an ability to receive and transmit that is enhanced in a specific (forward) direction and attenuated in one or more directions. See *front-to-back ratio* and *front-to-side ratio*.

Directional wattmeter — See *wattmeter*.

Director — A parasitic element of a *Yagi antenna* that focuses the radiated signal in the desired direction. See *reflector*.

Discharge — Extract energy from a *battery* or *cell*. **Self-discharge** refers to the internal loss of energy without an external *circuit*.

Discriminator — See *frequency discriminator*.

Dish — A curved *directional antenna* that uses a *reflector* to focus radio waves.

Distress call — A transmission made in order to attract attention in an emergency. (See *MAYDAY* and *SOS*)

Doping — Adding impurities to *semiconductor* material to change its conductive properties. **N-type** material is created if adding the impurity results in more *electrons* being available to flow as *current*. **P-type** material results if fewer *electrons* are available.

Doppler shift — A change in observed frequency of a signal caused by relative motion between the transmitter and receiver. Also called the **Doppler effect**.

Doubling — Two or more operators transmitting at the same time on the same frequency.

Downlink — Transmitted signals or the range of frequencies for transmissions from a satellite to Earth. See *uplink*.

Drain — See *transistor*.

Driven element — An antenna *element* supplied directly with power from the transmitter.

Driver — The *amplifier* stage immediately preceding a *power amplifier* in a transmitter.

Dual-band antenna — An antenna designed for use on two different amateur bands.

Dummy antenna or **dummy load** — A station accessory that dissipates a transmitted signal as heat to allow testing or adjustment of transmitting equipment without radiating a signal on the air.

Duplex — (1) Transmitting on one frequency and receiving on another, such as for *repeater* operation. (2) A mode of communications (also known as **full duplex**) in which a user transmits on one frequency and receives on another frequency simultaneously. This is in contrast to **half duplex** in which the user transmits at one time and receives at other times.

Duplexer — A device that allows bidirectional communication on closely spaced frequencies or *channels*. In a *repeater*, the duplexer also allows the transmitter and receiver to share a single antenna. See *diplexer*.

Duty cycle — The percentage of time that a signal or device, such as a transmitter, is active. **Duty factor** is the same as duty cycle, but expressed as a fraction instead of percent.

DX — Distance, distant stations, foreign countries.

DXpedition — An expedition for the purpose of making contacts from a rare or unusual location.

E region — The second lowest *ionospheric* region, the E region (or layer) exists only during the day. Under certain conditions, it may refract radio waves enough to return them to Earth.

Earth station — An amateur station located on or within 50 km of the Earth's surface, intended for communications with space stations or with other Earth stations by means of one or more other objects in space.

Earth-Moon-Earth (EME) or **moonbounce** — A method of communicating with other stations by reflecting radio signals off the Moon's surface.

Echolink — A system of linking repeaters and computer-based users by using the Voice-Over-Internet Protocol.

Electric field — A region of space in which electrical energy is stored and in which a stationary electrically charged object will feel a force. The **electric potential** between two points in the electric field is the amount of energy required to move a single electron between those two points.

Electrode — The general term for an electrical contact or connection point.

Electromagnetic wave — Energy composed of a continuously varying *electric field* and *magnetic field* moving through space or a *transmission line*.

Electromotive force (EMF) — The force that causes *electrons* or other charged objects to move.

Electron — A tiny, negatively charged particle, normally found in the volume surrounding the nucleus of an atom. Moving electrons make up an electrical *current*.

Electronic current — See *current*.

Element — (1) The conducting part or parts of an antenna designed to radiate or receive *radio waves*. (2) An examination for an FCC license in the amateur service. (3) A dot or dash in the *Morse code*.

Elmer — A ham radio mentor or teacher.

Emcomm — An abbreviation for *emergency communications*

Emergency — A situation where there is an immediate threat to the safety of human life or property.

Emergency communications — Communications conducted under adverse conditions where normal channels of communications are not available.

Emergency traffic — Messages with life and death urgency or requests for medical help and supplies that leave an area shortly after an emergency.

Emission — The transmitted signal from an *amateur station*.

Emission privilege — Permission to use a particular *emission type* (such as Morse code or voice).

Emission types — Term for the different modes authorized for use on the Amateur Radio bands. Examples are CW, SSB, RTTY and FM.

Emitter — See *transistor.*

Encoding — Changing the form of a signal into one suitable for storage or transmission. **Decoding** is the process of returning the signal to its original form.

Encryption — Changing the form of a signal into a privately-known format intended to obscure the meaning of the signal. **Decryption** is the process of reversing the encryption.

Energy — The ability to do work; the ability to exert a force to move some object.

Envelope — The outline of an RF signal formed by the peaks of the individual RF cycles.

Extended-coverage receiver — A receiver that tunes frequencies from around 30 MHz to several hundred MHz or into the GHz range. Also known as a **wide-range receiver**.

F region — A combination of the two highest *ionospheric* regions (or layers), the F1 and F2 regions. The F region refracts radio waves and returns them to Earth. Its height varies greatly depending on the time of day, season of the year and amount of sunspot activity.

Farad (F) — The basic unit of *capacitance*.

Federal Communications Commission (FCC) — Federal agency in the United States that regulates use and allocation of the frequency spectrum among many different services, including Amateur Radio.

Federal Registration Number (FRN) — An identification number assigned to an individual by the FCC to use when performing license modification or renewal.

Feed line — See *transmission line*.

Feed line loss — The fraction of power dissipated as heat as it travels through a feed line.

Feed point — The point at which a *transmission line* is electrically connected to an antenna.

Feed point impedance — The ratio of RF voltage to current at the *feed point* of an antenna.

Ferrite — A ceramic material with magnetic properties used in *inductors*. Ferrite is often formed into beads or cores so that it may be placed on cables, forming an *RF choke*.

FET — See *transistor*.

Filter — A circuit or system whose effect on a signal depends on its frequency or other characteristics. A **passive filter** is constructed entirely from unpowered devices such as *resistors*, *capacitors* and *inductors*. An **active filter** also uses powered devices such as *amplifiers* or *transistors*. A **digital filter** performs the filtering functions by operating on digital data that represents a signal.

Form 605 — An FCC form that serves as the application for your Amateur Radio license, or for modifications to an existing license.

Forward power — Power in a *transmission line* traveling from a transmitter toward a *load* or *antenna*.

Fox hunting — Exercises in which participants look for a hidden transmitter (the fox) to test *radio direction-finding* skills. Also called a **bunny hunt**.

Frequency — The number of complete cycles per second of an *ac current* or *ac voltage*.

Frequency band — A continuous range of frequencies in which one type of communications is authorized. See *band*.

Frequency coordination — Allocating *repeater* input and output frequencies to minimize interference between repeaters and to other users of the band.

Frequency coordinator — An elected individual or group that recommends repeater frequencies to reduce or eliminate interference between *repeaters* operating on or near the same frequency in the same geographical area.

Frequency discriminator — A *detector* used for FM signals.

Frequency modulation (FM) — The process of adding information to an RF signal or *carrier* by varying its frequency. FM broadcast stations and most professional communications (police, fire, taxi) use FM. **FM phone** is used on most *repeaters*.

Frequency privilege — Authorization to use a particular group of frequencies.

Frequency-shift keying (FSK) — A method of digital *modulation* that shifts the transmitter frequency to represent the bits of digital data.

Front-end overload — Interference to a receiver caused by a strong signal that causes the receiver's sensitive input circuitry ("front end") to be overloaded. Front-end overload results in distortion of the desired signal and the generation of unwanted spurious signals within the receiver. See **receiver overload**.

Front-to-back ratio (F/B) — The ratio of an antenna's *gain* in the forward direction to that in the opposite or rear direction.

Front-to-side ratio (F/S) — The ratio of an antenna's *gain* in the forward direction to that at right angles to the forward direction.

FRS — Family Radio Service. An unlicensed radio service that uses low-power radios operating near 460 MHz and intended for short-range communications by family members.

Fundamental — The frequency of which all *harmonics* are integer multiples.

Fundamental overload — *Radio frequency interference (RFI)* caused when a strong RF signal exceeds a receiver's ability to reject it.

Fuse — A thin metal strip mounted in a holder. When excessive *current* passes through the fuse, the metal strip melts and opens the *circuit* to protect it against further current overload.

Gain — (1) Enhancing an antenna's ability to receive or radiate signals in a specific direction. (2) The ability of a component, circuit, or piece of equipment to *amplify* a signal. (3) **Mic Gain** — sensitivity of the microphone amplifier circuit. (4) **RF Gain** — sensitivity of the receiver to incoming signals. (5) **AF Gain** — receiver output volume.

Gate — See *transistor*.

Gateway — A station that serves to connect one network of stations with the Internet or another network of stations.

General-coverage receiver — A receiver used to listen to a wide range of frequencies, not just specific bands. Most general-coverage receivers tune from frequencies below the AM broadcast band (550 – 1700 kHz) to around 30 MHz. (See *extended-coverage receiver*.)

Generator — A device that uses a motor to convert mechanical energy into ac or dc electrical energy. See also *signal generator*.

GFI (also GFCI) — Ground-fault interrupting *circuit breaker* that opens a circuit when an imbalance of current flow is detected between the hot and neutral wires of an ac power circuit. An **AFCI** or arc-fault circuit interrupter opens a circuit when an arc is detected.

Giga (or lower case G) — The metric prefix for 10^9 or multiplication by 1,000,000,000.

GMRS — General Mobile Radio Service. A licensed radio service operating 460 MHz intended for family businesses and members to communicate within a city or region.

Go-kit — A pre-packaged collection of equipment or supplies kept at hand to allow an operator to quickly report where needed in time of need.

Grace period — The time allowed by the FCC following the expiration of an amateur license to renew that license without having to retake an examination. Those who hold an expired license may not operate an *amateur station* until the license is reinstated.

Grant — Authorization given by the FCC

Grid square — A locator in the Maidenhead Locator System.

Ground — The electric *potential* of the Earth's surface. Also called **earth ground** or **ground potential**.

Ground loss — RF energy that is converted to heat while reflecting from or traveling through or along the Earth's surface.

Ground rod — A metallic rod that is driven into the Earth to make a **ground connection**.

Ground-plane — A conducting surface of continuous metal or discrete wires that acts to create an electrical image of an antenna. **Ground-plane antennas** require a ground-plane in order to operate properly.

Ground-wave propagation — *Propagation* in which radio waves travel along the Earth's surface.

Ham-band receiver — A receiver designed to receive only frequencies in the amateur bands.

Hamfest — A flea-market for ham radio, electronic and computer equipment and accessories.

Half-wave dipole — A popular antenna that is ½-wavelength long at the desired operating frequency. Dipoles usually consist of a single length of wire or tubing with a feed point at the center. See *dipole*.

Harmful interference — Interference that seriously degrades, obstructs or repeatedly interrupts a radio communication service operating in accordance with the Radio Regulations. [Part 97.3 (a) (22)]

Harmonic — A signal that is an integer multiple (2×, 3×, 4×, etc) of a *fundamental* frequency.

Hand-held radio — A VHF or UHF transceiver that can be carried in the hand or pocket. Also known as an **HT**.

Header — The first part of a digital message containing routing and control information about the message. See *preamble*.

Headphones — A pair of speakers held against or inserted into each ear. A **headset** or **boomset** combines headphones with a microphone for additional convenience.

Health and Welfare traffic — Messages about the well-being of individuals in a disaster area. Such messages must wait for **Emergency** and **Priority traffic** to clear, and results in advisories to those outside the disaster area awaiting news from family and friends.

Henry (H) — The basic unit of *inductance*.

Hertz (Hz) —The basic unit of frequency. 1 Hz = 1 cycle per second.

High frequency (HF) — The term used for the frequency range from 3 MHz to 30 MHz.

High-pass filter (HPF) — A filter designed to pass signals above a specified *cutoff frequency*, while attenuating lower-frequency signals.

Hop — See *sky-wave propagation*.

Impedance — The ratio of ac voltage to ac current, including *phase*. The combination of *reactance* and *resistance* that constitutes opposition to *ac current*. Impedance is measured in *ohms*.

Impedance match — To adjust *impedances* to be equal or the case in which two impedances are equal. Usually refers to the point at which a feed line is connected to an antenna or to transmitting equipment. If the impedances are different, that is a **mismatch**. See *matched*.

Impedance matching network — A device that transforms one impedance to another, such as an antenna system input *impedance* to match that of a transmitter or receiver. Also called an *antenna-matching network* or **transmatch**.

Indicator — (1) A device used to signal status audibly or visually. (2) Characters added before or after a call sign signifying a change in license class or that the station or operator is transmitting away from the registered location.

Inductance — A measure of the ability to store energy in a *magnetic field*. Inductance is measured in *henries*.

Inductor — An electrical *component* that stores energy in a *magnetic field*. An inductor is usually composed of a coil of wire wound around a central core.

Input frequency — A *repeater's* receiving frequency.

Insulator — A material in which *electrons* do not move easily in response to an applied *voltage*. Insulation is used to prevent *current* flow between points at different voltages.

Integrated circuit (IC or chip) — An electronic *component* made up of many individual components in a single package.

Intermediate frequency (IF) — The stages in a *superheterodyne* receiver that follow a *mixer* circuit and that operate at a fixed frequency. Most of the receiver's *gain* and *selectivity* are achieved in the IF stages.

Intermodulation (intermod or IMD) — Spurious signals created by the combination of other signals. Usually related to the overload of circuits by strong signals.

International Amateur Radio Union (IARU) — The international organization of national Amateur Radio societies.

International Telecommunication Union (ITU) — The organization of the United Nations responsible for coordinating international telecommunications agreements.

Internet Radio Linking Project (IRLP) — A system of linking repeaters by using the Voice-Over-Internet Protocol.

Inverted V — A *dipole* antenna supported in the middle with each half sloping downward.

Inverter — A circuit that converts dc power into ac power.

Ion — An electrically-charged atom or molecule.

Ionosphere — A region of electrically charged (ionized) gases high in the atmosphere that affects the *propagation* of radio waves through it. See *sky-wave propagation*.

Isotropic antenna — An *antenna* that radiates and receives equally in all directions, both vertical and horizontal.

Jack — A *connector* designed to have a mating assembly inserted into it, usually mounted on a piece of equipment.

Keplerian elements — Numeric parameters describing a satellite's orbit that can be used to compute the position of the satellite at any point in time.

Kerchunk — The sound made when a brief transmission activates a repeater.

Key — A manually operated switch that turns a transmitter on and off to send *Morse code*.

Key click — Spurious signals generated as a transmitter is turned on and off that are heard as clicks by stations on nearby frequencies.

Keyboard-to-keyboard — A *digital mode* intended for operators to exchange text messages as the characters are entered.

Keyer or **electronic keyer** — A device that makes it easier to send well-formed Morse code. It sends a continuous string of either dots or dashes, depending on which lever of a connected *paddle* is pressed.

Kilo (k) — The metric prefix for 10^3 or multiplication by 1000.

Knife-edge — See *diffraction*.

Ladder line — See *open-wire line*.

Lag — In comparing two *waveforms*, refers to the waveform in which positive change occurs last.

Lead — (pronounced *"leed"*) (1) Refers to the wires or connection points on an electrical *component* or the probes and cables that are used to connect test instruments to the devices being measured. (2) In comparing two *waveforms*, refers to the waveform in which change in the positive direction occurs first.

Light-emitting diode (LED) — A *diode* that emits light when current flows through it.

Lightning protection — Methods to prevent lightning damage to your equipment (and your house), such as unplugging equipment, disconnecting antenna *feed lines* and using a **lightning arrestor**.

Line-of-sight propagation — The term used to describe VHF and UHF *propagation* in a straight line directly from one station to another.

Linear — (1) To act on a signal such that the result is a replica of the original signal at a different scale. (2) Equipment that amplifies the output of a transmitter, often to the full legal amateur power limit of 1500 W *peak envelope power (PEP)*.

Liquid-crystal display (LCD) — A device for displaying graphics or characters by passing light through a liquid crystal between patterns of *electrodes*.

Local control — Operation of a station with a *control operator* physically present at the transmitter.

Load — A device or system to which electrical *power* is delivered, such as a heating element or *antenna*. Also the amount of power consumed or that can be safely dissipated, such as a "50-watt load."

Loading — Increasing an antenna's apparent electrical length by inserting inductance or capacitance.

Lobe — A direction of maximum reception or transmission in an antenna's *radiation pattern*. The **main lobe** has the greatest strength for the entire pattern. A **side lobe** is a maximum located at an angle to the main lobe.

Log — The documents of a station that detail operation of the station. They can be used as supporting evidence and for troubleshooting interference-related problems or complaints.

Log periodic — a type of frequency independent antenna

Loop — (1) An antenna with *element(s)* constructed as continuous lengths of wire or tubing. (2) A point of maximum voltage or current on an antenna.

Lower sideband (LSB) — (1) In an *AM* or *single sideband* signal, the *sideband* located below the *carrier* frequency. (2) The common single sideband operating mode on the 40, 80 and 160 meter amateur bands.

Lowest usable frequency (LUF) — The lowest frequency that can be used for communication using *sky-wave propagation* along a specific path.

Low-pass filter (LPF) — A filter designed to pass signals below a specified *cutoff frequency*, while attenuating higher-frequency signals.

Machine — Slang for *repeater*.

Magnetic field — A region of space in which magnetic energy is stored and in which a moving electrically charged object will feel a force.

Matched — A transmission or feed line that is terminated by a load that has the same *impedance* as the feed line's *characteristic impedance*.

Maximum usable frequency (MUF) — The highest frequency that can be used for communication using *sky-wave propagation* along a specific path.

Maximum permissible exposure (MPE) — The maximum intensity of RF radiation to which a human being may safely be exposed. FCC Rules establish maximum permissible exposure values for humans to RF radiation. [Part 1.1310 and Part 97.13 (c)]

MAYDAY — From the French *m'aidez* (help me), MAYDAY is used when calling for emergency assistance in voice modes.

Mega (M) — The metric prefix for 10^6 or multiplication by 1,000,000.

Memory channel — Frequency and mode information stored by a radio and referenced by a number or alphanumeric designator.

Meteor scatter — Communication by signals reflected by the ionized meteor trails in the upper atmosphere.

Meter (instrument) — A device that displays a numeric value as a number or as the position of an indicator on a numeric scale.

Metric prefixes — A series of terms used in the *metric system* of measurement.

Metric system — A system of measurement that uses a set of prefixes to indicate multiples of 10 of a basic unit.

Medium frequency (MF) — The term used for the frequency range from 300 kHz to 3 MHz.

Micro (μ) — The metric prefix for 10^{-6} or division by 1,000,000.

Microphone (mic) — A device that converts sound waves into electrical energy.

Microwave — The conventional term for frequencies greater than 1000 MHz (1 GHz).

Milli (m) — The metric prefix for 10^{-3} or division by 1000.

Mixer — A circuit that combines two RF signals and generates **products** at both the signal's sum and difference frequencies. An **audio mixer** adds multiple signals together into a single signal.

Mobile station — A radio transmitter designed to be mounted in a vehicle. Any station that can be operated on the move, typically in a car, but also on a boat, a motorcycle, truck or RV.

Mobile flutter — Rapid amplitude variation of a signal from a moving vehicle experiencing *multipath interference*. Also called **picket fencing**.

Mode — (1) The combination of a type of information and a method of transmission. For example, FM radiotelephony or *FM phone* consists of using FM modulation to carry voice information. (2) The combination of a satellite's *uplink* and *downlink* bands.

Mode-restricted — Portions of the amateur bands in which only certain *emission types* are allowed.

Modem — Short for **mo**dulator/**dem**odulator. A modem changes data into audio signals that can be transmitted by radio and *demodulates* a received signal to recover transmitted data.

Modulate — The process of adding information to an RF signal or *carrier* by varying its *amplitude*, *frequency*, or *phase*.

Morse code — The system of encoding characters as dots and dashes invented by Samuel Morse. See *continuous wave*.

MUF — See *maximum usable frequency*.

Multiband antenna — An antenna capable of operating on more than one amateur band, usually using a single *transmission line*.

Multi-hop — Long-distance radio *propagation* using several skips or hops between the Earth and the *ionosphere*. See *sky-wave propagation*.

Multimeter — An electronic test instrument used to measure *current*, *voltage* and *resistance*. Alternate names are **volt-ohm-milliammeter (VOM)** and **vacuum-tube voltmeter (VTVM)**. If the numeric display is digital, the instrument may also be called a **digital multimeter (DMM)** or **digital voltmeter (DVM)**.

Multimode radio — Transceiver capable of SSB, CW and FM operation.

Multipath propagation — *Propagation* by means of multiple reflections. When the reflected signals partially cancel, it is referred to as **multipath interference**.

Multiple Protocol Controller (MPC) — A piece of equipment that can act as a *modem* or *TNC* for several *protocols*.

N or type N connector — A type of *RF connector* that can be used through *microwave* frequencies.

N-type — See *doping*.

Nano (n) — The metric prefix for 10^{-9} or division by 1,000,000,000.

National Electrical Code (NEC) — A set of guidelines governing electrical safety, including antennas.

National Incident Management System (NIMS) — The method by which emergency situations are managed by US public safety agencies.

Net — A formal system of operation in order to exchange or manage information.

Net control station (NCS) — The station in charge of a *net*.

Network — (1) A term used to describe several digital stations linked together to relay data over long distances. (2) A general term for electrical circuits.

Node — (1) One station in a digital network. (2) A point of minimum voltage or current on an *antenna*.

Noise blanker — A circuit that mutes the receiver output during noise pulses.

Noise reduction — Removing random noise from a receiver's audio output.

Notch filter — A *filter* that removes a very narrow range of frequencies, usually from a receiver's audio output audio.

Null — (1) Tune or adjust for a minimum response (2) A direction of minimum reception or transmission in an antenna's *radiation pattern*.

Offset frequency — The difference between a *repeater's* transmitter and receiver frequencies. Also known as the repeater's **split** or **offset**.

Ohm — The basic unit of electrical *resistance* and represented by the symbol Ω.

Ohm's Law — A basic electrical law stating that the *current* (I) through a *circuit* is directly proportional to the *voltage* (E) across the circuit and inversely proportional to the *resistance* (R) of the circuit: I = E / R. Ohm's Law is equivalently stated as E = I × R or R = E / I.

Ohmmeter — A device used to measure *resistance*.

Omnidirectional — An *antenna* that radiates and receives equally in all horizontal directions.

One-way communications — Radio signals not directed to a specific station, or for which no reply is expected. The FCC Rules provide for limited types of one-way communications on the amateur bands. [Part 97.111 (b)]

Open circuit — A break in an electrical *circuit* that prevents *current* from flowing.

Open repeater — A *repeater* available for use by all hams.

Open-wire line — A *transmission line* made from two parallel wires separated by insulation. Also known as **ladder line**, **parallel-conductor feed line**, **twin-lead**, or **window line**.

Operator/primary station license — An amateur license actually consists of two licenses. The **operator license** is that portion of an Amateur Radio license that gives permission to operate an *amateur station*. The **primary station license** is that portion of an Amateur Radio license that authorizes an *amateur station* at a specific location. The station license also lists the *call sign* of that station.

OSCAR — Orbiting Satellite Carrying Amateur Radio

Oscillate — To vibrate continuously at a single *frequency*. An **oscillator** is a device or circuit that generates a signal at a single frequency.

Output frequency — A *repeater's* transmitting frequency.

P-type — See *doping*.

Packet radio — A system of digital communication using the AX.25 protocol whereby information is broken into data groups called **packets** that also contain addressing and error-detection information.

Paddle —A pair of contacts operated by one or two levers used to control an electronic *keyer* that generates *Morse code* automatically.

Parallel circuit — An electrical *circuit* in which the *electrons* may follow more than one path.

Parallel-conductor line — See *open-wire line.*

Parasitic element — An antenna *element* that affects the antenna performance by receiving and re-radiating energy from a *driven element* without being connected directly to the feed line.

Parity — An error detection method for *digital data* that counts the number of 1 bits in each data character. One bit added to each character — the parity bit — is used to indicate whether the correct number of 1 bits is odd or even.

Part 15 — The section of the FCC's rules that deal with unlicensed devices likely to transmit or receive RF signals.

Part 97 — The section of the FCC's rules that regulate Amateur Radio.

Passive filter — See *filter.*

Peak envelope power (PEP) — The average *power* during one RF cycle of a radio signal at the crest of the modulated *waveform*

Pecuniary — Payment of any type, whether money or other goods or services.

Perigee — The point in a satellite's orbit at which it is closest to the Earth. See *apogee.*

Period — The time it takes for one complete cycle of a repeating *waveform*. The reciprocal of *frequency.*

Phase — A measure of position in time within a repeating *waveform*, such as a *sine wave*. Phase is measured in *degrees* or *radians.*

Phase modulation (PM) — The process of adding information to a signal by varying its *phase*. Phase modulation is very similar to FM. PM signals can be received by FM receivers.

Phase-shift keying (PSK) — A method of digital *modulation* that shifts the transmitted signal's *phase* to represent the bits of digital data.

Phone — Another name for voice communications. An abbreviation for *radiotelephone.*

Phone emission — The FCC's name for voice transmissions.

Phone patch — Conducting a telephone call via radio communications.

Phonetic alphabet — A standardized list of words used on voice modes to make it easier to understand letters of the alphabet, such as those in call signs. The call sign KA6LMN stated phonetically is *Kilo Alfa Six Lima Mike November.*

Pico (p) — The metric prefix for 10^{-12} or division by 1,000,000,000,000.

PL (see CTCSS) — An abbreviation for Private Line, a trademark of Motorola.

Plug — An electrical *connector* designed to be inserted into a jack.

PN junction — The interface between *N-type* and *P-type* material.

Polarity — The orientation or direction of a *voltage* or *current* with respect to a standard or convention that assigns positive and negative.

Polarization — The orientation of the *electric field* of a *radio wave*. A radio wave can be horizontally, vertically, or circularly polarized.

Pole — In a *switch*, refers to a controlled *current* path or *circuit*.

Portable designator — Additional identifying information added to a *call sign* specifying the station's location.

Portable device — Generally considered to be a radio transmitting device designed to be transported easily and set up for operation independently of normal infrastructure. For purposes of RF exposure regulations, a portable device is one designed to have a transmitting antenna that is generally within 20 centimeters of a human body.

Potential — see *voltage*.

Potentiometer — (pronounced *po-ten-chee-AH-me-ter*) Another name for a variable *resistor* in which the resistance value can be changed without removing it from a circuit. Also called a **pot**.

Power — The rate of energy consumption or expenditure. To calculate power in an electrical *circuit* multiply the *voltage* applied to the circuit by the *current* through the circuit (P = I × E).

Power amplifier — See *linear*.

Power density — The strength of a *radio wave* measured as power per unit of area.

Power supply — A device that converts ac power from a utility or other service to ac or dc power used by equipment.

Preamble — The information at the beginning of a *radiogram* that contains routing and other information about the message. See *header*.

Preamplifier — An amplifier used to increase the strength of a received signal. Preamplifier circuits are often included in a receiver and may be turned on or off.

Prefix — The leading letters and numbers of a call sign that indicate the country in which the *call sign* was assigned.

Primary service — When a frequency band is shared among two or more different radio services, the primary service is preferred. Stations in a **secondary service** must not cause harmful interference to, and must accept interference from stations in the primary service. [Part 97.303]

Primary station license — See *operator/primary station license*.

Priority traffic — Emergency-related messages, but not as important as *emergency traffic*.

Privileges — The frequencies and modes of communication that are permitted in an FCC telecommunications service

Procedural signals (prosign) — For *Morse code* communications, one or two letters sent as a single character to indicate the operator's intention or to control the communication. For *phone* communications, prosigns consist of single words, such as "Break" or "Over."

Propagation — The method by which *radio waves* travel.

Protocol — A method of *encoding*, packaging, and exchanging digital data.

Push to talk (PTT) — Turning a transmitter on and off manually with a switch, usually thumb- or foot-activated.

Q signals — Three-letter symbols beginning with Q used in *Morse code* to save time and to improve communication. Some examples are QRS (send slower) and QTH (location).

Q system — A method of providing signal quality reports on a scale of 1 ("Q1") to 5 ("Q5").

QSL card — A postcard that serves as a confirmation of communication between two hams. QSL is a *Q-signal* meaning "received and understood."

QSO — A conversation between two radio amateurs. QSO is a *Q signal* meaning "I am in contact."

Quad antenna — A *directional antenna* with *elements* in the shape of four-sided loops, one *wavelength* in circumference.

Quarter-wave vertical — A *ground-plane* antenna constructed of a ¼-wavelength radiating *element*, usually oriented perpendicularly to the Earth or ground-plane.

Question pool — The set of questions from which an amateur license exam is constructed. There is one pool for each license class.

Radial — A wire forming part of a *ground plane*, attached at an antenna's base and running radially away from the antenna.

Radian — A measure of angle or *phase*. Each radian equals $360/2\pi$ or 57.3 degrees.

Radiation — To emit or give off energy, such as a radio wave. **Ionizing radiation** has sufficient energy to cause an electron to escape from an atom, creating a charged *ion*. RF energy used for radio communication is much less energetic and is called **non-ionizing radiation**.

Radiation pattern — A graph showing how an *antenna* radiates and receives in different directions. An **azimuthal pattern** shows radiation in horizontal directions. An **elevation pattern** shows radiation at different vertical angles.

Radio Amateur Civil Emergency Service (RACES) — A part of the Amateur Service that provides radio communications for civil defense organizations during local, regional or national civil emergencies.

Radio direction finding (RDF) — The method of locating a transmitter by determining the bearings of received signals.

Radio frequency (RF) exposure — FCC Rules establish *maximum permissible exposure* (MPE) values for humans to RF radiation. [Part 1.1310 and Part 97.13 (c)]

Radio frequency (RF) signals — RF or **radio signals** are generally considered to be any electrical signals with a frequency higher than 20,000 Hz, up to 300 GHz.

Radio-frequency interference (RFI) — Disturbance to electronic equipment or to radio communication caused by radio-frequency signals.

Radiogram — A formal message exchanged via radio.

Radio horizon — The most distant point to which radio signals can be sent without *ionospheric* or *tropospheric propagation*. See *sky-wave propagation*.

Radioteletype (RTTY) — A *data mode* that used the Baudot code to encode characters.

Radio wave — An *electromagnetic wave* with a frequency greater than 20 kHz.

Ragchew — An informal conversation.

Range — The longest distance over which radio signals can be exchanged.

Reactance — The property of opposition to ac current. Capacitors exhibit **capacitive reactance** and inductors exhibit **inductive reactance**. Reactance is measured in *ohms*.

Receiver (RVCR) — A device that converts radio waves into signals we can hear, see, or be read by a computer.

Receiver overload — Interference to a receiver caused by a RF signal too strong for the receiver input circuits. A signal that overloads the receiver RF input causes *front-end overload*. Receiver overload is sometimes called **RF overload**.

Receiver incremental tuning (RIT) — A transceiver control to adjust the receive frequency without affecting the transmit frequency.

Receiving converter — A device that shifts the frequency of incoming signals so that a receiver can be used on another band.

Recharge — See *charge*.

Reciprocal operating authority — Permission for amateur radio operators from another country to operate in the US using their home license. This permission is based on various treaties between the US government and the governments of other countries.

Rectifier — A diode intended for use in *power supplies* and power conversion circuits.

Rectify — Convert *ac* to *dc*.

Reflected power — Power in a *transmission line* returning to the transmitter from the load or antenna.

Reflector — (1) A parasitic element of a *Yagi antenna* that cancels the radiated signal in the undesired direction. (2) A conducting surface that acts as an electrical mirror to reflect radio waves.

Refract — Bending of an *electromagnetic wave* as it travels through materials with different properties. Radio waves are refracted as they travel through the *ionosphere*.

Region — Administrative areas defined by the *International Telecommunication Union (ITU)*.

Regulation — The ability of a *power supply* to control output voltage.

Relay — A *switch* operated by an electromagnet.

Remote control — Operation of a station in which the control functions of the station are operated by a *control operator* over a *control link*.

Remote receiver — A receiver at a separate location from a transmitter. Used by *repeater* systems to extend listening range or by individual stations to improve reception capabilities.

Repeater — A station that retransmits the signals of other stations to give them greater range.

Resistance — The property of opposing an electric *current*. Resistance is measured in *ohms*.

Resistor — An electronic *component* with a specific value of *resistance*, used to oppose or control current through a *circuit*. Resistors can be either fixed or variable. (See *potentiometer*.)

Resonance — The condition in an electrical circuit or antenna in which *reactance* is zero.

Resonant circuit — A circuit that exhibits *resonance* at one or more frequencies.

Resonant frequency — The frequency at which a circuit or antenna is resonant. See *tuned circuit*.

RF burn — A burn produced by coming in contact with RF voltages.

RF choke — An *inductor* or other impedance used to prevent or reduce the flow of RF current.

RF connector — A type of electrical *connector* designed specifically for use with RF signals.

RF feedback — Distortion of transmitted speech caused by RF signals being picked up by the microphone input circuits.

RF ground — The technique of maintaining the enclosures of radio equipment at a common RF voltage. See *bonding*.

RF overload — See *receiver overload*.

RF safety — Preventing injury or illness to humans from the effects of radio-frequency energy.

Rig — The radio amateur's term for a transmitter, receiver or transceiver.

Round-table — A contact in which several station take turns transmitting.

RST — A system of numbers used for signal reports: R is readability, S is strength and T is tone. (On phone, only R and S reports are used.)

Rubber duck antenna — A flexible rubber-coated antenna used mainly with hand-held VHF or UHF transceivers.

S meter — A meter that provides an indication of the relative strength of received signals in **S-units**.

Safety interlock — A switch that automatically turns off power to a piece of equipment when the enclosure is opened.

Safety ground — A *ground* connection intended to prevent shock hazards.

Scanning — Rapidly switching between a list of frequencies to listen for an active *channel*. **Tone scanning** determines what CTCSS access tones are present in specific signal.

Scattering — Radio wave *propagation* by means of multiple reflections in the layers of the atmosphere or from an obstruction.

Schematic diagram — A drawing that describes the electrical connections in a piece of electric or electronic equipment by using symbols to represent the electrical *components*.

Schematic symbol — A standardized symbol used to represent an electrical or electronic *component* on a *schematic diagram*.

Secondary service — See *primary service*.

Selectivity — The ability of a receiver to distinguish between signals.

Semiconductor — (1) A material with conductivity between that of a *conductor* and an *insulator*. (2) An electrical *component* constructed from semiconductor material.

Sensitivity — The ability of a receiver to detect signals.

Series circuit — An electrical *circuit* in which there is only one path for the *current* to follow.

Service — A set of regulations by the FCC that defines a certain type of telecommunications activity.

Shack — The room or location in which an *amateur station* is constructed.

Shield — (1) A cable's metallic layer or coating intended to prevent external signals from being picked up by an internal conductor or to prevent signals from being radiated from the internal conductor. (2) A metal wall or case that blocks RF signals.

Shielding — Surrounding an electronic circuit with conductive material to block RF signals from being radiated or received.

Short circuit — An electrical connection that causes *current* to bypass the intended path. Short-circuit often refers to an accidental connection that results in improper operation of equipment or circuits.

Sideband — An RF signal that results from modulating the amplitude or frequency of a *carrier*. An AM sideband can be either higher in frequency (**upper sideband** or **USB**) or lower in frequency (**lower sideband** or **LSB**) than the carrier. FM sidebands are produced on both sides of the carrier frequency.

Signal generator — A device that produces a low-level signal that can be set to a desired frequency.

Signal report — An evaluation of the transmitting station's signal and reception quality. See *Q system* and *RST*.

Simplex — Receiving and transmitting on the same frequency. See *duplex* and *half-duplex*.

Sine wave — A *waveform* with an amplitude equal to the sine of frequency × time.

Single sideband (SSB) — SSB is a form of *amplitude modulation* in which one *sideband* and the *carrier* are removed.

Skip — See *sky-wave propagation*.

Skip zone — An area of poor radio communication, too distant for *ground-wave propagation* and too close for *sky-wave propagation*.

Skyhook — Slang for antenna.

Sky-wave propagation — The method of *propagation* by which radio waves travel through the *ionosphere* and back to Earth. Also referred to as *skip*. Travel from the Earth's surface to the ionosphere and back is called a *hop*.

Slow-scan television (SSTV) — A television system used by amateurs to transmit pictures within the *bandwidth* required for a voice signal.

SMA — A type of RF connector used at *microwave* frequencies.

SOS — A *Morse code* call for emergency assistance.

Source — See *transistor*.

Space station — An *amateur station* located more than 50 km above the Earth's surface.

Speaker — A device that turns an *audio frequency* electrical signal into sound.

Specific absorption rate (SAR) — A term that describes the rate at which RF energy is absorbed by the human body. *Maximum permissible exposure (MPE)* limits are based on whole-body SAR values.

Spectrum — The range of electromagnetic signals. The radio spectrum includes signals between audio frequencies and infrared light.

Speech compression or **processing** — Increasing the average *power* of a voice signal by amplifying low-level *components* of the signal more than high-level *components*.

Splatter — A type of interference to stations on nearby frequencies that occurs when a transmitter is *overmodulated*.

Sporadic E (Es or E-skip) — A form of *propagation* that occurs when radio signals are reflected from small, densely ionized regions in the *E region* of the *ionosphere*. Sporadic E has been observed from the 15 meter through 1.25 meter bands.

Spurious emissions — Signals from a transmitter on frequencies other than the operating frequency.

Squelch — Circuitry that mutes the audio output of a receiver when no signal is received. **Carrier squelch** operates only on the presence of a signal carrier. **Tone squelch** requires a specific *CTCSS* tone to be present before allowing receiver audio to be heard. **Digital Code Squelch (DCS)** requires a continuous sequence of tones.

Squelch tail — The burst of noise heard from an FM receiver between when a station stops transmitting and when the receiver's *squelch* circuit mutes the receiver.

Stage — One of several circuits or devices that act on a signal in sequence.

Standard frequency offset — The standard transmitter/receiver frequency offset used by a *repeater* on a particular *amateur band*. For example, the standard offset on 2 meters is 600 kHz. Also see *offset frequency*.

Standing-wave ratio (SWR) —A measure of the impedance match between the transmission line's *characteristic impedance* and that of the *load* (usually an *antenna* or *antenna system*). **VSWR** is the ratio of maximum voltage to minimum voltage along the transmission line formed by the standing waves that result from power being reflected by the antenna or load. SWR is also the ratio of feed point impedance or load impedance to the feed line's *characteristic impedance*.

Station license — See *operator/primary station license*.

Stratosphere — The part of the Earth's *atmosphere* between the troposphere and ionosphere, extending from about 7 miles to 30 miles above the Earth.

Sub-audible tone — See *CTCSS*.

Suffix — The letters that follow a *call sign* prefix identifying a specific amateur.

Sunspot cycle — The number of sunspots increases and decreases in a predictable cycle that lasts about 11 years.

Sunspots — Dark spots on the surface of the Sun caused by magnetic fields.

Superheterodyne — A type of receiver that shifts signals to a fixed *intermediate frequency (IF)* for *amplification* and *demodulation*. Each frequency shift is termed a **conversion** and the superheterodyne is described as being a single-, double-, or triple-conversion.

Surge protector — A device that is used to prevent temporary or *transient* excessive voltages from damaging sensitive electronic equipment.

Switch — A *component* used by an operator to connect or disconnect electrical *circuits*.

SWR meter — A measuring instrument that can indicate when an antenna system is working well. A device used to measure SWR. See *standing wave ratio*.

Tactical call signs — Names used to identify a station's location or function during emergency communications.

Tactical communications — A first-response communications under emergency conditions that involves a few people in a small area.

Telecommand — A one-way radio transmission to start, change or end functions of a device at a distance.

Telemetry — Information about a device sent to a receiving station by radio.

Television interference (TVI) — Disruption of television reception caused by another signal.

Temporary state of communications emergency — When a disaster disrupts normal communications in a particular area, the FCC can declare this type of emergency. Certain rules may apply for the duration of the emergency.

Terminal Node Controller (TNC) — A device that acts as an interface between a computer and a radio for implementing a *data mode*.

Termination — A load or antenna connected to a *transmission line*.

Third-party — An unlicensed person on whose behalf communications is passed by amateur radio.

Third-party communications — Messages passed from one amateur to another on behalf of a third person.

Third-party communications agreement — An official agreement between the United States and another country that allows amateurs in both countries to participate in third-party communications.

Third-party participation — An unlicensed person participating in amateur communications. A *control operator* must ensure compliance with FCC rules.

Throw — In a *switch*, refers to the number of alternative *current* paths for a controlled *circuit*.

Ticket — Slang for an Amateur Radio license.

Time-out timer — A device that limits the amount of time a *repeater* can transmit without a pause by the input signal.

Tolerance — The allowed variation in the dimensions or value of an electrical or mechanical *component*, usually expressed in percent or as a range of values.

Track — To follow a satellite as it travels around the Earth. **Tracking software** uses the satellite's *Keplerian elements* to determine its location and when it is visible from a specific location.

Traffic — Formal messages exchanged via radio. *Traffic handling* is the process of exchanging traffic. A *traffic net* is a net specially created and managed to handle traffic.

Transformer — An electrical *component* that transfers ac power from one circuit to another by means of a magnetic field shared by two or more *inductors*.

Transient — A short pulse of electrical energy.

Transceiver (XCVR) — A radio transmitter and receiver combined in one unit.

Transistor — A *semiconductor* device used as a *switch* or *amplifier*. A **bipolar junction transistor (BJT)** is made from a pair of back-to-back *PN junctions*, and is controlled by a *current*. A BJT has three *electrodes*: **base**, **collector**, and **emitter**. A **field-effect transistor (FET)** uses an *electric field* to control *current* flow through a conducting *channel*. An FET has three electrodes: **gate**, **drain**, and **source**.

Transmatch — see *impedance matching network*.

Transmission line — Cable used to connect a transmitter, receiver or transceiver to an *antenna* or *load*.

Transmit-receive (TR) switch — A circuit or device that switches an *antenna* between *transmitter* and *receiver* circuits or equipment.

Transmitter (XMTR) — A device that produces radio frequency signals with sufficient power to be useful for communications.

Transponder — A device usually used on satellites that retransmit all signals in a range of frequencies.

Transverter — A device that converts signals so that a transceiver can operate on another band.

Trip — Activate when a threshold is exceeded or an event is detected. A *circuit breaker* trips, opening a circuit, when excessive *current* flow occurs, for example.

Troposphere — The region in Earth's atmosphere between the Earth's surface and the *stratosphere*.

Tropospheric bending — When radio waves are bent or *refracted* in the *troposphere*, they return to Earth farther away than the visible horizon.

Tropospheric ducting — A type of VHF propagation that can occur when warm air overruns cold air (a **temperature inversion**).

Tropospheric propagation (tropo) — Any method of propagation by means of atmospheric phenomena in the *troposphere*.

Tuned circuit — A circuit with a *resonant frequency* that can be adjusted, usually through the use of adjustable *capacitors* or *inductors*.

Tuning — Adjusting a radio or *circuit* that is frequency-sensitive.

Twin-lead — See *open-wire line*.

UHF connector — A type of *RF connector* usually used below 500 MHz.

Ultra high frequency (UHF) — The term used for the frequency range from 300 MHz to 3000 MHz (3 GHz).

Ultraviolet (UV) — *Electromagnetic waves* with frequencies greater than visible light. Literally, "above violet," which is the high-frequency end of the visible range.

Unbalanced line — *Transmission line* with one conductor connected to *ground*, such as *coaxial cable*.

Uncontrolled environment — Any area in which an RF signal may cause radiation exposure to people who may not be aware of the radiated electric and magnetic fields. The FCC generally considers members of the general public and an amateur's neighbors to be in an uncontrolled RF radiation exposure environment to determine the *maximum permissible exposure* levels. See *controlled environment.*

Unidentified communications or signals — Signals or radio communications in which the transmitting station's *call sign* is not transmitted.

Unintentional radiator — A device that radiates RF signals not required for its normal operation.

Universal Licensing System (ULS) — FCC database for all FCC radio services and licensees.

Uplink — Transmitted signals or the range of frequencies for transmissions from Earth to a satellite. See *downlink.*

Upper sideband (USB) — (1) In an AM signal, the *sideband* located above the carrier frequency. (2) The common single-sideband operating mode on the 60, 20, 17, 15, 12 and 10 meter HF amateur bands, and all the VHF and UHF bands.

Vacuum tube — An electronic *component* that operates by controlling *electron* flow between two or more *electrodes* in a vacuum.

Vanity call — A *call sign* selected by the amateur instead of one sequentially assigned by the FCC.

Variable-frequency oscillator (VFO) — An *oscillator* with an adjustable frequency. A VFO is used in receivers and transmitters to control the operating frequency.

Vertical antenna — An antenna with a single vertical radiating *element*. See *ground-plane antenna.*

Very high frequency (VHF) — The term used for the frequency range from 30 MHz to 300 MHz.

Visible horizon — The most distant point one can see by line of sight.

Voice — Any of the several methods used by amateurs to transmit speech.

Voice communications — Hams can use several voice modes, including FM and SSB.

Voice Over Internet Protocol (VOIP) — A method of sending voice and other audio over the Internet as digital data.

Volt (V) — The basic unit of electric potential or *electromotive force.*

Voltage — A measure of *electric potential* between two points.

Voltmeter — A test instrument used to measure *voltage.*

Volunteer Examiner (VE) — A licensed amateur who is accredited by a Volunteer Examiner Coordinator (VEC) to administer amateur license examinations.

Volunteer Examiner Coordinator (VEC) — An organization that has entered into an agreement with the FCC to coordinate amateur license examinations.

Voice-Operated Transmission (VOX) — Turning a transmitter on and off under control of the operator's voice.

Waterfall display — Used with digital modes, this type of display consists of a sequence of horizontal lines showing signal strength as a change of brightness with frequency represented by position on the line. Older lines move down the display so that the history of the signal's strength and frequency form a "waterfall-like" picture.

Watt (W) — The unit of *power* in the metric system.

Wattmeter — Also called a *power meter*, a test instrument used to measure the power output (in watts) of a transmitter. A **directional wattmeter** can measure power flowing in either direction in a feed line.

Waveform — The amplitude of an ac signal over time.

Wavelength —The distance a *radio wave* travels during one *cycle*. The wavelength relates to frequency in that higher frequency waves have shorter wavelengths. Represented by the symbol λ.

Weak-signal — (1) Refers to the use of SSB or CW on the VHF and UHF bands because they provide better communications at low signal levels than FM signals. (2) Any mode of operation that involves very low signal levels, such as *Earth-Moon-Earth*.

Whip antenna — An antenna with an *element* made of a single, flexible rod or tube.

Willful interference — Intentional, deliberate obstruction of radio communications.

Window line — See *open-wire line*.

Winlink — A system of email transmission and distribution using Amateur Radio for the connection between individual amateurs and mailbox stations known as **Radio Message Servers (RMS)**.

WSJT — A suite of software programs for *weak signal* and *meteor scatter* communications.

WWV/WWVH — Radio stations run by the National Institute of Standards and Technology (NIST) to provide accurate time and frequencies.

XCVR — Transceiver

XMTR — Transmitter

Yagi antenna — The most popular type of *directional antenna* or *beam*. It has one *driven element* and one or more *parasitic elements*.

73 — Ham abbreviation for "best regards." Used on both phone and CW toward the end of a contact.

Chapter 11

Question Pool

Technician Class Syllabus

Effective July 1, 2014

SUBELEMENT T1 — FCC Rules, descriptions and definitions for the Amateur Radio Service, operator and station license responsibilities

[6 Exam Questions — 6 Groups]

T1A Amateur Radio Service: purpose and permissible use of the Amateur Radio Service; operator/primary station license grant; where FCC rules are codified; basis and purpose of FCC rules; meanings of basic terms used in FCC rules; interference; spectrum management

T1B Authorized frequencies: frequency allocations; ITU regions; emission modes; restricted sub-bands; spectrum sharing; transmissions near band edges

T1C Operator licensing: operator classes; sequential, special event, and vanity call sign systems; international communications; reciprocal operation; station license and licensee; places where the amateur service is regulated by the FCC; name and address on FCC license database; license term; renewal; grace period

T1D Authorized and prohibited transmission: communications with other countries; music; exchange of information with other services; indecent language; compensation for use of station; retransmission of other amateur signals; codes and ciphers; sale of equipment; unidentified transmissions; broadcasting

T1E Control operator and control types: control operator required; eligibility; designation of control operator; privileges and duties; control point; local, automatic and remote control; location of control operator

T1F Station identification; repeaters; third party communications; club stations; FCC inspection

SUBELEMENT T2 — Operating Procedures

[3 Exam Questions — 3 Groups]

T2A Station operation: choosing an operating frequency; calling another station; test transmissions; procedural signs; use of minimum power; choosing an operating frequency; band plans; calling frequencies; repeater offsets

T2B VHF/UHF operating practices: SSB phone; FM repeater; simplex; splits and shifts; CTCSS; DTMF; tone squelch; carrier squelch; phonetics; operational problem resolution; Q signals

T2C Public service: emergency and non-emergency operations; applicability of FCC rules; RACES and ARES; net and traffic procedures; emergency restrictions

SUBELEMENT T3 — Radio wave characteristics: properties of radio waves; propagation modes

[3 Exam Questions — 3 Groups]

T3A Radio wave characteristics: how a radio signal travels; fading; multipath; wavelength vs. penetration; antenna orientation

T3B Radio and electromagnetic wave properties: the electromagnetic spectrum; wavelength vs. frequency; velocity of electromagnetic waves; calculating wavelength

T3C Propagation modes: line of sight; sporadic E; meteor and auroral scatter and reflections; tropospheric ducting; F layer skip; radio horizon

SUBELEMENT T4 — Amateur radio practices and station set up

[2 Exam Questions — 2 Groups]

T4A Station setup: connecting microphones; reducing unwanted emissions; power source; connecting a computer; RF grounding; connecting digital equipment; connecting an SWR meter

T4B Operating controls: tuning; use of filters; squelch function; AGC; repeater offset; memory channels

SUBELEMENT T5 — Electrical principles: math for electronics; electronic principles; Ohm's Law

[4 Exam Questions — 4 Groups]

T5A   Electrical principles, units, and terms: current and voltage; conductors and insulators; alternating and direct current

T5B   Math for electronics: conversion of electrical units; decibels; the metric system

T5C   Electronic principles: capacitance; inductance; current flow in circuits; alternating current; definition of RF; DC power calculations; impedance

T5D   Ohm's Law: formulas and usage

SUBELEMENT T6 — Electrical components: semiconductors; circuit diagrams; component functions

[4 Exam Questions — 4 Groups]

T6A   Electrical components: fixed and variable resistors; capacitors and inductors; fuses; switches; batteries

T6B   Semiconductors: basic principles and applications of solid state devices; diodes and transistors

T6C   Circuit diagrams; schematic symbols

T6D   Component functions: rectification; switches; indicators; power supply components; resonant circuit; shielding; power transformers; integrated circuits

SUBELEMENT T7 — Station equipment: common transmitter and receiver problems; antenna measurements; troubleshooting; basic repair and testing

[4 Exam Questions — 4 Groups]

T7A   Station equipment: receivers; transmitters; transceivers; modulation; transverters; low power and weak signal operation; transmit and receive amplifiers

T7B   Common transmitter and receiver problems: symptoms of overload and overdrive; distortion; causes of interference; interference and consumer electronics; part 15 devices; over and under modulation; RF feedback; off frequency signals; fading and noise; problems with digital communications interfaces

T7C   Antenna measurements and troubleshooting: measuring SWR; dummy loads; coaxial cables; feed line failure modes

T7D   Basic repair and testing: soldering; using basic test instruments; connecting a voltmeter, ammeter, or ohmmeter

SUBELEMENT T8 — Modulation modes: amateur satellite operation; operating activities; non-voice communications

[4 Exam Questions — 4 Groups]

T8A   Modulation modes: bandwidth of various signals; choice of emission type

T8B   Amateur satellite operation; Doppler shift, basic orbits, operating protocols; control operator, transmitter power considerations; satellite tracking; digital modes

T8C   Operating activities: radio direction finding; radio control; contests; linking over the Internet; grid locators

T8D   Non-voice communications: image signals; digital modes; CW; packet; PSK31; APRS; error detection and correction; NTSC

SUBELEMENT T9 — Antennas and feed lines

[2 Exam Questions   2 Groups]

T9A   Antennas: vertical and horizontal polarization; concept of gain; common portable and mobile antennas; relationships between antenna length and frequency

T9B   Feed lines: types of feed lines; attenuation vs. frequency; SWR concepts; matching; weather protection; choosing RF connectors and feed lines

SUBELEMENT T0 — Electrical safety: AC and DC power circuits; antenna installation; RF hazards

[3 Exam Questions — 3 Groups]

T0A   Power circuits and hazards: hazardous voltages; fuses and circuit breakers; grounding; lightning protection; battery safety; electrical code compliance

T0B   Antenna safety: tower safety; erecting an antenna support; overhead power lines; installing an antenna

T0C   RF hazards: radiation exposure; proximity to antennas; recognized safe power levels; exposure to others; radiation types; duty cycle

Technician Class Question Pool
Effective July 1, 2014

There are significant differences between the order of topics in the Question Pool subelements and the arrangement of material in the text. An alternate arrangement of the questions that follows the text more closely is available at on this book's website at **www.arrl.org/ham-radio-license-manual**. Follow the Study Guide link to download the material as a PDF file you can print or view.

SUBELEMENT T1 — FCC Rules, descriptions and definitions for the Amateur Radio Service, operator and station license responsibilities
[6 Exam Questions — 6 Groups]

T1A — Amateur Radio Service: purpose and permissible use of the Amateur Radio Service; operator/primary station license grant; where FCC rules are codified; basis and purpose of FCC rules; meanings of basic terms used in FCC rules; interference; spectrum management

T1A01
Which of the following is a purpose of the Amateur Radio Service as stated in the FCC rules and regulations?
A. Providing personal radio communications for as many citizens as possible
B. Providing communications for international non-profit organizations
C. Advancing skills in the technical and communication phases of the radio art
D. All of these choices are correct

T1A02
Which agency regulates and enforces the rules for the Amateur Radio Service in the United States?
A. FEMA
B. The ITU
C. The FCC
D. Homeland Security

T1A03
Which part of the FCC regulations contains the rules governing the Amateur Radio Service?
A. Part 73
B. Part 95
C. Part 90
D. Part 97

T1A01
(C)
[97.1]
Page 7-2

T1A02
(C)
[97.1]
Page 7-2

T1A03
(D)
Page 7-1

T1A04
(C)
[97.3(a)(23)]
Page 8-7

T1A04

Which of the following meets the FCC definition of harmful interference?

A. Radio transmissions that annoy users of a repeater
B. Unwanted radio transmissions that cause costly harm to radio station apparatus
C. That which seriously degrades, obstructs, or repeatedly interrupts a radio communication service operating in accordance with the Radio Regulations
D. Static from lightning storms

T1A05
(A)
[97.1 (e)]
Page 7-2

T1A05

Which of the following is a purpose of the Amateur Radio Service rules and regulations as defined by the FCC?

A. Enhancing international goodwill
B. Providing inexpensive communication for local emergency organizations
C. Training of operators in military radio operating procedures
D. All of these choices are correct

T1A06
(D)
[97.101(d), 97.303(o)(2)]
Page 8-7

T1A06

Which of the following services are protected from interference by amateur signals under all circumstances?

A. Citizens Radio Service
B. Broadcast Service
C. Land Mobile Radio Service
D. Radionavigation Service

T1A07
(C)
[97.3(a)(46)]
Page 6-33

T1A07

What is the FCC Part 97 definition of telemetry?

A. An information bulletin issued by the FCC
B. A one-way transmission to initiate, modify or terminate functions of a device at a distance
C. A one-way transmission of measurements at a distance from the measuring instrument
D. An information bulletin from a VEC

T1A08
(B)
[97.3(a)(22)]
Page 7-16

T1A08

Which of the following entities recommends transmit/receive channels and other parameters for auxiliary and repeater stations?

A. Frequency Spectrum Manager
B. Frequency Coordinator
C. FCC Regional Field Office
D. International Telecommunication Union

T1A09
(C)
[97.3(a)(22)]
Page 7-17

T1A09

Who selects a Frequency Coordinator?

A. The FCC Office of Spectrum Management and Coordination Policy
B. The local chapter of the Office of National Council of Independent Frequency Coordinators
C. Amateur operators in a local or regional area whose stations are eligible to be auxiliary or repeater stations
D. FCC Regional Field Office

T1A10

What is the FCC Part 97 definition of an amateur station?

A. A station in the Amateur Radio Service consisting of the apparatus necessary for carrying on radio communications
B. A building where Amateur Radio receivers, transmitters, and RF power amplifiers are installed
C. Any radio station operated by a non-professional
D. Any radio station for hobby use

T1A11

When is willful interference to other amateur radio stations permitted?

A. Only if the station being interfered with is expressing extreme religious or political views
B. At no time
C. Only during a contest
D. At any time, amateurs are not protected from willful interference

T1A12

Which of the following is a permissible use of the Amateur Radio Service?

A. Broadcasting music and videos to friends
B. Providing a way for amateur radio operators to earn additional income by using their stations to pass messages
C. Providing low-cost communications for start-up businesses
D. Allowing a person to conduct radio experiments and to communicate with other licensed hams around the world

T1A13

What is the FCC Part 97 definition of telecommand?

A. An instruction bulletin issued by the FCC
B. A one-way radio transmission of measurements at a distance from the measuring instrument
C. A one-way transmission to initiate, modify or terminate functions of a device at a distance
D. An instruction from a VEC

T1A14

What must you do if you are operating on the 23 cm band and learn that you are interfering with a radiolocation station outside the United States?

A. Stop operating or take steps to eliminate the harmful interference
B. Nothing, because this band is allocated exclusively to the amateur service
C. Establish contact with the radiolocation station and ask them to change frequency
D. Change to CW mode, because this would not likely cause interference

T1B — Authorized frequencies: frequency allocations; ITU regions; emission modes; restricted sub-bands; spectrum sharing; transmissions near band edges

T1B01

What is the ITU?

A. An agency of the United States Department of Telecommunications Management
B. A United Nations agency for information and communication technology issues
C. An independent frequency coordination agency
D. A department of the FCC

T1A10
(A)
[97.3(a)(5)]
Page 7-3

T1A11
(B)
[97.101(d)]
Page 8-8

T1A12
(D)
Page 7-2

T1A13
(C)
[97.3(a)(45)]
Page 6-33

T1A14
(A)
[97.303(d)]
Page 7-16

T1B01
(B)
Page 7-17

T1B02
(A)
[97.301]
Page 7-18

T1B02

Why are the frequency assignments for some U.S. Territories different from those in the 50 U.S. States?

A. Some U. S. Territories are located in ITU regions other than region 2
B. Territorial governments are allowed to select their own frequency allocations
C. Territorial frequency allocations must also include those of adjacent countries
D. Any territory that was in existence before the ratification of the Communications Act of 1934 is exempt from FCC frequency regulations

T1B03
(B)
[97.301(a)]
Page 7-12

T1B03

Which frequency is within the 6 meter band?

A. 49.00 MHz
B. 52.525 MHz
C. 28.50 MHz
D. 222.15 MHz

T1B04
(A)
[97.301(a)]
Page 7-12

T1B04

Which amateur band are you using when your station is transmitting on 146.52 MHz?

A. 2 meter band
B. 20 meter band
C. 14 meter band
D. 6 meter band

T1B05
(C)
[97.301(a)]
Page 7-12

T1B05

Which 70 cm frequency is authorized to a Technician Class license holder operating in ITU Region 2?

A. 53.350 MHz
B. 146.520 MHz
C. 443.350 MHz
D. 222.520 MHz

T1B06
(B)
[97.301(a)]
Page 7-12

T1B06

Which 23 cm frequency is authorized to a Technician Class licensee?

A. 2315 MHz
B. 1296 MHz
C. 3390 MHz
D. 146.52 MHz

T1B07
(D)
[97.301(a)]
Page 7-12

T1B07

What amateur band are you using if you are transmitting on 223.50 MHz?

A. 15 meter band
B. 10 meter band
C. 2 meter band
D. 1.25 meter band

T1B08
(A)
[97.303]
Page 7-15

T1B08

Which of the following is a result of the fact that the amateur service is secondary in some portions of the 70 cm band?

A. U.S. amateurs may find non-amateur stations in the bands, and must avoid interfering with them
B. U.S. amateurs must give foreign amateur stations priority in those portions
C. International communications are not permitted on 70 cm
D. Digital transmissions are not permitted on 70 cm

T1B09

Why should you not set your transmit frequency to be exactly at the edge of an amateur band or sub-band?
A. To allow for calibration error in the transmitter frequency display
B. So that modulation sidebands do not extend beyond the band edge
C. To allow for transmitter frequency drift
D. All of these choices are correct

T1B10

Which of the bands above 30 MHz that are available to Technician Class operators have mode-restricted sub-bands?
A. The 6 meter, 2 meter, and 70 cm bands
B. The 2 meter and 13 cm bands
C. The 6 meter, 2 meter, and 1.25 meter bands
D. The 2 meter and 70 cm bands

T1B11

What emission modes are permitted in the mode-restricted sub-bands at 50.0 to 50.1 MHz and 144.0 to 144.1 MHz?
A. CW only
B. CW and RTTY
C. SSB only
D. CW and SSB

T1B12

Why are frequency assignments for U.S. stations operating maritime mobile not the same everywhere in the world?
A. Amateur maritime mobile stations in international waters must conform to the frequency assignments of the country nearest to their vessel
B. Amateur frequency assignments can vary among the three ITU regions
C. Frequency assignments are determined by the captain of the vessel
D. Amateur frequency assignments are different in each of the 90 ITU zones

T1B13

Which emission may be used between 219 and 220 MHz?
A. Spread spectrum
B. Data
C. SSB voice
D. Fast-scan television

T1C — Operator licensing: operator classes; sequential, special event, and vanity call sign systems; international communications; reciprocal operation; station license and licensee; places where the amateur service is regulated by the FCC; name and address on FCC license database; license term; renewal; grace period

T1C01

Which type of call sign has a single letter in both its prefix and suffix?
A. Vanity
B. Sequential
C. Special event
D. In-memoriam

T1B09
(D)
[97.101(a),
97.301(a-e)]
Page 2-10

T1B10
(C)
[97.301(e),
97.305(c)]
Page 7-13

T1B11
(A)
[97.301(a),
97.305(a),(c)]
Page 7-13

T1B12
(B)
[97.301]
Page 7-18

T1B13
(B)
[97.305(c)]
Page 7-13

T1C01
(C)
[97.3(a)(11)
(iii)] 7-22

T1C02
(B)
Page 7-20

T1C02

Which of the following is a valid US amateur radio station call sign?

A. KMA3505
B. W3ABC
C. KDKA
D. 11Q1176

T1C03
(A)
[97.117]
Page 7-19

T1C03

What types of international communications are permitted by an FCC-licensed amateur station?

A. Communications incidental to the purposes of the amateur service and remarks of a personal character
B. Communications incidental to conducting business or remarks of a personal nature
C. Only communications incidental to contest exchanges, all other communications are prohibited
D. Any communications that would be permitted by an international broadcast station

T1C04
(A)
[97.107]
Page 7-18

T1C04

When are you allowed to operate your amateur station in a foreign country?

A. When the foreign country authorizes it
B. When there is a mutual agreement allowing third party communications
C. When authorization permits amateur communications in a foreign language
D. When you are communicating with non-licensed individuals in another country

T1C05
(A)
Page 7-22

T1C05

Which of the following is a vanity call sign which a technician class amateur operator might select if available?

A. K1XXX
B. KA1X
C. W1XX
D. All of these choices are correct

T1C06
(D)
[97.5(a)(2)]
Page 7-18

T1C06

From which of the following locations may an FCC-licensed amateur station transmit, in addition to places where the FCC regulates communications?

A. From within any country that belongs to the International Telecommunication Union
B. From within any country that is a member of the United Nations
C. From anywhere within in ITU Regions 2 and 3
D. From any vessel or craft located in international waters and documented or registered in the United States

T1C07
(B)
[97.23]
Page 7-9

T1C07

What may result when correspondence from the FCC is returned as undeliverable because the grantee failed to provide the correct mailing address?

A. Fine or imprisonment
B. Revocation of the station license or suspension of the operator license
C. Require the licensee to be re-examined
D. A reduction of one rank in operator class

T1C08

What is the normal term for an FCC-issued primary station/operator amateur radio license grant?

A. Five years
B. Life
C. Ten years
D. Twenty years

T1C09

What is the grace period following the expiration of an amateur license within which the license may be renewed?

A. Two years
B. Three years
C. Five years
D. Ten years

T1C10

How soon after passing the examination for your first amateur radio license may you operate a transmitter on an amateur service frequency?

A. Immediately
B. 30 days after the test date
C. As soon as your operator/station license grant appears in the FCC's license database
D. You must wait until you receive your license in the mail from the FCC

T1C11

If your license has expired and is still within the allowable grace period, may you continue to operate a transmitter on amateur service frequencies?

A. No, transmitting is not allowed until the FCC license database shows that the license has been renewed
B. Yes, but only if you identify using the suffix GP
C. Yes, but only during authorized nets
D. Yes, for up to two years

T1C12

Who may select a desired call sign under the vanity call sign rules?

A. Only licensed amateurs with general or extra class licenses
B. Only licensed amateurs with an extra class license
C. Only an amateur licensee who has been licensed continuously for more than 10 years
D. Any licensed amateur

T1C13

For which license classes are new licenses currently available from the FCC?

A. Novice, Technician, General, Advanced
B. Technician, Technician Plus, General, Advanced
C. Novice, Technician Plus, General, Advanced
D. Technician, General, Amateur Extra

T1C14

Who may select a vanity call sign for a club station?

A. Any Extra Class member of the club
B. Any member of the club
C. Any officer of the club
D. Only the person named as trustee on the club station license grant

T1C08
(C)
[97.25]
Page 7-8

T1C09
(A)
[97.21(b)]
Page 7-8

T1C10
(C)
[97.5(a)]
Page 7-6

T1C11
(A)
[97.21(b)]
Page 7-8

T1C12
(D)
[97.19]
Page 7-22

T1C13
(D)
[97.9(a),
97.17(a)]
Page 7-3

T1C14
(D)
[97.21(a) (1)]
Page 7-22

T1D — Authorized and prohibited transmission: communications with other countries; music; exchange of information with other services; indecent language; compensation for use of station; retransmission of other amateur signals; codes and ciphers; sale of equipment; unidentified transmissions; broadcasting

T1D01
(A)
[97.111(a)
(1)]
Page 7-19

T1D01

With which countries are FCC-licensed amateur stations prohibited from exchanging communications?
A. Any country whose administration has notified the ITU that it objects to such communications
B. Any country whose administration has notified the ARRL that it objects to such communications
C. Any country engaged in hostilities with another country
D. Any country in violation of the War Powers Act of 1934

T1D02
(A)
[97.111(a)
(5)]
Page 8-13

T1D02

On which of the following occasions may an FCC-licensed amateur station exchange messages with a U.S. military station?
A. During an Armed Forces Day Communications Test
B. During a Memorial Day Celebration
C. During an Independence Day celebration
D. During a propagation test

T1D03
(C)
[97.211(b),
97.215(b)]
Page 8-12

T1D03

When is the transmission of codes or ciphers that hide the meaning of a message allowed by an amateur station?
A. Only during contests
B. Only when operating mobile
C. Only when transmitting control commands to space stations or radio control craft
D. Only when frequencies above 1280 MHz are used

T1D04
(A)
[97.113(a)
(4),
97.113(c)]
Page 8-13

T1D04

What is the only time an amateur station is authorized to transmit music?
A. When incidental to an authorized retransmission of manned spacecraft communications
B. When the music produces no spurious emissions
C. When the purpose is to interfere with an illegal transmission
D. When the music is transmitted above 1280 MHz

T1D05
(A)
[97.113(a)
(3)(ii)]
Page 8-12

T1D05

When may amateur radio operators use their stations to notify other amateurs of the availability of equipment for sale or trade?
A. When the equipment is normally used in an amateur station and such activity is not conducted on a regular basis
B. When the asking price is $100.00 or less
C. When the asking price is less than its appraised value
D. When the equipment is not the personal property of either the station licensee or the control operator or their close relatives

T1D06

What, if any, are the restrictions concerning transmission of language that may be considered indecent or obscene?

A. The FCC maintains a list of words that are not permitted to be used on amateur frequencies
B. Any such language is prohibited
C. The ITU maintains a list of words that are not permitted to be used on amateur frequencies
D. There is no such prohibition

T1D07

What types of amateur stations can automatically retransmit the signals of other amateur stations?

A. Auxiliary, beacon, or Earth stations
B. Auxiliary, repeater, or space stations
C. Beacon, repeater, or space stations
D. Earth, repeater, or space stations

T1D08

In which of the following circumstances may the control operator of an amateur station receive compensation for operating the station?

A. When engaging in communications on behalf of their employer
B. When the communication is incidental to classroom instruction at an educational institution
C. When re-broadcasting weather alerts during a RACES net
D. When notifying other amateur operators of the availability for sale or trade of apparatus

T1D09

Under which of the following circumstances are amateur stations authorized to transmit signals related to broadcasting, program production, or news gathering, assuming no other means is available?

A. Only where such communications directly relate to the immediate safety of human life or protection of property
B. Only when broadcasting communications to or from the space shuttle
C. Only where noncommercial programming is gathered and supplied exclusively to the National Public Radio network
D. Only when using amateur repeaters linked to the Internet

T1D10

What is the meaning of the term "broadcasting" in the FCC rules for the amateur services?

A. Two-way transmissions by amateur stations
B. Transmission of music
C. Transmission of messages directed only to amateur operators
D. Transmissions intended for reception by the general public

T1D11

When may an amateur station transmit without identifying?

A. When the transmissions are of a brief nature to make station adjustments
B. When the transmissions are unmodulated
C. When the transmitted power level is below 1 watt
D. When transmitting signals to control a model craft

T1D06
(B)
[97.113(a)(4)]
Page 8-11

T1D07
(B)
[97.113(d)]
Page 8-13

T1D08
(B)
[97.113(a)(3)(iii)]
Page 8-12

T1D09
(A)
[97.113(5)(b)]
Page 8-13

T1D10
(D)
[97.3(a)(10)]
Page 8-13

T1D11
(D)
[97.119(a), 97.215(a)]
Page 8-3

T1D12
(B)
[97.111(b)
(4,5,6)]
Page 8-13

T1D12

Under which of the following circumstances may an amateur radio station engage in broadcasting?
A. Under no circumstances
B. When transmitting code practice, information bulletins, or transmissions necessary to provide emergency communications
C. At any time as long as no music is transmitted
D. At any time as long as the material being transmitted did not originate from a commercial broadcast station

T1E — Control operator and control types: control operator required; eligibility; designation of control operator; privileges and duties; control point; local, automatic and remote control; location of control operator

T1E01
(D)
[97.7]
Page 8-1

T1E01

When is an amateur station permitted to transmit without a control operator?
A. When using automatic control, such as in the case of a repeater
B. When the station licensee is away and another licensed amateur is using the station
C. When the transmitting station is an auxiliary station
D. Never

T1E02
(D)
[97.7(a)(b)]
Page 8-2

T1E02

Who may a station licensee designate to be the control operator of an amateur station?
A. Any U.S. citizen or registered alien
B. Any family member of the station licensee
C. Any person over the age of 18
D. Only a person for whom an amateur operator/primary station license grant appears in the FCC database or who is authorized for alien reciprocal operation

T1E03
(A)
[97.103(b)]
Page 8-1

T1E03

Who must designate the station control operator?
A. The station licensee
B. The FCC
C. The frequency coordinator
D. The ITU

T1E04
(D)
[97.105(b)]
Page 8-2

T1E04

What determines the transmitting privileges of an amateur station?
A. The frequency authorized by the frequency coordinator
B. The class of operator license held by the station licensee
C. The highest class of operator license held by anyone on the premises
D. The class of operator license held by the control operator

T1E05
(C)
[97.3(a)(14)]
Page 8-2

T1E05

What is an amateur station control point?
A. The location of the station's transmitting antenna
B. The location of the station transmitting apparatus
C. The location at which the control operator function is performed
D. The mailing address of the station licensee

T1E06

Under what type of control do APRS network digipeaters operate?

A. Automatic
B. Remote
C. Local
D. Manual

T1E06
(A)
[97.109(d)]
Page 8-11

T1E07

When the control operator is not the station licensee, who is responsible for the proper operation of the station?

A. All licensed amateurs who are present at the operation
B. Only the station licensee
C. Only the control operator
D. The control operator and the station licensee are equally responsible

T1E07
(D)
[97.103(a)]
Page 8-2

T1E08

Which of the following is an example of automatic control?

A. Repeater operation
B. Controlling the station over the Internet
C. Using a computer or other device to automatically send CW
D. Using a computer or other device to automatically identify

T1E08
(A)
[97.3(a)(6),
97.205(d)]
Page 8-11

T1E09

What type of control is being used when the control operator is at the control point?

A. Radio control
B. Unattended control
C. Automatic control
D. Local control

T1E09
(D)
[97.109(b)]
Page 8-10

T1E10

Which of the following is an example of remote control as defined in Part 97?

A. Repeater operation
B. Operating a station over the Internet
C. Controlling a model aircraft, boat or car by amateur radio
D. All of these choices are correct

T1E10
(B)
[97.3(a)(39)]
Page 8-10

T1E11

Who does the FCC presume to be the control operator of an amateur station, unless documentation to the contrary is in the station records?

A. The station custodian
B. The third party participant
C. The person operating the station equipment
D. The station licensee

T1E11
(D)
[97.103(a)]
Page 8-2

T1E12

When, under normal circumstances, may a Technician Class licensee be the control operator of a station operating in an exclusive Extra Class operator segment of the amateur bands?

A. At no time
B. When operating a special event station
C. As part of a multi-operator contest team
D. When using a club station whose trustee is an Extra Class operator licensee

T1E12
(A)
[97.119(e)]
Page 8-2

T1F01
(A)
Page 8-4

T1F01

What type of identification is being used when identifying a station on the air as Race Headquarters?

A. Tactical call sign
B. An official call sign reserved for RACES drills
C. SSID
D. Broadcast station

T1F02
(C)
[97.119(a)]
Page 8-4

T1F02

When using tactical identifiers such as "Race Headquarters" during a community service net operation, how often must your station transmit the station's FCC-assigned call sign?

A. Never, the tactical call is sufficient
B. Once during every hour
C. At the end of each communication and every ten minutes during a communication
D. At the end of every transmission

T1F03
(D)
[97.119(a)]
Page 8-3

T1F03

When is an amateur station required to transmit its assigned call sign?

A. At the beginning of each contact, and every 10 minutes thereafter
B. At least once during each transmission
C. At least every 15 minutes during and at the end of a communication
D. At least every 10 minutes during and at the end of a communication

T1F04
(C)
[97.119(b)(2)]
Page 8-4

T1F04

Which of the following is an acceptable language to use for station identification when operating in a phone sub-band?

A. Any language recognized by the United Nations
B. Any language recognized by the ITU
C. The English language
D. English, French, or Spanish

T1F05
(B)
[97.119(b)(2)]
Page 8-4

T1F05

What method of call sign identification is required for a station transmitting phone signals?

A. Send the call sign followed by the indicator RPT
B. Send the call sign using CW or phone emission
C. Send the call sign followed by the indicator R
D. Send the call sign using only phone emission

T1F06
(D)
[97.119(c)]
Page 8-4

T1F06

Which of the following formats of a self-assigned indicator is acceptable when identifying using a phone transmission?

A. KL7CC stroke W3
B. KL7CC slant W3
C. KL7CC slash W3
D. All of these choices are correct

T1F07

Which of the following restrictions apply when a non-licensed person is allowed to speak to a foreign station using a station under the control of a Technician Class control operator?
A. The person must be a U.S. citizen
B. The foreign station must be one with which the U.S. has a third party agreement
C. The licensed control operator must do the station identification
D. All of these choices are correct

T1F07
(B)
[97.115(a)(2)]
Page 8-10

T1F08

Which indicator is required by the FCC to be transmitted after a station call sign?
A. /M when operating mobile
B. /R when operating a repeater
C. / followed the FCC Region number when operating out of the region in which the license was issued
D. /KT, /AE or /AG when using new license privileges earned by CSCE while waiting for an upgrade to a previously issued license to appear in the FCC license database

T1F08
(D)
[97.119(f)]
Page 8-5

T1F09

What type of amateur station simultaneously retransmits the signal of another amateur station on a different channel or channels?
A. Beacon station
B. Earth station
C. Repeater station
D. Message forwarding station

T1F09
(C)
[97.3(a)(40)]
Page 2-12

T1F10

Who is accountable should a repeater inadvertently retransmit communications that violate the FCC rules?
A. The control operator of the originating station
B. The control operator of the repeater
C. The owner of the repeater
D. Both the originating station and the repeater owner

T1F10
(A)
[97.205(g)]
Page 8-11

T1F11

To which foreign stations do the FCC rules authorize the transmission of non-emergency third party communications?
A. Any station whose government permits such communications
B. Those in ITU Region 2 only
C. Those in ITU Regions 2 and 3 only
D. Those in ITU Region 3 only

T1F11
(A)
[97.115(a)]
Page 8-9

T1F12

How many persons are required to be members of a club for a club station license to be issued by the FCC?
A. At least 5
B. At least 4
C. A trustee and 2 officers
D. At least 2

T1F12
(B)
[97.5(b)(2)]
Page 7-22

T1F13
(B)
[97.103(c)]
Page 7-9

T1F13

When must the station licensee make the station and its records available for FCC inspection?

A. At any time ten days after notification by the FCC of such an inspection
B. At any time upon request by an FCC representative
C. Only after failing to comply with an FCC notice of violation
D. Only when presented with a valid warrant by an FCC official or government agent

SUBELEMENT T2 — Operating Procedures
[3 Exam Questions — 3 Groups]

T2A — Station operation: choosing an operating frequency; calling another station; test transmissions; procedural signs; use of minimum power; choosing an operating frequency; band plans; calling frequencies; repeater offsets

T2A01
What is the most common repeater frequency offset in the 2 meter band?
A. Plus 500 kHz
B. Plus or minus 600 kHz
C. Minus 500 kHz
D. Only plus 600 kHz

T2A01
(B)
Page 6-16

T2A02
What is the national calling frequency for FM simplex operations in the 70 cm band?
A. 146.520 MHz
B. 145.000 MHz
C. 432.100 MHz
D. 446.000 MHz

T2A02
(D)
Page 6-14

T2A03
What is a common repeater frequency offset in the 70 cm band?
A. Plus or minus 5 MHz
B. Plus or minus 600 kHz
C. Minus 600 kHz
D. Plus 600 kHz

T2A03
(A)
Page 6-16

T2A04
What is an appropriate way to call another station on a repeater if you know the other station's call sign?
A. Say break, break then say the station's call sign
B. Say the station's call sign then identify with your call sign
C. Say CQ three times then the other station's call sign
D. Wait for the station to call CQ then answer it

T2A04
(B)
Page 6-12

T2A05
How should you respond to a station calling CQ?
A. Transmit CQ followed by the other station's call sign
B. Transmit your call sign followed by the other station's call sign
C. Transmit the other station's call sign followed by your call sign
D. Transmit a signal report followed by your call sign

T2A05
(C)
Page 6-13

T2A06
What must an amateur operator do when making on-air transmissions to test equipment or antennas?
A. Properly identify the transmitting station
B. Make test transmissions only after 10:00 p.m. local time
C. Notify the FCC of the test transmission
D. State the purpose of the test during the test procedure

T2A06
(A)
Page 8-6

T2A07
(D)
Page 8-6

T2A07

Which of the following is true when making a test transmission?

A. Station identification is not required if the transmission is less than 15 seconds

B. Station identification is not required if the transmission is less than 1 watt

C. Station identification is only required once an hour when the transmissions are for test purposes only

D. Station identification is required at least every ten minutes during the test and at the end of the test

T2A08
(D)
Page 6-13

T2A08

What is the meaning of the procedural signal "CQ"?

A. Call on the quarter hour

B. A new antenna is being tested (no station should answer)

C. Only the called station should transmit

D. Calling any station

T2A09
(B)
Page 6-11

T2A09

What brief statement is often transmitted in place of "CQ" to indicate that you are listening on a repeater?

A. The words "Hello test" followed by your call sign

B. Your call sign

C. The repeater call sign followed by your call sign

D. The letters "QSY" followed by your call sign

T2A10
(A)
Page 7-16

T2A10

What is a band plan, beyond the privileges established by the FCC?

A. A voluntary guideline for using different modes or activities within an amateur band

B. A mandated list of operating schedules

C. A list of scheduled net frequencies

D. A plan devised by a club to indicate frequency band usage

T2A11
(D)
[97.313(a)]
Page 7-15

T2A11

Which of the following is an FCC rule regarding power levels used in the amateur bands, under normal, non-distress circumstances?

A. There is no limit to power as long as there is no interference with other services

B. No more than 200 watts PEP may be used

C. Up to 1500 watts PEP may be used on any amateur frequency without restriction

D. While not exceeding the maximum power permitted on a given band, use the minimum power necessary to carry out the desired communication

T2A12
(D)
Page 6-13

T2A12

Which of the following is a guideline to use when choosing an operating frequency for calling CQ?

A. Listen first to be sure that no one else is using the frequency

B. Ask if the frequency is in use

C. Make sure you are in your assigned band

D. All of these choices are correct

T2B — VHF/UHF operating practices: SSB phone; FM repeater; simplex; splits and shifts; CTCSS; DTMF; tone squelch; carrier squelch; phonetics; operational problem resolution; Q signals

T2B01

What is the term used to describe an amateur station that is transmitting and receiving on the same frequency?

A. Full duplex communication
B. Diplex communication
C. Simplex communication
D. Multiplex communication

T2B02

What is the term used to describe the use of a sub-audible tone transmitted with normal voice audio to open the squelch of a receiver?

A. Carrier squelch
B. Tone burst
C. DTMF
D. CTCSS

T2B03

Which of the following describes the muting of receiver audio controlled solely by the presence or absence of an RF signal?

A. Tone squelch
B. Carrier squelch
C. CTCSS
D. Modulated carrier

T2B04

Which of the following common problems might cause you to be able to hear but not access a repeater even when transmitting with the proper offset?

A. The repeater receiver may require an audio tone burst for access
B. The repeater receiver may require a CTCSS tone for access
C. The repeater receiver may require a DCS tone sequence for access
D. All of these choices are correct

T2B05

What determines the amount of deviation of an FM (as opposed to PM) signal?

A. Both the frequency and amplitude of the modulating signal
B. The frequency of the modulating signal
C. The amplitude of the modulating signal
D. The relative phase of the modulating signal and the carrier

T2B06

What happens when the deviation of an FM transmitter is increased?

A. Its signal occupies more bandwidth
B. Its output power increases
C. Its output power and bandwidth increases
D. Asymmetric modulation occurs

T2B01
(C)
Page 6-9

T2B02
(D)
Page 6-16

T2B03
(B)
Page 5-7

T2B04
(D)
Page 6-17

T2B05
(C)
Page 2-10

T2B06
(A)
Page 2-9

T2B07
(A)
Page 2-9

T2B07

What could cause your FM signal to interfere with stations on nearby frequencies?
A. Microphone gain too high, causing over-deviation
B. SWR too high
C. Incorrect CTCSS Tone
D. All of these choices are correct

T2B08
(A)
Page 8-7

T2B08

Which of the following applies when two stations transmitting on the same frequency interfere with each other?
A. Common courtesy should prevail, but no one has absolute right to an amateur frequency
B. Whoever has the strongest signal has priority on the frequency
C. Whoever has been on the frequency the longest has priority on the frequency
D. The station which has the weakest signal has priority on the frequency

T2B09
(A)
[97.119(b)
(2)]
Page 8-4

T2B09

Which of the following methods is encouraged by the FCC when identifying your station when using phone?
A. Use of a phonetic alphabet
B. Send your call sign in CW as well as voice
C. Repeat your call sign three times
D. Increase your signal to full power when identifying

T2B10
(A)
Page 6-5

T2B10

Which Q signal indicates that you are receiving interference from other stations?
A. QRM
B. QRN
C. QTH
D. QSB

T2B11
(B)
Page 6-5

T2B11

Which Q signal indicates that you are changing frequency?
A. QRU
B. QSY
C. QSL
D. QRZ

T2B12
(A)
Page 6-14

T2B12

Under what circumstances should you consider communicating via simplex rather than a repeater?
A. When the stations can communicate directly without using a repeater
B. Only when you have an endorsement for simplex operation on your license
C. Only when third party traffic is not being passed
D. Only if you have simplex modulation capability

T2B13
(C)
Page 6-9

T2B13

Which of the following is true of the use of SSB phone in amateur bands above 50 MHz?
A. It is permitted only by holders of a General Class or higher license
B. It is permitted only on repeaters
C. It is permitted in at least some portion of all the amateur bands above 50 MHz
D. It is permitted only when power is limited to no more than 100 watts

T2C — Public service: emergency and non-emergency operations; applicability of FCC rules; RACES and ARES; net and traffic procedures; emergency restrictions

T2C01

When do the FCC rules NOT apply to the operation of an amateur station?

A. When operating a RACES station
B. When operating under special FEMA rules
C. When operating under special ARES rules
D. Never, FCC rules always apply

T2C02

What is one way to recharge a 12-volt lead-acid station battery if the commercial power is out?

A. Cool the battery in ice for several hours
B. Add acid to the battery
C. Connect the battery in parallel with a vehicle's battery and run the engine
D. All of these choices are correct

T2C03

What should be done to insure that voice message traffic containing proper names and unusual words are copied correctly by the receiving station?

A. The entire message should be repeated at least four times
B. Such messages must be limited to no more than 10 words
C. Such words and terms should be spelled out using a standard phonetic alphabet
D. All of these choices are correct

T2C04

What do RACES and ARES have in common?

A. They represent the two largest ham clubs in the United States
B. Both organizations broadcast road and weather information
C. Neither may handle emergency traffic supporting public service agencies
D. Both organizations may provide communications during emergencies

T2C05

Which of the following describes the Radio Amateur Civil Emergency Service (RACES)?

A. A radio service using amateur frequencies for emergency management or civil defense communications
B. A radio service using amateur stations for emergency management or civil defense communications
C. An emergency service using amateur operators certified by a civil defense organization as being enrolled in that organization
D. All of these choices are correct

T2C06

Which of the following is an accepted practice to get the immediate attention of a net control station when reporting an emergency?

A. Repeat the words SOS three times followed by the call sign of the reporting station
B. Press the push-to-talk button three times
C. Begin your transmission by saying "Priority" or "Emergency" followed by your call sign
D. Play a pre-recorded emergency alert tone followed by your call sign

T2C01
(D)
[97.103(a)]
Page 6-25

T2C02
(C)
Page 5-18

T2C03
(C)
Page 6-22

T2C04
(D)
Page 6-24

T2C05
(D)
[97.3(a)(38), 97.407]
Page 6-24

T2C06
(C)
Page 6-21

T2C07

Which of the following is an accepted practice for an amateur operator who has checked into an emergency traffic net?

A. Provided that the frequency is quiet, announce the station call sign and location every 5 minutes
B. Move 5 kHz away from the net's frequency and use high power to ask other hams to keep clear of the net frequency
C. Remain on frequency without transmitting until asked to do so by the net control station
D. All of the choices are correct

T2C08

Which of the following is a characteristic of good emergency traffic handling?

A. Passing messages exactly as received
B. Making decisions as to whether or not messages should be relayed or delivered
C. Communicating messages to the news media for broadcast outside the disaster area
D. All of these choices are correct

T2C09

Are amateur station control operators ever permitted to operate outside the frequency privileges of their license class?

A. No
B. Yes, but only when part of a FEMA emergency plan
C. Yes, but only when part of a RACES emergency plan
D. Yes, but only if necessary in situations involving the immediate safety of human life or protection of property

T2C10

What is the preamble in a formal traffic message?

A. The first paragraph of the message text
B. The message number
C. The priority handling indicator for the message
D. The information needed to track the message as it passes through the amateur radio traffic handling system

T2C11

What is meant by the term "check" in reference to a formal traffic message?

A. The check is a count of the number of words or word equivalents in the text portion of the message
B. The check is the value of a money order attached to the message
C. The check is a list of stations that have relayed the message
D. The check is a box on the message form that tells you the message was received

T2C12

What is the Amateur Radio Emergency Service (ARES)?

A. Licensed amateurs who have voluntarily registered their qualifications and equipment for communications duty in the public service
B. Licensed amateurs who are members of the military and who voluntarily agreed to provide message handling services in the case of an emergency
C. A training program that provides licensing courses for those interested in obtaining an amateur license to use during emergencies
D. A training program that certifies amateur operators for membership in the Radio Amateur Civil Emergency Service

SUBELEMENT T3 — Radio wave characteristics: properties of radio waves; propagation modes
[3 Exam Questions — 3 Groups]

T3A — Radio wave characteristics: how a radio signal travels; fading; multipath; wavelength vs. penetration; antenna orientation

T3A01

What should you do if another operator reports that your station's 2 meter signals were strong just a moment ago, but now they are weak or distorted?
A. Change the batteries in your radio to a different type
B. Turn on the CTCSS tone
C. Ask the other operator to adjust his squelch control
D. Try moving a few feet or changing the direction of your antenna if possible, as reflections may be causing multi-path distortion

T3A01
(D)
Page 4-2

T3A02

Why are UHF signals often more effective from inside buildings than VHF signals?
A. VHF signals lose power faster over distance
B. The shorter wavelength allows them to more easily penetrate the structure of buildings
C. This is incorrect; VHF works better than UHF inside buildings
D. UHF antennas are more efficient than VHF antennas

T3A02
(B)
Page 4-2

T3A03

What antenna polarization is normally used for long-distance weak-signal CW and SSB contacts using the VHF and UHF bands?
A. Right-hand circular
B. Left-hand circular
C. Horizontal
D. Vertical

T3A03
(C)
Page 4-15

T3A04

What can happen if the antennas at opposite ends of a VHF or UHF line of sight radio link are not using the same polarization?
A. The modulation sidebands might become inverted
B. Signals could be significantly weaker
C. Signals have an echo effect on voices
D. Nothing significant will happen

T3A04
(B)
Page 4-6

T3A05

When using a directional antenna, how might your station be able to access a distant repeater if buildings or obstructions are blocking the direct line of sight path?
A. Change from vertical to horizontal polarization
B. Try to find a path that reflects signals to the repeater
C. Try the long path
D. Increase the antenna SWR

T3A05
(B)
Page 4-14

T3A06
(B)
Page 4-2

T3A06

What term is commonly used to describe the rapid fluttering sound sometimes heard from mobile stations that are moving while transmitting?

A. Flip-flopping
B. Picket fencing
C. Frequency shifting
D. Pulsing

T3A07
(A)
Page 4-6

T3A07

What type of wave carries radio signals between transmitting and receiving stations?

A. Electromagnetic
B. Electrostatic
C. Surface acoustic
D. Magnetostrictive

T3A08
(C)
Page 4-2

T3A08

Which of the following is a likely cause of irregular fading of signals received by ionospheric reflection?

A. Frequency shift due to Faraday rotation
B. Interference from thunderstorms
C. Random combining of signals arriving via different paths
D. Intermodulation distortion

T3A09
(B)
Page 4-6

T3A09

Which of the following results from the fact that skip signals refracted from the ionosphere are elliptically polarized?

A. Digital modes are unusable
B. Either vertically or horizontally polarized antennas may be used for transmission or reception
C. FM voice is unusable
D. Both the transmitting and receiving antennas must be of the same polarization

T3A10
(D)
Page 4-2

T3A10

What may occur if data signals propagate over multiple paths?

A. Transmission rates can be increased by a factor equal to the number of separate paths observed
B. Transmission rates must be decreased by a factor equal to the number of separate paths observed
C. No significant changes will occur if the signals are transmitting using FM
D. Error rates are likely to increase

T3A11
(C)
Page 4-3

T3A11

Which part of the atmosphere enables the propagation of radio signals around the world?

A. The stratosphere
B. The troposphere
C. The ionosphere
D. The magnetosphere

T3B — Radio and electromagnetic wave properties: the electromagnetic spectrum; wavelength vs. frequency; velocity of electromagnetic waves; calculating wavelength

T3B01

What is the name for the distance a radio wave travels during one complete cycle?

A. Wave speed
B. Waveform
C. Wavelength
D. Wave spread

T3B01
(C)
Page 2-5

T3B02

What property of a radio wave is used to describe its polarization?

A. The orientation of the electric field
B. The orientation of the magnetic field
C. The ratio of the energy in the magnetic field to the energy in the electric field
D. The ratio of the velocity to the wavelength

T3B02
(A)
Page 4-6

T3B03

What are the two components of a radio wave?

A. AC and DC
B. Voltage and current
C. Electric and magnetic fields
D. Ionizing and non-ionizing radiation

T3B03
(C)
Page 4-6

T3B04

How fast does a radio wave travel through free space?

A. At the speed of light
B. At the speed of sound
C. Its speed is inversely proportional to its wavelength
D. Its speed increases as the frequency increases

T3B04
(A)
Page 2-5

T3B05

How does the wavelength of a radio wave relate to its frequency?

A. The wavelength gets longer as the frequency increases
B. The wavelength gets shorter as the frequency increases
C. There is no relationship between wavelength and frequency
D. The wavelength depends on the bandwidth of the signal

T3B05
(B)
Page 2-5

T3B06

What is the formula for converting frequency to approximate wavelength in meters?

A. Wavelength in meters equals frequency in hertz multiplied by 300
B. Wavelength in meters equals frequency in hertz divided by 300
C. Wavelength in meters equals frequency in megahertz divided by 300
D. Wavelength in meters equals 300 divided by frequency in megahertz

T3B06
(D)
Page 2-6

T3B07

What property of radio waves is often used to identify the different frequency bands?

A. The approximate wavelength
B. The magnetic intensity of waves
C. The time it takes for waves to travel one mile
D. The voltage standing wave ratio of waves

T3B07
(A)
Page 2-5

T3B08
(B)
Page 2-4

T3B08

What are the frequency limits of the VHF spectrum?
A. 30 to 300 kHz
B. 30 to 300 MHz
C. 300 to 3000 kHz
D. 300 to 3000 MHz

T3B09
(D)
Page 2-4

T3B09

What are the frequency limits of the UHF spectrum?
A. 30 to 300 kHz
B. 30 to 300 MHz
C. 300 to 3000 kHz
D. 300 to 3000 MHz

T3B10
(C)
Page 2-4

T3B10

What frequency range is referred to as HF?
A. 300 to 3000 MHz
B. 30 to 300 MHz
C. 3 to 30 MHz
D. 300 to 3000 kHz

T3B11
(B)
Page 2-5

T3B11

What is the approximate velocity of a radio wave as it travels through free space?
A. 3000 kilometers per second
B. 300,000,000 meters per second
C. 300,000 miles per hour
D. 186,000 miles per hour

T3C — Propagation modes: line of sight; sporadic E; meteor and auroral scatter and reflections; tropospheric ducting; F layer skip; radio horizon

T3C01
(C)
Page 4-3

T3C01

Why are direct (not via a repeater) UHF signals rarely heard from stations outside your local coverage area?
A. They are too weak to go very far
B. FCC regulations prohibit them from going more than 50 miles
C. UHF signals are usually not reflected by the ionosphere
D. They collide with trees and shrubbery and fade out

T3C02
(D)
Page 4-4

T3C02

Which of the following might be happening when VHF signals are being received from long distances?
A. Signals are being reflected from outer space
B. Signals are arriving by sub-surface ducting
C. Signals are being reflected by lightning storms in your area
D. Signals are being refracted from a sporadic E layer

T3C03
(B)
Page 4-4

T3C03

What is a characteristic of VHF signals received via auroral reflection?
A. Signals from distances of 10,000 or more miles are common
B. The signals exhibit rapid fluctuations of strength and often sound distorted
C. These types of signals occur only during winter nighttime hours
D. These types of signals are generally strongest when your antenna is aimed west

T3C04

Which of the following propagation types is most commonly associated with occasional strong over-the-horizon signals on the 10, 6, and 2 meter bands?
A. Backscatter
B. Sporadic E
C. D layer absorption
D. Gray-line propagation

T3C05

Which of the following effects might cause radio signals to be heard despite obstructions between the transmitting and receiving stations?
A. Knife-edge diffraction
B. Faraday rotation
C. Quantum tunneling
D. Doppler shift

T3C06

What mode is responsible for allowing over-the-horizon VHF and UHF communications to ranges of approximately 300 miles on a regular basis?
A. Tropospheric scatter
B. D layer refraction
C. F2 layer refraction
D. Faraday rotation

T3C07

What band is best suited for communicating via meteor scatter?
A. 10 meters
B. 6 meters
C. 2 meters
D. 70 cm

T3C08

What causes tropospheric ducting?
A. Discharges of lightning during electrical storms
B. Sunspots and solar flares
C. Updrafts from hurricanes and tornadoes
D. Temperature inversions in the atmosphere

T3C09

What is generally the best time for long-distance 10 meter band propagation via the F layer?
A. From dawn to shortly after sunset during periods of high sunspot activity
B. From shortly after sunset to dawn during periods of high sunspot activity
C. From dawn to shortly after sunset during periods of low sunspot activity
D. From shortly after sunset to dawn during periods of low sunspot activity

T3C10

What is the radio horizon?
A. The distance over which two stations can communicate by direct path
B. The distance from the ground to a horizontally mounted antenna
C. The farthest point you can see when standing at the base of your antenna tower
D. The shortest distance between two points on the Earth's surface

T3C04
(B)
Page 4-4

T3C05
(A)
Page 4-1

T3C06
(A)
Page 4-2

T3C07
(B)
Page 4-5

T3C08
(D)
Page 4-2

T3C09
(A)
Page 4-4

T3C10
(A)
Page 4-1

T3C11
(C)
Page 4-1

T3C11

Why do VHF and UHF radio signals usually travel somewhat farther than the visual line of sight distance between two stations?

A. Radio signals move somewhat faster than the speed of light
B. Radio waves are not blocked by dust particles
C. The Earth seems less curved to radio waves than to light
D. Radio waves are blocked by dust particles

T3C12
(A)
Page 4-4

T3C12

Which of the following bands may provide long distance communications during the peak of the sunspot cycle?

A. Six or ten meters
B. 23 centimeters
C. 70 centimeters or 1.25 meters
D. All of these choices are correct

SUBELEMENT T4 — Amateur radio practices and station set up
[2 Exam Questions — 2 Groups]

T4A — Station setup: connecting microphones; reducing unwanted emissions; power source; connecting a computer; RF grounding; connecting digital equipment; connecting an SWR meter

T4A01
Which of the following is true concerning the microphone connectors on amateur transceivers?
A. All transceivers use the same microphone connector type
B. Some connectors include push-to-talk and voltages for powering the microphone
C. All transceivers using the same connector type are wired identically
D. Un-keyed connectors allow any microphone to be connected

T4A02
How might a computer be used as part of an amateur radio station?
A. For logging contacts and contact information
B. For sending and/or receiving CW
C. For generating and decoding digital signals
D. All of these choices are correct

T4A03
Which is a good reason to use a regulated power supply for communications equipment?
A. It prevents voltage fluctuations from reaching sensitive circuits
B. A regulated power supply has FCC approval
C. A fuse or circuit breaker regulates the power
D. Power consumption is independent of load

T4A04
Where must a filter be installed to reduce harmonic emissions from your station?
A. Between the transmitter and the antenna
B. Between the receiver and the transmitter
C. At the station power supply
D. At the microphone

T4A05
Where should an in-line SWR meter be connected to monitor the standing wave ratio of the station antenna system?
A. In series with the feed line, between the transmitter and antenna
B. In series with the station's ground
C. In parallel with the push-to-talk line and the antenna
D. In series with the power supply cable, as close as possible to the radio

T4A06
Which of the following would be connected between a transceiver and computer in a packet radio station?
A. Transmatch
B. Mixer
C. Terminal node controller
D. Antenna

T4A01
(B)
Page 5-6

T4A02
(D)
Page 5-1

T4A03
(A)
Page 5-15

T4A04
(A)
Page 5-21

T4A05
(A)
Page 4-18

T4A06
(C)
Page 5-13

T4A07
(C)
Page 5-13

T4A07

How is a computer's sound card used when conducting digital communications using a computer?

A. The sound card communicates between the computer CPU and the video display
B. The sound card records the audio frequency for video display
C. The sound card provides audio to the microphone input and converts received audio to digital form
D. All of these choices are correct

T4A08
(D)
Page 5-25

T4A08

Which type of conductor is best to use for RF grounding?

A. Round stranded wire
B. Round copper-clad steel wire
C. Twisted-pair cable
D. Flat strap

T4A09
(D)
Page 5-20

T4A09

Which of the following could you use to cure distorted audio caused by RF current flowing on the shield of a microphone cable?

A. Band-pass filter
B. Low-pass filter
C. Preamplifier
D. Ferrite choke

T4A10
(B)
Page 5-16

T4A10

What is the source of a high-pitched whine that varies with engine speed in a mobile transceiver's receive audio?

A. The ignition system
B. The alternator
C. The electric fuel pump
D. Anti-lock braking system controllers

T4A11
(A)
Page 5-15

T4A11

Where should the negative return connection of a mobile transceiver's power cable be connected?

A. At the battery or engine block ground strap
B. At the antenna mount
C. To any metal part of the vehicle
D. Through the transceiver's mounting bracket

T4A12
(D)
Page 5-16

T4A12

What could be happening if another operator reports a variable high-pitched whine on the audio from your mobile transmitter?

A. Your microphone is picking up noise from an open window
B. You have the volume on your receiver set too high
C. You need to adjust your squelch control
D. Noise on the vehicle's electrical system is being transmitted along with your speech audio

T4B — Operating controls: tuning; use of filters; squelch function; AGC; repeater offset; memory channels

T4B01
What may happen if a transmitter is operated with the microphone gain set too high?
A. The output power might be too high
B. The output signal might become distorted
C. The frequency might vary
D. The SWR might increase

T4B01
(B)
Page 5-4

T4B02
Which of the following can be used to enter the operating frequency on a modern transceiver?
A. The keypad or VFO knob
B. The CTCSS or DTMF encoder
C. The Automatic Frequency Control
D. All of these choices are correct

T4B02
(A)
Page 5-2

T4B03
What is the purpose of the squelch control on a transceiver?
A. To set the highest level of volume desired
B. To set the transmitter power level
C. To adjust the automatic gain control
D. To mute receiver output noise when no signal is being received

T4B03
(D)
Page 5-7

T4B04
What is a way to enable quick access to a favorite frequency on your transceiver?
A. Enable the CTCSS tones
B. Store the frequency in a memory channel
C. Disable the CTCSS tones
D. Use the scan mode to select the desired frequency

T4B04
(B)
Page 5-3

T4B05
Which of the following would reduce ignition interference to a receiver?
A. Change frequency slightly
B. Decrease the squelch setting
C. Turn on the noise blanker
D. Use the RIT control

T4B05
(C)
Page 5-7

T4B06
Which of the following controls could be used if the voice pitch of a single-sideband signal seems too high or low?
A. The AGC or limiter
B. The bandwidth selection
C. The tone squelch
D. The receiver RIT or clarifier

T4B06
(D)
Page 5-7

T4B07
What does the term "RIT" mean?
A. Receiver Input Tone
B. Receiver Incremental Tuning
C. Rectifier Inverter Test
D. Remote Input Transmitter

T4B07
(B)
Page 5-7

T4B08
(B)
Page 5-7

T4B08

What is the advantage of having multiple receive bandwidth choices on a multimode transceiver?

A. Permits monitoring several modes at once
B. Permits noise or interference reduction by selecting a bandwidth matching the mode
C. Increases the number of frequencies that can be stored in memory
D. Increases the amount of offset between receive and transmit frequencies

T4B09
(C)
Page 5-7

T4B09

Which of the following is an appropriate receive filter bandwidth to select in order to minimize noise and interference for SSB reception?

A. 500 Hz
B. 1000 Hz
C. 2400 Hz
D. 5000 Hz

T4B10
(A)
Page 5-7

T4B10

Which of the following is an appropriate receive filter bandwidth to select in order to minimize noise and interference for CW reception?

A. 500 Hz
B. 1000 Hz
C. 2400 Hz
D. 5000 Hz

T4B11
(C)
Page 6-16

T4B11

Which of the following describes the common meaning of the term "repeater offset"?

A. The distance between the repeater's transmit and receive antennas
B. The time delay before the repeater timer resets
C. The difference between the repeater's transmit and receive frequencies
D. Matching the antenna impedance to the feed line impedance

T4B12
(A)
Page 5-7

T4B12

What is the function of automatic gain control or AGC?

A. To keep received audio relatively constant
B. To protect an antenna from lightning
C. To eliminate RF on the station cabling
D. An asymmetric goniometer control used for antenna matching

SUBELEMENT T5 — Electrical principles: math for electronics; electronic principles; Ohm's Law
[4 Exam Questions — 4 Groups]

T5A — Electrical principles, units, and terms: current and voltage; conductors and insulators; alternating and direct current

T5A01
Electrical current is measured in which of the following units?
A. Volts
B. Watts
C. Ohms
D. Amperes

T5A02
Electrical power is measured in which of the following units?
A. Volts
B. Watts
C. Ohms
D. Amperes

T5A03
What is the name for the flow of electrons in an electric circuit?
A. Voltage
B. Resistance
C. Capacitance
D. Current

T5A04
What is the name for a current that flows only in one direction?
A. Alternating current
B. Direct current
C. Normal current
D. Smooth current

T5A05
What is the electrical term for the electromotive force (EMF) that causes electron flow?
A. Voltage
B. Ampere-hours
C. Capacitance
D. Inductance

T5A06
How much voltage does a mobile transceiver usually require?
A. About 12 volts
B. About 30 volts
C. About 120 volts
D. About 240 volts

T5A01
(D)
Page 3-1

T5A02
(B)
Page 3-5

T5A03
(D)
Page 3-1

T5A04
(B)
Page 3-6

T5A05
(A)
Page 3-1

T5A06
(A)
Page 5-15

T5A07
(C)
Page 3-4

T5A07
Which of the following is a good electrical conductor?
A. Glass
B. Wood
C. Copper
D. Rubber

T5A08
(B)
Page 3-4

T5A08
Which of the following is a good electrical insulator?
A. Copper
B. Glass
C. Aluminum
D. Mercury

T5A09
(A)
Page 3-6

T5A09
What is the name for a current that reverses direction on a regular basis?
A. Alternating current
B. Direct current
C. Circular current
D. Vertical current

T5A10
(C)
Page 3-5

T5A10
Which term describes the rate at which electrical energy is used?
A. Resistance
B. Current
C. Power
D. Voltage

T5A11
(A)
Page 3-1

T5A11
What is the basic unit of electromotive force?
A. The volt
B. The watt
C. The ampere
D. The ohm

T5A12
(D)
Page 2-1

T5A12
What term describes the number of times per second that an alternating current reverses direction?
A. Pulse rate
B. Speed
C. Wavelength
D. Frequency

T5B — Math for electronics: conversion of electrical units; decibels; the metric system

T5B01
(C)
Page 2-2

T5B01
How many milliamperes is 1.5 amperes?
A. 15 milliamperes
B. 150 milliamperes
C. 1,500 milliamperes
D. 15,000 milliamperes

T5B02

What is another way to specify a radio signal frequency of 1,500,000 hertz?

A. 1500 kHz
B. 1500 MHz
C. 15 GHz
D. 150 kHz

T5B02
(A)
Page 2-2

T5B03

How many volts are equal to one kilovolt?

A. One one-thousandth of a volt
B. One hundred volts
C. One thousand volts
D. One million volts

T5B03
(C)
Page 2-2

T5B04

How many volts are equal to one microvolt?

A. One one-millionth of a volt
B. One million volts
C. One thousand kilovolts
D. One one-thousandth of a volt

T5B04
(A)
Page 2-2

T5B05

Which of the following is equivalent to 500 milliwatts?

A. 0.02 watts
B. 0.5 watts
C. 5 watts
D. 50 watts

T5B05
(B)
Page 2-2

T5B06

If an ammeter calibrated in amperes is used to measure a 3000-milliampere current, what reading would it show?

A. 0.003 amperes
B. 0.3 amperes
C. 3 amperes
D. 3,000,000 amperes

T5B06
(C)
Page 2-2

T5B07

If a frequency readout calibrated in megahertz shows a reading of 3.525 MHz, what would it show if it were calibrated in kilohertz?

A. 0.003525 kHz
B. 35.25 kHz
C. 3525 kHz
D. 3,525,000 kHz

T5B07
(C)
Page 2-2

T5B08

How many microfarads are 1,000,000 picofarads?

A. 0.001 microfarads
B. 1 microfarad
C. 1000 microfarads
D. 1,000,000,000 microfarads

T5B08
(B)
Page 2-2

T5B09

What is the approximate amount of change, measured in decibels (dB), of a power increase from 5 watts to 10 watts?

A. 2 dB
B. 3 dB
C. 5 dB
D. 10 dB

T5B10

What is the approximate amount of change, measured in decibels (dB), of a power decrease from 12 watts to 3 watts?

A. −1 dB
B. −3 dB
C. −6 dB
D. −9 dB

T5B11

What is the approximate amount of change, measured in decibels (dB), of a power increase from 20 watts to 200 watts?

A. 10 dB
B. 12 dB
C. 18 dB
D. 28 dB

T5B12

Which of the following frequencies is equal to 28,400 kHz?

A. 28.400 MHz
B. 2.800 MHz
C. 284.00 MHz
D. 28.400 kHz

T5B13

If a frequency readout shows a reading of 2425 MHz, what frequency is that in GHz?

A. 0.002425 GHz
B. 24.25 GHz
C. 2.425 GHz
D. 2425 GHz

T5C — Electronic principles: capacitance; inductance; current flow in circuits; alternating current; definition of RF; DC power calculations; impedance

T5C01

What is the ability to store energy in an electric field called?

A. Inductance
B. Resistance
C. Tolerance
D. Capacitance

T5C02

What is the basic unit of capacitance?
A. The farad
B. The ohm
C. The volt
D. The henry

T5C03

What is the ability to store energy in a magnetic field called?
A. Admittance
B. Capacitance
C. Resistance
D. Inductance

T5C04

What is the basic unit of inductance?
A. The coulomb
B. The farad
C. The henry
D. The ohm

T5C05

What is the unit of frequency?
A. Hertz
B. Henry
C. Farad
D. Tesla

T5C06

What does the abbreviation "RF" refer to?
A. Radio frequency signals of all types
B. The resonant frequency of a tuned circuit
C. The real frequency transmitted as opposed to the apparent frequency
D. Reflective force in antenna transmission lines

T5C07

What is a usual name for electromagnetic waves that travel through space?
A. Gravity waves
B. Sound waves
C. Radio waves
D. Pressure waves

T5C08

What is the formula used to calculate electrical power in a DC circuit?
A. Power (P) equals voltage (E) multiplied by current (I)
B. Power (P) equals voltage (E) divided by current (I)
C. Power (P) equals voltage (E) minus current (I)
D. Power (P) equals voltage (E) plus current (I)

T5C02
(A)
Page 3-7

T5C03
(D)
Page 3-7

T5C04
(C)
Page 3-7

T5C05
(A)
Page 2-3

T5C06
(A)
Page 2-3

T5C07
(C)
Page 4-6

T5C08
(A)
Page 3-5

T5C09

How much power is being used in a circuit when the applied voltage is 13.8 volts DC and the current is 10 amperes?
A. 138 watts
B. 0.7 watts
C. 23.8 watts
D. 3.8 watts

T5C10

How much power is being used in a circuit when the applied voltage is 12 volts DC and the current is 2.5 amperes?
A. 4.8 watts
B. 30 watts
C. 14.5 watts
D. 0.208 watts

T5C11

How many amperes are flowing in a circuit when the applied voltage is 12 volts DC and the load is 120 watts?
A. 0.1 amperes
B. 10 amperes
C. 12 amperes
D. 132 amperes

T5C12

What is meant by the term impedance?
A. It is a measure of the opposition to AC current flow in a circuit
B. It is the inverse of resistance
C. It is a measure of the Q or Quality Factor of a component
D. It is a measure of the power handling capability of a component

T5C13

What are the units of impedance?
A. Volts
B. Amperes
C. Coulombs
D. Ohms

T5D — Ohm's Law: formulas and usage

T5D01

What formula is used to calculate current in a circuit?
A. Current (I) equals voltage (E) multiplied by resistance (R)
B. Current (I) equals voltage (E) divided by resistance (R)
C. Current (I) equals voltage (E) added to resistance (R)
D. Current (I) equals voltage (E) minus resistance (R)

T5D02

What formula is used to calculate voltage in a circuit?
A. Voltage (E) equals current (I) multiplied by resistance (R)
B. Voltage (E) equals current (I) divided by resistance (R)
C. Voltage (E) equals current (I) added to resistance (R)
D. Voltage (E) equals current (I) minus resistance (R)

T5D03

What formula is used to calculate resistance in a circuit?
A. Resistance (R) equals voltage (E) multiplied by current (I)
B. Resistance (R) equals voltage (E) divided by current (I)
C. Resistance (R) equals voltage (E) added to current (I)
D. Resistance (R) equals voltage (E) minus current (I)

T5D04

What is the resistance of a circuit in which a current of 3 amperes flows through a resistor connected to 90 volts?
A. 3 ohms
B. 30 ohms
C. 93 ohms
D. 270 ohms

T5D05

What is the resistance in a circuit for which the applied voltage is 12 volts and the current flow is 1.5 amperes?
A. 18 ohms
B. 0.125 ohms
C. 8 ohms
D. 13.5 ohms

T5D06

What is the resistance of a circuit that draws 4 amperes from a 12-volt source?
A. 3 ohms
B. 16 ohms
C. 48 ohms
D. 8 ohms

T5D07

What is the current flow in a circuit with an applied voltage of 120 volts and a resistance of 80 ohms?
A. 9600 amperes
B. 200 amperes
C. 0.667 amperes
D. 1.5 amperes

T5D08

What is the current flowing through a 100-ohm resistor connected across 200 volts?
A. 20,000 amperes
B. 0.5 amperes
C. 2 amperes
D. 100 amperes

T5D02
(A)
Page 3-4

T5D03
(B)
Page 3-4

T5D04
(B)
Page 3-5

T5D05
(C)
Page 3-5

T5D06
(A)
Page 3-5

T5D07
(D)
Page 3-5

T5D08
(C)
Page 3-5

T5D09
(C)
Page 3-5

T5D09
What is the current flowing through a 24-ohm resistor connected across 240 volts?
A. 24,000 amperes
B. 0.1 amperes
C. 10 amperes
D. 216 amperes

T5D10
(A)
Page 3-5

T5D10
What is the voltage across a 2-ohm resistor if a current of 0.5 amperes flows through it?
A. 1 volt
B. 0.25 volts
C. 2.5 volts
D. 1.5 volts

T5D11
(B)
Page 3-5

T5D11
What is the voltage across a 10-ohm resistor if a current of 1 ampere flows through it?
A. 1 volt
B. 10 volts
C. 11 volts
D. 9 volts

T5D12
(D)
Page 3-5

T5D12
What is the voltage across a 10-ohm resistor if a current of 2 amperes flows through it?
A. 8 volts
B. 0.2 volts
C. 12 volts
D. 20 volts

SUBELEMENT T6 — Electrical components: semiconductors; circuit diagrams; component functions
[4 Exam Questions — 4 Groups]

T6A — Electrical components: fixed and variable resistors; capacitors and inductors; fuses; switches; batteries

T6A01
What electrical component is used to oppose the flow of current in a DC circuit?
A. Inductor
B. Resistor
C. Voltmeter
D. Transformer

T6A01
(B)
Page 3-7

T6A02
What type of component is often used as an adjustable volume control?
A. Fixed resistor
B. Power resistor
C. Potentiometer
D. Transformer

T6A02
(C)
Page 3-9

T6A03
What electrical parameter is controlled by a potentiometer?
A. Inductance
B. Resistance
C. Capacitance
D. Field strength

T6A03
(B)
Page 3-9

T6A04
What electrical component stores energy in an electric field?
A. Resistor
B. Capacitor
C. Inductor
D. Diode

T6A04
(B)
Page 3-7

T6A05
What type of electrical component consists of two or more conductive surfaces separated by an insulator?
A. Resistor
B. Potentiometer
C. Oscillator
D. Capacitor

T6A05
(D)
Page 3-7

T6A06
What type of electrical component stores energy in a magnetic field?
A. Resistor
B. Capacitor
C. Inductor
D. Diode

T6A06
(C)
Page 3-7

T6A07

What electrical component is usually composed of a coil of wire?

A. Switch
B. Capacitor
C. Diode
D. Inductor

T6A08

What electrical component is used to connect or disconnect electrical circuits?

A. Magnetron
B. Switch
C. Thermistor
D. All of these choices are correct

T6A09

What electrical component is used to protect other circuit components from current overloads?

A. Fuse
B. Capacitor
C. Inductor
D. All of these choices are correct

T6A10

Which of the following battery types is rechargeable?

A. Nickel-metal hydride
B. Lithium-ion
C. Lead-acid gel-cell
D. All of these choices are correct

T6A11

Which of the following battery types is not rechargeable?

A. Nickel-cadmium
B. Carbon-zinc
C. Lead-acid
D. Lithium-ion

T6B — Semiconductors: basic principles and applications of solid state devices; diodes and transistors

T6B01

What class of electronic components is capable of using a voltage or current signal to control current flow?

A. Capacitors
B. Inductors
C. Resistors
D. Transistors

T6B02

What electronic component allows current to flow in only one direction?

A. Resistor
B. Fuse
C. Diode
D. Driven Element

T6B03

Which of these components can be used as an electronic switch or amplifier?
A. Oscillator
B. Potentiometer
C. Transistor
D. Voltmeter

T6B04

Which of the following components can be made of three layers of semiconductor material?
A. Alternator
B. Transistor
C. Triode
D. Pentagrid converter

T6B05

Which of the following electronic components can amplify signals?
A. Transistor
B. Variable resistor
C. Electrolytic capacitor
D. Multi-cell battery

T6B06

How is the cathode lead of a semiconductor diode usually identified?
A. With the word cathode
B. With a stripe
C. With the letter C
D. All of these choices are correct

T6B07

What does the abbreviation LED stand for?
A. Low Emission Diode
B. Light Emitting Diode
C. Liquid Emission Detector
D. Long Echo Delay

T6B08

What does the abbreviation FET stand for?
A. Field Effect Transistor
B. Fast Electron Transistor
C. Free Electron Transition
D. Field Emission Thickness

T6B09

What are the names of the two electrodes of a diode?
A. Plus and minus
B. Source and drain
C. Anode and cathode
D. Gate and base

T6B03
(C)
Page 3-11

T6B04
(B)
Page 3-11

T6B05
(A)
Page 3-11

T6B06
(B)
Page 3-10

T6B07
(B)
Page 3-10

T6B08
(A)
Page 3-11

T6B09
(C)
Page 3-10

T6B10
(A)
Page 3-11

T6B10
What are the three electrodes of a PNP or NPN transistor?
A. Emitter, base, and collector
B. Source, gate, and drain
C. Cathode, grid, and plate
D. Cathode, drift cavity, and collector

T6B11
(B)
Page 3-11

T6B11
What are the three electrodes of a field effect transistor?
A. Emitter, base, and collector
B. Source, gate, and drain
C. Cathode, grid, and plate
D. Cathode, gate, and anode

T6B12
(A)
Page 3-11

T6B12
What is the term that describes a transistor's ability to amplify a signal?
A. Gain
B. Forward resistance
C. Forward voltage drop
D. On resistance

T6C — Circuit diagrams; schematic symbols

T6C01
(C)
Page 3-13

T6C01
What is the name for standardized representations of components in an electrical wiring diagram?
A. Electrical depictions
B. Grey sketch
C. Schematic symbols
D. Component callouts

T6C02
(A)
Page 3-13

T6C02
What is component 1 in figure T1?
A. Resistor
B. Transistor
C. Battery
D. Connector

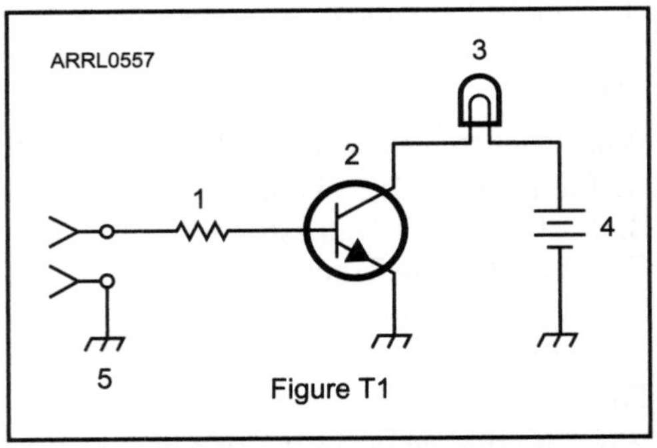

Figure T1 — Refer to this figure for questions T2C02 through T2C05 and T6D10.

T6C03
(B)
Page 3-13

T6C03
What is component 2 in figure T1?
A. Resistor
B. Transistor
C. Indicator lamp
D. Connector

T6C04
(C)
Page 3-13

T6C04
What is component 3 in figure T1?
A. Resistor
B. Transistor
C. Lamp
D. Ground symbol

T6C05

What is component 4 in figure T1?
A. Resistor
B. Transistor
C. Battery
D. Ground symbol

T6C06

What is component 6 in figure T2?
A. Resistor
B. Capacitor
C. Regulator IC
D. Transistor

T6C07

What is component 8 in figure T2?
A. Resistor
B. Inductor
C. Regulator IC
D. Light emitting diode

T6C08

What is component 9 in figure T2?
A. Variable capacitor
B. Variable inductor
C. Variable resistor
D. Variable transformer

T6C09

What is component 4 in figure T2?
A. Variable inductor
B. Double-pole switch
C. Potentiometer
D. Transformer

T6C10

What is component 3 in figure T3?
A. Connector
B. Meter
C. Variable capacitor
D. Variable inductor

T6C11

What is component 4 in figure T3?
A. Antenna
B. Transmitter
C. Dummy load
D. Ground

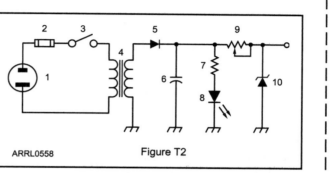

ARRL0558   Figure T2

Figure T2 — Refer to this figure for questions T2C06 through T2C09 and T6D03.

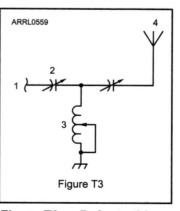

ARRL0559   Figure T3

Figure T3 — Refer to this figure for questions T2C10 and T6C11.

T6C05
(C)
Page 3-13

T6C06
(B)
Page 3-13

T6C07
(D)
Page 3-13

T6C08
(C)
Page 3-13

T6C09
(D)
Page 3-13

T6C10
(D)
Page 3-13

T6C11
(A)
Page 3-13

T6C12
(A)
Page 3-13

T6C12

What do the symbols on an electrical circuit schematic diagram represent?

A. Electrical components
B. Logic states
C. Digital codes
D. Traffic nodes

T6C13
(C)
Page 3-14

T6C13

Which of the following is accurately represented in electrical circuit schematic diagrams?

A. Wire lengths
B. Physical appearance of components
C. The way components are interconnected
D. All of these choices are correct

T6D — Component functions: rectification; switches; indicators; power supply components; resonant circuit; shielding; power transformers; integrated circuits

T6D01
(B)
Page 3-10

T6D01

Which of the following devices or circuits changes an alternating current into a varying direct current signal?

A. Transformer
B. Rectifier
C. Amplifier
D. Reflector

T6D02
(A)
Page 3-12

T6D02

What best describes a relay?

A. A switch controlled by an electromagnet
B. A current controlled amplifier
C. An optical sensor
D. A pass transistor

[Refer to Figure T2 on page 11-45]

T6D03
(A)
Page 3-13

T6D03

What type of switch is represented by component 3 in figure T2?

A. Single-pole single-throw
B. Single-pole double-throw
C. Double-pole single-throw
D. Double-pole double-throw

T6D04
(C)
Page 3-13

T6D04

Which of the following can be used to display signal strength on a numeric scale?

A. Potentiometer
B. Transistor
C. Meter
D. Relay

T6D05

What type of circuit controls the amount of voltage from a power supply?
A. Regulator
B. Oscillator
C. Filter
D. Phase inverter

T6D05
(A)
Page 5-15

T6D06

What component is commonly used to change 120V AC house current to a lower AC voltage for other uses?
A. Variable capacitor
B. Transformer
C. Transistor
D. Diode

T6D06
(B)
Page 3-9

T6D07

Which of the following is commonly used as a visual indicator?
A. LED
B. FET
C. Zener diode
D. Bipolar transistor

T6D07
(A)
Page 3-11

T6D08

Which of the following is used together with an inductor to make a tuned circuit?
A. Resistor
B. Zener diode
C. Potentiometer
D. Capacitor

T6D08
(D)
Page 3-9

T6D09

What is the name of a device that combines several semiconductors and other components into one package?
A. Transducer
B. Multi-pole relay
C. Integrated circuit
D. Transformer

T6D09
(C)
Page 3-11

[Refer to Figure T1 on page 11-44]

T6D10

What is the function of component 2 in Figure T1?
A. Give off light when current flows through it
B. Supply electrical energy
C. Control the flow of current
D. Convert electrical energy into radio waves

T6D10
(C)
Page 3-11

T6D11

What is a simple resonant or tuned circuit?
A. An inductor and a capacitor connected in series or parallel to form a filter
B. A type of voltage regulator
C. A resistor circuit used for reducing standing wave ratio
D. A circuit designed to provide high fidelity audio

T6D11
(A)
Page 3-9

T6D12
(C)
Page 5-22

T6D12

Which of the following is a common reason to use shielded wire?

A. To decrease the resistance of DC power connections
B. To increase the current carrying capability of the wire
C. To prevent coupling of unwanted signals to or from the wire
D. To couple the wire to other signals

SUBELEMENT T7 — Station equipment: common transmitter and receiver problems; antenna measurements; troubleshooting; basic repair and testing
[4 Exam Questions — 4 Groups]

T7A — Station equipment: receivers; transmitters; transceivers; modulation; transverters; low power and weak signal operation; transmit and receive amplifiers

T7A01

Which term describes the ability of a receiver to detect the presence of a signal?
A. Linearity
B. Sensitivity
C. Selectivity
D. Total Harmonic Distortion

T7A02

What is a transceiver?
A. A type of antenna switch
B. A unit combining the functions of a transmitter and a receiver
C. A component in a repeater which filters out unwanted interference
D. A type of antenna matching network

T7A03

Which of the following is used to convert a radio signal from one frequency to another?
A. Phase splitter
B. Mixer
C. Inverter
D. Amplifier

T7A04

Which term describes the ability of a receiver to discriminate between multiple signals?
A. Discrimination ratio
B. Sensitivity
C. Selectivity
D. Harmonic Distortion

T7A05

What is the name of a circuit that generates a signal of a desired frequency?
A. Reactance modulator
B. Product detector
C. Low-pass filter
D. Oscillator

T7A01
(B)
Page 3-18

T7A02
(B)
Page 2-12

T7A03
(B)
Page 3-18

T7A04
(C)
Page 3-18

T7A05
(D)
Page 3-16

T7A06
(C)
Page 3-19

T7A06
What device takes the output of a low-powered 28 MHz SSB exciter and produces a 222 MHz output signal?
A. High-pass filter
B. Low-pass filter
C. Transverter
D. Phase converter

T7A07
(D)
Page 5-6

T7A07
What is meant by the term "PTT"?
A. Pre-transmission tuning to reduce transmitter harmonic emission
B. Precise tone transmissions used to limit repeater access to only certain signals
C. A primary transformer tuner use to match antennas
D. The push to talk function which switches between receive and transmit

T7A08
(C)
Page 3-17

T7A08
Which of the following describes combining speech with an RF carrier signal?-
A. Impedance matching
B. Oscillation
C. Modulation
D. Low-pass filtering

T7A09
(B)
Page 6-28

T7A09
Which of the following devices is most useful for VHF weak-signal communication?
A. A quarter-wave vertical antenna
B. A multi-mode VHF transceiver
C. An omni-directional antenna
D. A mobile VHF FM transceiver

T7A10
(B)
Page 5-8

T7A10
What device increases the low-power output from a handheld transceiver?
A. A voltage divider
B. An RF power amplifier
C. An impedance network
D. All of these choices are correct

T7A11
(A)
Page 3-18

T7A11
Where is an RF preamplifier installed?
A. Between the antenna and receiver
B. At the output of the transmitter's power amplifier
C. Between a transmitter and antenna tuner
D. At the receiver's audio output

T7B — Common transmitter and receiver problems: symptoms of overload and overdrive; distortion; causes of interference; interference and consumer electronics; part 15 devices; over and under modulation; RF feedback; off frequency signals; fading and noise; problems with digital communications interfaces

T7B01

What can you do if you are told your FM handheld or mobile transceiver is over-deviating?

A. Talk louder into the microphone
B. Let the transceiver cool off
C. Change to a higher power level
D. Talk farther away from the microphone

T7B01
(D)
Page 5-4

T7B02

What would cause a broadcast AM or FM radio to receive an amateur radio transmission unintentionally?

A. The receiver is unable to reject strong signals outside the AM or FM band
B. The microphone gain of the transmitter is turned up too high
C. The audio amplifier of the transmitter is overloaded
D. The deviation of an FM transmitter is set too low

T7B02
(A)
Page 5-21

T7B03

Which of the following may be a cause of radio frequency interference?

A. Fundamental overload
B. Harmonics
C. Spurious emissions
D. All of these choices are correct

T7B03
(D)
Page 5-19

T7B04

Which of the following is a way to reduce or eliminate interference by an amateur transmitter to a nearby telephone?

A. Put a filter on the amateur transmitter
B. Reduce the microphone gain
C. Reduce the SWR on the transmitter transmission line
D. Put a RF filter on the telephone

T7B04
(D)
Page 5-21

T7B05

How can overload of a non-amateur radio or TV receiver by an amateur signal be reduced or eliminated?

A. Block the amateur signal with a filter at the antenna input of the affected receiver
B. Block the interfering signal with a filter on the amateur transmitter
C. Switch the transmitter from FM to SSB
D. Switch the transmitter to a narrow-band mode

T7B05
(A)
Page 5-21

T7B06

Which of the following actions should you take if a neighbor tells you that your station's transmissions are interfering with their radio or TV reception?

A. Make sure that your station is functioning properly and that it does not cause interference to your own radio or television when it is tuned to the same channel
B. Immediately turn off your transmitter and contact the nearest FCC office for assistance
C. Tell them that your license gives you the right to transmit and nothing can be done to reduce the interference
D. Install a harmonic doubler on the output of your transmitter and tune it until the interference is eliminated

T7B06
(A)
Page 5-22

T7B07
(D)
Page 5-19

T7B07
Which of the following may be useful in correcting a radio frequency interference problem?
A. Snap-on ferrite chokes
B. Low-pass and high-pass filters
C. Band-reject and band-pass filters
D. All of these choices are correct

T7B08
(D)
Page 5-23

T7B08
What should you do if something in a neighbor's home is causing harmful interference to your amateur station?
A. Work with your neighbor to identify the offending device
B. Politely inform your neighbor about the rules that prohibit the use of devices which cause interference
C. Check your station and make sure it meets the standards of good amateur practice
D. All of these choices are correct

T7B09
(A)
Page 5-23

T7B09
What is a Part 15 device?
A. An unlicensed device that may emit low powered radio signals on frequencies used by a licensed service
B. A type of amateur radio that can legally be used in the citizen's band
C. A device for long distance communications using special codes sanctioned by the International Amateur Radio Union
D. A type of test set used to determine whether a transmitter is in compliance with FCC regulation 91.15

T7B10
(D)
Page 6-12

T7B10
What might be the problem if you receive a report that your audio signal through the repeater is distorted or unintelligible?
A. Your transmitter may be slightly off frequency
B. Your batteries may be running low
C. You could be in a bad location
D. All of these choices are correct

T7B11
(C)
Page 5-24

T7B11
What is a symptom of RF feedback in a transmitter or transceiver?
A. Excessive SWR at the antenna connection
B. The transmitter will not stay on the desired frequency
C. Reports of garbled, distorted, or unintelligible transmissions
D. Frequent blowing of power supply fuses

T7B12
(D)
Page 5-21

T7B12
What might be the first step to resolve cable TV interference from your ham radio transmission?
A. Add a low pass filter to the TV antenna input
B. Add a high pass filter to the TV antenna input
C. Add a preamplifier to the TV antenna input
D. Be sure all TV coaxial connectors are installed properly

T7C — Antenna measurements and troubleshooting: measuring SWR; dummy loads; coaxial cables; feed line failure modes

T7C01

What is the primary purpose of a dummy load?
A. To prevent the radiation of signals when making tests
B. To prevent over-modulation of your transmitter
C. To improve the radiation from your antenna
D. To improve the signal to noise ratio of your receiver

T7C01
(A)
Page 5-4

T7C02

Which of the following instruments can be used to determine if an antenna is resonant at the desired operating frequency?
A. A VTVM
B. An antenna analyzer
C. A Q meter
D. A frequency counter

T7C02
(B)
Page 4-19

T7C03

What, in general terms, is standing wave ratio (SWR)?
A. A measure of how well a load is matched to a transmission line
B. The ratio of high to low impedance in a feed line
C. The transmitter efficiency ratio
D. An indication of the quality of your station's ground connection

T7C03
(A)
Page 4-10

T7C04

What reading on an SWR meter indicates a perfect impedance match between the antenna and the feed line?
A. 2 to 1
B. 1 to 3
C. 1 to 1
D. 10 to 1

T7C04
(C)
Page 4-10

T7C05

What is the approximate SWR value above which the protection circuits in most solid-state transmitters begin to reduce transmitter power?
A. 2 to 1
B. 1 to 2
C. 6 to 1
D. 10 to 1

T7C05
(A)
Page 4-10

T7C06

What does an SWR reading of 4:1 indicate?
A. Loss of –4 dB
B. Good impedance match
C. Gain of +4 dB
D. Impedance mismatch

T7C06
(D)
Page 4-10

T7C07
(C)
Page 4-8

T7C07
What happens to power lost in a feed line?
A. It increases the SWR
B. It comes back into your transmitter and could cause damage
C. It is converted into heat
D. It can cause distortion of your signal

T7C08
(D)
Page 4-18

T7C08
What instrument other than an SWR meter could you use to determine if a feed line and antenna are properly matched?
A. Voltmeter
B. Ohmmeter
C. Iambic pentameter
D. Directional wattmeter

T7C09
(A)
Page 4-16

T7C09
Which of the following is the most common cause for failure of coaxial cables?
A. Moisture contamination
B. Gamma rays
C. The velocity factor exceeds 1.0
D. Overloading

T7C10
(D)
Page 4-16

T7C10
Why should the outer jacket of coaxial cable be resistant to ultraviolet light?
A. Ultraviolet resistant jackets prevent harmonic radiation
B. Ultraviolet light can increase losses in the cable's jacket
C. Ultraviolet and RF signals can mix together, causing interference
D. Ultraviolet light can damage the jacket and allow water to enter the cable

T7C11
(C)
Page 4-17

T7C11
What is a disadvantage of air core coaxial cable when compared to foam or solid dielectric types?
A. It has more loss per foot
B. It cannot be used for VHF or UHF antennas
C. It requires special techniques to prevent water absorption
D. It cannot be used at below freezing temperatures

T7C12
(B)
Page 4-9

T7C12
Which of the following is a common use of coaxial cable?
A. Carrying dc power from a vehicle battery to a mobile radio
B. Carrying RF signals between a radio and antenna
C. Securing masts, tubing, and other cylindrical objects on towers
D. Connecting data signals from a TNC to a computer

T7C13
(B)
Page 5-4

T7C13
What does a dummy load consist of?
A. A high-gain amplifier and a TR switch
B. A non-inductive resistor and a heat sink
C. A low voltage power supply and a DC relay
D. A 50 ohm reactance used to terminate a transmission line

T7D — Basic repair and testing: soldering; using basic test instruments; connecting a voltmeter, ammeter, or ohmmeter

T7D01

Which instrument would you use to measure electric potential or electromotive force?

A. An ammeter
B. A voltmeter
C. A wavemeter
D. An ohmmeter

T7D01
(B)
Page 3-1

T7D02

What is the correct way to connect a voltmeter to a circuit?

A. In series with the circuit
B. In parallel with the circuit
C. In quadrature with the circuit
D. In phase with the circuit

T7D02
(B)
Page 3-3

T7D03

How is an ammeter usually connected to a circuit?

A. In series with the circuit
B. In parallel with the circuit
C. In quadrature with the circuit
D. In phase with the circuit

T7D03
(A)
Page 3-3

T7D04

Which instrument is used to measure electric current?

A. An ohmmeter
B. A wavemeter
C. A voltmeter
D. An ammeter

T7D04
(D)
Page 3-1

T7D05

What instrument is used to measure resistance?

A. An oscilloscope
B. A spectrum analyzer
C. A noise bridge
D. An ohmmeter

T7D05
(D)
Page 3-4

T7D06

Which of the following might damage a multimeter?

A. Measuring a voltage too small for the chosen scale
B. Leaving the meter in the milliamps position overnight
C. Attempting to measure voltage when using the resistance setting
D. Not allowing it to warm up properly

T7D06
(C)
Page 3-3

T7D07

Which of the following measurements are commonly made using a multimeter?

A. SWR and RF power
B. Signal strength and noise
C. Impedance and reactance
D. Voltage and resistance

T7D07
(D)
Page 3-3

T7D08
(C)
Page 4-17

T7D08

Which of the following types of solder is best for radio and electronic use?

A. Acid-core solder
B. Silver solder
C. Rosin-core solder
D. Aluminum solder

T7D09
(C)
Page 4-17

T7D09

What is the characteristic appearance of a cold solder joint?

A. Dark black spots
B. A bright or shiny surface
C. A grainy or dull surface
D. A greenish tint

T7D10
(B)
Page 3-3

T7D10

What is probably happening when an ohmmeter, connected across an unpowered circuit, initially indicates a low resistance and then shows increasing resistance with time?

A. The ohmmeter is defective
B. The circuit contains a large capacitor
C. The circuit contains a large inductor
D. The circuit is a relaxation oscillator

T7D11
(B)
Page 3-3

T7D11

Which of the following precautions should be taken when measuring circuit resistance with an ohmmeter?

A. Ensure that the applied voltages are correct
B. Ensure that the circuit is not powered
C. Ensure that the circuit is grounded
D. Ensure that the circuit is operating at the correct frequency

T7D12
(B)
Page 3-3

T7D12

Which of the following precautions should be taken when measuring high voltages with a voltmeter?

A. Ensure that the voltmeter has very low impedance
B. Ensure that the voltmeter and leads are rated for use at the voltages to be measured
C. Ensure that the circuit is grounded through the voltmeter
D. Ensure that the voltmeter is set to the correct frequency

SUBELEMENT T8 — Modulation modes: amateur satellite operation; operating activities; non-voice communications
[4 Exam Questions — 4 Groups]

T8A — Modulation modes: bandwidth of various signals; choice of emission type

T8A01
Which of the following is a form of amplitude modulation?
A. Spread-spectrum
B. Packet radio
C. Single sideband
D. Phase shift keying

T8A01
(C)
Page 2-9

T8A02
What type of modulation is most commonly used for VHF packet radio transmissions?
A. FM
B. SSB
C. AM
D. Spread Spectrum

T8A02
(A)
Page 2-10

T8A03
Which type of voice mode is most often used for long-distance (weak signal) contacts on the VHF and UHF bands?
A. FM
B. DRM
C. SSB
D. PM

T8A03
(C)
Page 2-11

T8A04
Which type of modulation is most commonly used for VHF and UHF voice repeaters?
A. AM
B. SSB
C. PSK
D. FM

T8A04
(D)
Page 2-10

T8A05
Which of the following types of emission has the narrowest bandwidth?
A. FM voice
B. SSB voice
C. CW
D. Slow-scan TV

T8A05
(C)
Page 2-10

T8A06
Which sideband is normally used for 10 meter HF, VHF and UHF single-sideband communications?
A. Upper sideband
B. Lower sideband
C. Suppressed sideband
D. Inverted sideband

T8A06
(A)
Page 2-11

T8A07
(C)
Page 2-11

T8A07
What is the primary advantage of single sideband over FM for voice transmissions?
A. SSB signals are easier to tune
B. SSB signals are less susceptible to interference
C. SSB signals have narrower bandwidth
D. All of these choices are correct

T8A08
(B)
Page 2-10

T8A08
What is the approximate bandwidth of a single sideband voice signal?
A. 1 kHz
B. 3 kHz
C. 6 kHz
D. 15 kHz

T8A09
(C)
Page 2-10

T8A09
What is the approximate bandwidth of a VHF repeater FM phone signal?
A. Less than 500 Hz
B. About 150 kHz
C. Between 10 and 15 kHz
D. Between 50 and 125 kHz

T8A10
(B)
Page 2-10

T8A10
What is the typical bandwidth of analog fast-scan TV transmissions on the 70 cm band?
A. More than 10 MHz
B. About 6 MHz
C. About 3 MHz
D. About 1 MHz

T8A11
(B)
Page 2-10

T8A11
What is the approximate maximum bandwidth required to transmit a CW signal?
A. 2.4 kHz
B. 150 Hz
C. 1000 Hz
D. 15 kHz

T8B — Amateur satellite operation; Doppler shift, basic orbits, operating protocols; control operator, transmitter power considerations; satellite tracking; digital modes

T8B01
(D)
[97.301, 97.207(c)]
Page 6-30

T8B01
Who may be the control operator of a station communicating through an amateur satellite or space station?
A. Only an Amateur Extra Class operator
B. A General Class licensee or higher licensee who has a satellite operator certification
C. Only an Amateur Extra Class operator who is also an AMSAT member
D. Any amateur whose license privileges allow them to transmit on the satellite uplink frequency

T8B02

How much transmitter power should be used on the uplink frequency of an amateur satellite or space station?

A. The maximum power of your transmitter
B. The minimum amount of power needed to complete the contact
C. No more than half the rating of your linear amplifier
D. Never more than 1 watt

T8B02
(B)
[97.313]
Page 6-31

T8B03

Which of the following are provided by satellite tracking programs?

A. Maps showing the real-time position of the satellite track over the earth
B. The time, azimuth, and elevation of the start, maximum altitude, and end of a pass
C. The apparent frequency of the satellite transmission, including effects of Doppler shift
D. All of these answers are correct

T8B03
(D)
Page 6-31

T8B04

Which amateur stations may make contact with an amateur station on the International Space Station using 2 meter and 70 cm band amateur radio frequencies?

A. Only members of amateur radio clubs at NASA facilities
B. Any amateur holding a Technician or higher class license
C. Only the astronaut's family members who are hams
D. You cannot talk to the ISS on amateur radio frequencies

T8B04
(B)
[97.301,
97.207(c)]
Page 6-30

T8B05

What is a satellite beacon?

A. The primary transmit antenna on the satellite
B. An indicator light that shows where to point your antenna
C. A reflective surface on the satellite
D. A transmission from a space station that contains information about a satellite

T8B05
(D)
Page 6-30

T8B06

Which of the following are inputs to a satellite tracking program?

A. The weight of the satellite
B. The Keplerian elements
C. The last observed time of zero Doppler shift
D. All of these answers are correct

T8B06
(B)
Page 6-31

T8B07

With regard to satellite communications, what is Doppler shift?

A. A change in the satellite orbit
B. A mode where the satellite receives signals on one band and transmits on another
C. An observed change in signal frequency caused by relative motion between the satellite and the earth station
D. A special digital communications mode for some satellites

T8B07
(C)
Page 6-30

T8B08

What is meant by the statement that a satellite is operating in mode U/V?

A. The satellite uplink is in the 15 meter band and the downlink is in the 10 meter band
B. The satellite uplink is in the 70 cm band and the downlink is in the 2 meter band
C. The satellite operates using ultraviolet frequencies
D. The satellite frequencies are usually variable

T8B08
(B)
Page 6-31

T8B09
(B)
Page 6-31

T8B09

What causes spin fading when referring to satellite signals?
A. Circular polarized noise interference radiated from the sun
B. Rotation of the satellite and its antennas
C. Doppler shift of the received signal
D. Interfering signals within the satellite uplink band

T8B10
(C)
Page 6-30

T8B10

What do the initials LEO tell you about an amateur satellite?
A. The satellite battery is in Low Energy Operation mode
B. The satellite is performing a Lunar Ejection Orbit maneuver
C. The satellite is in a Low Earth Orbit
D. The satellite uses Light Emitting Optics

T8B11
(C)
Page 6-31

T8B11

What is a commonly used method of sending signals to and from a digital satellite?
A. USB AFSK
B. PSK31
C. FM Packet
D. WSJT

T8C — Operating activities: radio direction finding; radio control; contests; linking over the Internet; grid locators

T8C01
(C)
Page 6-29

T8C01

Which of the following methods is used to locate sources of noise interference or jamming?
A. Echolocation
B. Doppler radar
C. Radio direction finding
D. Phase locking

T8C02
(B)
Page 6-29

T8C02

Which of these items would be useful for a hidden transmitter hunt?
A. Calibrated SWR meter
B. A directional antenna
C. A calibrated noise bridge
D. All of these choices are correct

T8C03
(A)
Page 6-28

T8C03

What popular operating activity involves contacting as many stations as possible during a specified period of time?
A. Contesting
B. Net operations
C. Public service events
D. Simulated emergency exercises

T8C04
(C)
Page 6-28

T8C04

Which of the following is good procedure when contacting another station in a radio contest?
A. Be sure to sign only the last two letters of your call if there is a pileup calling the station
B. Work the station twice to be sure that you are in his log
C. Send only the minimum information needed for proper identification and the contest exchange
D. All of these choices are correct

T8C05

What is a grid locator?

A. A letter-number designator assigned to a geographic location
B. A letter-number designator assigned to an azimuth and elevation
C. An instrument for neutralizing a final amplifier
D. An instrument for radio direction finding

T8C05
(A)
Page 6-4

T8C06

How is access to an IRLP node accomplished?

A. By obtaining a password which is sent via voice to the node
B. By using DTMF signals
C. By entering the proper Internet password
D. By using CTCSS tone codes

T8C06
(B)
Page 6-19

T8C07

What is the maximum power allowed when transmitting telecommand signals to radio controlled models?

A. 500 milliwatts
B. 1 watt
C. 25 watts
D. 1500 watts

T8C07
(B)
[97.215(c)]
Page 6-33

T8C08

What is required in place of on-air station identification when sending signals to a radio control model using amateur frequencies?

A. Voice identification must be transmitted every 10 minutes
B. Morse code ID must be sent once per hour
C. A label indicating the licensee's name, call sign and address must be affixed to the transmitter
D. A flag must be affixed to the transmitter antenna with the station call sign in 1 inch high letters or larger

T8C08
(C)
[97.215(a)]
Page 6-33

T8C09

How might you obtain a list of active nodes that use VoIP?

A. From the FCC Rulebook
B. From your local emergency coordinator
C. From a repeater directory
D. From the local repeater frequency coordinator

T8C09
(C)
Page 6-19

T8C10

How do you select a specific IRLP node when using a portable transceiver?

A. Choose a specific CTCSS tone
B. Choose the correct DSC tone
C. Access the repeater autopatch
D. Use the keypad to transmit the IRLP node ID

T8C10
(D)
Page 6-19

T8C11

What name is given to an amateur radio station that is used to connect other amateur stations to the Internet?

A. A gateway
B. A repeater
C. A digipeater
D. A beacon

T8C11
(A)
Page 5-14

T8C12
(D)
Page 6-19

T8C12

What is meant by Voice Over Internet Protocol (VoIP) as used in amateur radio?

A. A set of rules specifying how to identify your station when linked over the Internet to another station
B. A set of guidelines for working DX during contests using Internet access
C. A technique for measuring the modulation quality of a transmitter using remote sites monitored via the Internet
D. A method of delivering voice communications over the Internet using digital techniques

T8C13
(A)
Page 6-18

T8C13

What is the Internet Radio Linking Project (IRLP)?

A. A technique to connect amateur radio systems, such as repeaters, via the Internet using Voice Over Internet Protocol
B. A system for providing access to websites via amateur radio
C. A system for informing amateurs in real time of the frequency of active DX stations
D. A technique for measuring signal strength of an amateur transmitter via the Internet

T8D — Non-voice communications: image signals; digital modes; CW; packet; PSK31; APRS; error detection and correction; NTSC

T8D01
(D)
Page 5-9

T8D01

Which of the following is an example of a digital communications method?

A. Packet
B. PSK31
C. MFSK
D. All of these choices are correct

T8D02
(A)
Page 5-11

T8D02

What does the term "APRS" mean?

A. Automatic Packet Reporting System
B. Associated Public Radio Station
C. Auto Planning Radio Set-up
D. Advanced Polar Radio System

T8D03
(D)
Page 5-11

T8D03

Which of the following devices provides data to the transmitter when sending automatic position reports from a mobile amateur radio station?

A. The vehicle speedometer
B. A WWV receiver
C. A connection to a broadcast FM sub-carrier receiver
D. A Global Positioning System receiver

T8D04
(C)
Page 6-32

T8D04

What type of transmission is indicated by the term NTSC?

A. A Normal Transmission mode in Static Circuit
B. A special mode for earth satellite uplink
C. An analog fast scan color TV signal
D. A frame compression scheme for TV signals

T8D05

Which of the following is an application of APRS (Automatic Packet Reporting System)?
A. Providing real time tactical digital communications in conjunction with a map showing the locations of stations
B. Showing automatically the number of packets transmitted via PACTOR during a specific time interval
C. Providing voice over Internet connection between repeaters
D. Providing information on the number of stations signed into a repeater

T8D06

What does the abbreviation PSK mean?
A. Pulse Shift Keying
B. Phase Shift Keying
C. Packet Short Keying
D. Phased Slide Keying

T8D07

What is PSK31?
A. A high-rate data transmission mode
B. A method of reducing noise interference to FM signals
C. A method of compressing digital television signals
D. A low-rate data transmission mode

T8D08

Which of the following may be included in packet transmissions?
A. A check sum which permits error detection
B. A header which contains the call sign of the station to which the information is being sent
C. Automatic repeat request in case of error
D. All of these choices are correct

T8D09

What code is used when sending CW in the amateur bands?
A. Baudot
B. Hamming
C. International Morse
D. Gray

T8D10

Which of the following can be used to transmit CW in the amateur bands?
A. Straight Key
B. Electronic Keyer
C. Computer Keyboard
D. All of these choices are correct

T8D11

What is an ARQ transmission system?
A. A special transmission format limited to video signals
B. A system used to encrypt command signals to an amateur radio satellite
C. A digital scheme whereby the receiving station detects errors and sends a request to the sending station to retransmit the information
D. A method of compressing the data in a message so more information can be sent in a shorter time

T8D05
(A)
Page 5-11

T8D06
(B)
Page 5-11

T8D07
(D)
Page 5-11

T8D08
(D)
Page 5-10

T8D09
(C)
Page 5-9

T8D10
(D)
Page 5-6

T8D11
(C)
Page 5-10

SUBELEMENT T9 — Antennas and feed lines
[2 Exam Questions — 2 Groups]

T9A — Antennas: vertical and horizontal polarization; concept of gain; common portable and mobile antennas; relationships between antenna length and frequency

T9A01
(C)
Page 4-14

T9A01

What is a beam antenna?
A. An antenna built from aluminum I-beams
B. An omnidirectional antenna invented by Clarence Beam
C. An antenna that concentrates signals in one direction
D. An antenna that reverses the phase of received signals

T9A02
(B)
Page 4-6

T9A02

Which of the following is true regarding vertical antennas?
A. The magnetic field is perpendicular to the Earth
B. The electric field is perpendicular to the Earth
C. The phase is inverted
D. The phase is reversed

T9A03
(B)
Page 4-11

T9A03

Which of the following describes a simple dipole mounted so the conductor is parallel to the Earth's surface?
A. A ground wave antenna
B. A horizontally polarized antenna
C. A rhombic antenna
D. A vertically polarized antenna

T9A04
(A)
Page 4-13

T9A04

What is a disadvantage of the "rubber duck" antenna supplied with most handheld radio transceivers?
A. It does not transmit or receive as effectively as a full-sized antenna
B. It transmits a circularly polarized signal
C. If the rubber end cap is lost it will unravel very quickly
D. All of these choices are correct

T9A05
(C)
Page 4-12

T9A05

How would you change a dipole antenna to make it resonant on a higher frequency?
A. Lengthen it
B. Insert coils in series with radiating wires
C. Shorten it
D. Add capacitive loading to the ends of the radiating wires

T9A06
(C)
Page 4-15

T9A06

What type of antennas are the quad, Yagi, and dish?
A. Non-resonant antennas
B. Loop antennas
C. Directional antennas
D. Isotropic antennas

T9A07

What is a good reason not to use a "rubber duck" antenna inside your car?
A. Signals can be significantly weaker than when it is outside of the vehicle
B. It might cause your radio to overheat
C. The SWR might decrease, decreasing the signal strength
D. All of these choices are correct

T9A07
(A)
Page 4-13

T9A08

What is the approximate length, in inches, of a quarter-wavelength vertical antenna for 146 MHz?
A. 112
B. 50
C. 19
D. 12

T9A08
(C)
Page 4-11

T9A09

What is the approximate length, in inches, of a 6 meter ½-wavelength wire dipole antenna?
A. 6
B. 50
C. 112
D. 236

T9A09
(C)
Page 4-11

T9A10

In which direction is the radiation strongest from a half-wave dipole antenna in free space?
A. Equally in all directions
B. Off the ends of the antenna
C. Broadside to the antenna
D. In the direction of the feed line

T9A10
(C)
Page 4-11

T9A11

What is meant by the gain of an antenna?
A. The additional power that is added to the transmitter power
B. The additional power that is lost in the antenna when transmitting on a higher frequency
C. The increase in signal strength in a specified direction when compared to a reference antenna
D. The increase in impedance on receive or transmit compared to a reference antenna

T9A11
(C)
Page 4-6

T9A12

What is a reason to use a properly mounted ⅝ wavelength antenna for VHF or UHF mobile service?
A. It offers a lower angle of radiation and more gain than a ¼ wavelength antenna and usually provides improved coverage
B. It features a very high angle of radiation and is better for communicating via a repeater
C. The ⅝ wavelength antenna completely eliminates distortion caused by reflected signals
D. The ⅝ wavelength antenna offers a 10-times power gain over a ¼ wavelength design

T9A12
(A)
Page 4-13

T9A13

Why are VHF or UHF mobile antennas often mounted in the center of the vehicle roof?
A. Roof mounts have the lowest possible SWR of any mounting configuration
B. Only roof mounting can guarantee a vertically polarized signal
C. A roof mounted antenna normally provides the most uniform radiation pattern
D. Roof mounted antennas are always the easiest to install

T9A13
(C)
Page 4-13

T9A14

Which of the following terms describes a type of loading when referring to an antenna?
A. Inserting an inductor in the radiating portion of the antenna to make it electrically longer
B. Inserting a resistor in the radiating portion of the antenna to make it resonant
C. Installing a spring at the base of the antenna to absorb the effects of collisions with other objects
D. Making the antenna heavier so it will resist wind effects when in motion

T9B — Feed lines: types of feed lines; attenuation vs. frequency; SWR concepts; matching; weather protection; choosing RF connectors and feed lines

T9B01

Why is it important to have a low SWR in an antenna system that uses coaxial cable feed line?
A. To reduce television interference
B. To allow the efficient transfer of power and reduce losses
C. To prolong antenna life
D. All of these choices are correct

T9B02

What is the impedance of the most commonly used coaxial cable in typical amateur radio installations?
A. 8 ohms
B. 50 ohms
C. 600 ohms
D. 12 ohms

T9B03

Why is coaxial cable used more often than any other feed line for amateur radio antenna systems?
A. It is easy to use and requires few special installation considerations
B. It has less loss than any other type of feed line
C. It can handle more power than any other type of feed line
D. It is less expensive than any other types of feed line

T9B04

What does an antenna tuner do?
A. It matches the antenna system impedance to the transceiver's output impedance
B. It helps a receiver automatically tune in weak stations
C. It allows an antenna to be used on both transmit and receive
D. It automatically selects the proper antenna for the frequency band being used

T9B05

What generally happens as the frequency of a signal passing through coaxial cable is increased?
A. The apparent SWR increases
B. The reflected power increases
C. The characteristic impedance increases
D. The loss increases

T9B06

Which of the following connectors is most suitable for frequencies above 400 MHz?

A. A UHF (PL-259/SO-239) connector
B. A Type N connector
C. An RS-213 connector
D. A DB-25 connector

T9B06
(B)
Page 4-17

T9B07

Which of the following is true of PL-259 type coax connectors?

A. They are preferred for microwave operation
B. They are water tight
C. They are commonly used at HF frequencies
D. They are a bayonet type connector

T9B07
(C)
Page 4-17

T9B08

Why should coax connectors exposed to the weather be sealed against water intrusion?

A. To prevent an increase in feed line loss
B. To prevent interference to telephones
C. To keep the jacket from becoming loose
D. All of these choices are correct

T9B08
(A)
Page 4-17

T9B09

What might cause erratic changes in SWR readings?

A. The transmitter is being modulated
B. A loose connection in an antenna or a feed line
C. The transmitter is being over-modulated
D. Interference from other stations is distorting your signal

T9B09
(B)
Page 4-10

T9B10

What electrical difference exists between the smaller RG-58 and larger RG-8 coaxial cables?

A. There is no significant difference between the two types
B. RG-58 cable has less loss at a given frequency
C. RG-8 cable has less loss at a given frequency
D. RG-58 cable can handle higher power levels

T9B10
(C)
Page 4-16

T9B11

Which of the following types of feed line has the lowest loss at VHF and UHF?

A. 50-ohm flexible coax
B. Multi-conductor unbalanced cable
C. Air-insulated hard line
D. 75-ohm flexible coax

T9B11
(C)
Page 4-9

SUBELEMENT T0 — Electrical safety: AC and DC power circuits; antenna installation; RF hazards
[3 Exam Questions — 3 Groups]

T0A — Power circuits and hazards: hazardous voltages; fuses and circuit breakers; grounding; lightning protection; battery safety; electrical code compliance

T0A01
(B)
Page 9-3

T0A01

Which of the following is a safety hazard of a 12-volt storage battery?

A. Touching both terminals with the hands can cause electrical shock
B. Shorting the terminals can cause burns, fire, or an explosion
C. RF emissions from the battery
D. All of these choices are correct

T0A02
(D)
Page 9-2

T0A02

How does current flowing through the body cause a health hazard?

A. By heating tissue
B. It disrupts the electrical functions of cells
C. It causes involuntary muscle contractions
D. All of these choices are correct

T0A03
(C)
Page 9-4

T0A03

What is connected to the green wire in a three-wire electrical AC plug?

A. Neutral
B. Hot
C. Safety ground
D. The white wire

T0A04
(B)
Page 3-12

T0A04

What is the purpose of a fuse in an electrical circuit?

A. To prevent power supply ripple from damaging a circuit
B. To interrupt power in case of overload
C. To limit current to prevent shocks
D. All of these choices are correct

T0A05
(C)
Page 3-12

T0A05

Why is it unwise to install a 20-ampere fuse in the place of a 5-ampere fuse?

A. The larger fuse would be likely to blow because it is rated for higher current
B. The power supply ripple would greatly increase
C. Excessive current could cause a fire
D. All of these choices are correct

T0A06
(D)
Page 9-3

T0A06

What is a good way to guard against electrical shock at your station?

A. Use three-wire cords and plugs for all AC powered equipment
B. Connect all AC powered station equipment to a common safety ground
C. Use a circuit protected by a ground-fault interrupter
D. All of these choices are correct

T0A07

Which of these precautions should be taken when installing devices for lightning protection in a coaxial cable feed line?

A. Include a parallel bypass switch for each protector so that it can be switched out of the circuit when running high power
B. Include a series switch in the ground line of each protector to prevent RF overload from inadvertently damaging the protector
C. Keep the ground wires from each protector separate and connected to station ground
D. Ground all of the protectors to a common plate which is in turn connected to an external ground

T0A07
(D)
Page 9-5

T0A08

What safety equipment should always be included in home-built equipment that is powered from 120V AC power circuits?

A. A fuse or circuit breaker in series with the AC hot conductor
B. An AC voltmeter across the incoming power source
C. An inductor in series with the AC power source
D. A capacitor across the AC power source

T0A08
(A)
Page 9-4

T0A09

What kind of hazard is presented by a conventional 12-volt storage battery?

A. It emits ozone which can be harmful to the atmosphere
B. Shock hazard due to high voltage
C. Explosive gas can collect if not properly vented
D. All of these choices are correct

T0A09
(C)
Page 5-18

T0A10

What can happen if a lead-acid storage battery is charged or discharged too quickly?

A. The battery could overheat and give off flammable gas or explode
B. The voltage can become reversed
C. The memory effect will reduce the capacity of the battery
D. All of these choices are correct

T0A10
(A)
Page 5-18

T0A11

What kind of hazard might exist in a power supply when it is turned off and disconnected?

A. Static electricity could damage the grounding system
B. Circulating currents inside the transformer might cause damage
C. The fuse might blow if you remove the cover
D. You might receive an electric shock from the charge stored in large capacitors

T0A11
(D)
Page 9-3

T0B — Antenna safety: tower safety; erecting an antenna support; overhead power lines; installing an antenna

T0B01

When should members of a tower work team wear a hard hat and safety glasses?

A. At all times except when climbing the tower
B. At all times except when belted firmly to the tower
C. At all times when any work is being done on the tower
D. Only when the tower exceeds 30 feet in height

T0B01
(C)
Page 9-13

T0B02
(C)
Page 9-13

T0B02

What is a good precaution to observe before climbing an antenna tower?

A. Make sure that you wear a grounded wrist strap
B. Remove all tower grounding connections
C. Put on a climbing harness and safety glasses
D. All of the these choices are correct

T0B03
(D)
Page 9-14

T0B03

Under what circumstances is it safe to climb a tower without a helper or observer?

A. When no electrical work is being performed
B. When no mechanical work is being performed
C. When the work being done is not more than 20 feet above the ground
D. Never

T0B04
(C)
Page 9-12

T0B04

Which of the following is an important safety precaution to observe when putting up an antenna tower?

A. Wear a ground strap connected to your wrist at all times
B. Insulate the base of the tower to avoid lightning strikes
C. Look for and stay clear of any overhead electrical wires
D. All of these choices are correct

T0B05
(C)
Page 9-14

T0B05

What is the purpose of a gin pole?

A. To temporarily replace guy wires
B. To be used in place of a safety harness
C. To lift tower sections or antennas
D. To provide a temporary ground

T0B06
(D)
Page 9-12

T0B06

What is the minimum safe distance from a power line to allow when installing an antenna?

A. Half the width of your property
B. The height of the power line above ground
C. ½ wavelength at the operating frequency
D. So that if the antenna falls unexpectedly, no part of it can come closer than 10 feet to the power wires

T0B07
(C)
Page 9-14

T0B07

Which of the following is an important safety rule to remember when using a crank-up tower?

A. This type of tower must never be painted
B. This type of tower must never be grounded
C. This type of tower must never be climbed unless it is in the fully retracted position
D. All of these choices are correct

T0B08
(C)
Page 9-13

T0B08

What is considered to be a proper grounding method for a tower?

A. A single four-foot ground rod, driven into the ground no more than 12 inches from the base
B. A ferrite-core RF choke connected between the tower and ground
C. Separate eight-foot long ground rods for each tower leg, bonded to the tower and each other
D. A connection between the tower base and a cold water pipe

T0B09

Why should you avoid attaching an antenna to a utility pole?

A. The antenna will not work properly because of induced voltages
B. The utility company will charge you an extra monthly fee
C. The antenna could contact high-voltage power wires
D. All of these choices are correct

T0B09
(C)
Page 9-12

T0B10

Which of the following is true concerning grounding conductors used for lightning protection?

A. Only non-insulated wire must be used
B. Wires must be carefully routed with precise right-angle bends
C. Sharp bends must be avoided
D. Common grounds must be avoided

T0B10
(C)
Page 9-4

T0B11

Which of the following establishes grounding requirements for an amateur radio tower or antenna?

A. FCC Part 97 Rules
B. Local electrical codes
C. FAA tower lighting regulations
D. Underwriters Laboratories' recommended practices

T0B11
(B)
Page 9-4

T0B12

Which of the following is good practice when installing ground wires on a tower for lightning protection?

A. Put a loop in the ground connection to prevent water damage to the ground system
B. Make sure that all bends in the ground wires are clean, right angle bends
C. Ensure that connections are short and direct
D. All of these choices are correct

T0B12
(C)
Page 9-4

T0C — RF hazards: radiation exposure; proximity to antennas; recognized safe power levels; exposure to others; radiation types; duty cycle

T0C01

What type of radiation are VHF and UHF radio signals?

A. Gamma radiation
B. Ionizing radiation
C. Alpha radiation
D. Non-ionizing radiation

T0C01
(D)
Page 9-5

T0C02

Which of the following frequencies has the lowest value for Maximum Permissible Exposure limit?

A. 3.5 MHz
B. 50 MHz
C. 440 MHz
D. 1296 MHz

T0C02
(B)
Page 9-7

T0C03

What is the maximum power level that an amateur radio station may use at VHF frequencies before an RF exposure evaluation is required?
A. 1500 watts PEP transmitter output
B. 1 watt forward power
C. 50 watts PEP at the antenna
D. 50 watts PEP reflected power

T0C04

What factors affect the RF exposure of people near an amateur station antenna?
A. Frequency and power level of the RF field
B. Distance from the antenna to a person
C. Radiation pattern of the antenna
D. All of these choices are correct

T0C05

Why do exposure limits vary with frequency?
A. Lower frequency RF fields have more energy than higher frequency fields
B. Lower frequency RF fields do not penetrate the human body
C. Higher frequency RF fields are transient in nature
D. The human body absorbs more RF energy at some frequencies than at others

T0C06

Which of the following is an acceptable method to determine that your station complies with FCC RF exposure regulations?
A. By calculation based on FCC OET Bulletin 65
B. By calculation based on computer modeling
C. By measurement of field strength using calibrated equipment
D. All of these choices are correct

T0C07

What could happen if a person accidentally touched your antenna while you were transmitting?
A. Touching the antenna could cause television interference
B. They might receive a painful RF burn
C. They might develop radiation poisoning
D. All of these choices are correct

T0C08

Which of the following actions might amateur operators take to prevent exposure to RF radiation in excess of FCC-supplied limits?
A. Relocate antennas
B. Relocate the transmitter
C. Increase the duty cycle
D. All of these choices are correct

T0C09

How can you make sure your station stays in compliance with RF safety regulations?
A. By informing the FCC of any changes made in your station
B. By re-evaluating the station whenever an item of equipment is changed
C. By making sure your antennas have low SWR
D. All of these choices are correct

T0C10

Why is duty cycle one of the factors used to determine safe RF radiation exposure levels?
A. It affects the average exposure of people to radiation
B. It affects the peak exposure of people to radiation
C. It takes into account the antenna feed line loss
D. It takes into account the thermal effects of the final amplifier

T0C10
(A)
Page 9-7

T0C11

What is the definition of duty cycle during the averaging time for RF exposure?
A. The difference between the lowest power output and the highest power output of a transmitter
B. The difference between the PEP and average power output of a transmitter
C. The percentage of time that a transmitter is transmitting
D. The percentage of time that a transmitter is not transmitting

T0C11
(C)
Page 9-7

T0C12

How does RF radiation differ from ionizing radiation (radioactivity)?
A. RF radiation does not have sufficient energy to cause genetic damage
B. RF radiation can only be detected with an RF dosimeter
C. RF radiation is limited in range to a few feet
D. RF radiation is perfectly safe

T0C12
(A)
Page 9-5

T0C13

If the averaging time for exposure is 6 minutes, how much power density is permitted if the signal is present for 3 minutes and absent for 3 minutes rather than being present for the entire 6 minutes?
A. 3 times as much
B. ½ as much
C. 2 times as much
D. There is no adjustment allowed for shorter exposure times

T0C13
(C)
Page 9-7

Choosing a Ham Radio

Your guide to selecting the right equipment

Lead Author—Ward Silver, NØAX;
Co-authors—Greg Widin, KØGW and David Haycock, KI6AWR

- **About This Publication**
- **Types of Operation**
- **VHF/UHF Equipment**
- **HF Equipment**
- **Manufacturer's Directory**

WHO NEEDS THIS PUBLICATION AND WHY?

Hello and welcome to this handy guide to selecting a radio. Choosing just one from the variety of radio models is a challenge! The good news is that most commercially manufactured Amateur Radio equipment performs the basics very well, so you shouldn't be overly concerned about a "wrong" choice of brands or models. This guide is intended to help you make sense of common features and decide which are most important to you. We provide explanations and definitions, along with what a particular feature might mean to you on the air.

This publication is aimed at the new Technician licensee ready to acquire a first radio, a licensee recently upgraded to General Class and wanting to explore HF, or someone getting back into ham radio after a period of inactivity. A technical background is not needed to understand the material.

ABOUT THIS PUBLICATION

After this introduction and a "Quick Start" guide, there are two main sections; one covering gear for the VHF and UHF bands and one for HF band equipment. You'll encounter a number of terms and abbreviations—watch for *italicized* words—so two glossaries are provided; one for the VHF/UHF section and one for the HF section. You'll be comfortable with these terms by the time you've finished reading!

We assume that you'll be buying commercial equipment and accessories as new gear. Used equipment is cheaper, of course, but may have faults or defects with which you might be unfamiliar, leading to problems. Teaming up with an experienced ham or a reputable dealer is the way to evaluate used equipment. Unless you are experienced with electronics, kits and homebuilt rigs are not recommended as a first radio, either. Websites of some radio manufacturers are listed at the end of this document.

What This Publication Is Not

This isn't a traditional "buyer's guide" with feature lists and prices for many radios. Manufacturer's websites and catalogs from radio stores have plenty of information on the latest models and features. You won't find operating instructions or technical specifications here—download brochures and manuals directly from the manufacturer!

Your Best Resource

Your best resource is a knowledgeable friend. Better yet, how about a group of friends? This is where a ham radio club or team can really help. Find local clubs via the ARRL's home page: **www.arrl.org**—enter "Clubs" in the site's search window. A club can help you with every aspect of choosing a radio; from explaining a feature to filling you in on what works best in your area. These helpful *Elmers* (ham radio mentors) may be willing to loan or demonstrate a radio so you can experience different styles of operating before making buying decisions. If the group is supporting a public-service activity, such as a walkathon, parade, or race, volunteer to help so you can see how the radios are actually used.

The "Choosing a Ham Radio" Website

It's not possible to give you all the details in one publication, so a supporting web page has been created at **www.arrl.org/buying-your-first-radio**. It is referenced on a web page for new hams **www.arrl.org/get-on-the-air** that includes handy references to download and print, supplements that will be updated, and links to a more complete glossary and other online resources and books to help you learn even more.

NEW TO HAM RADIO?

If you are new to ham radio with a background in other types of radio communications, you should read the supplement "Ham Ways – A Primer" available on the Buying Your First Radio web page listed in the preceding paragraph. It will explain some of the ways hams communicate that may be unfamiliar. This will help you understand more about ham radio features and why they are important to hams.

ARE YOU READY?

Finally, savor the experience—you'll never buy another "first" radio, so have fun! If things don't turn out exactly as you expect, you can easily sell or trade for different equipment. Most hams try many radios and you probably will, too. So relax and prepare to enjoy the ride! Once you're finished, others can use this publication, too. Give it away or tell them about this guide's website where they can download and print their own copy!

CONTRIBUTORS

The following Elmers contributed by reviewing this publication to make it as useful as possible, their way of giving back to ham radio: Norm, K6YXH; Ken, WA3KD; Dave, KA1HDG; Mike, N4FOZ; James, KG8DZ; Jason, KI6PCN; Barb, N6DNI; Patrick, KI6PCS; Ken, WD9DPK; Mary, KI6TOS; Paula, KI6SAK; Jack, AD7NK; Katie, W1KRB; Sharon, KE7HBZ; Matt, N8MS; Marty, N6VI; Diane, KE7PCS; Charles, KE4SKY and Chris, KB7YOU.

WHAT DO YOU WANT TO DO?

Answering that question starts with the type of activities that interest you, such as emergency communications, casual conversation with friends or club members, or communicating using digital data. You'll need to then consider the range over which you expect to make contacts and the type of station you expect to use.

Where Are The Hams You Want To Contact?

Considering how radio waves of different frequencies *propagate* (travel) helps answer the question of what bands your radio will need. Different bands support different ranges of contacts as shown in **Table 1**. Evaluate the activities in which you want to participate to see where you will need to make contacts. (Glossaries in the *Ham Radio License Manual* or on-line at **www.arrl.org** explain these and many other terms.)

Table 1

Bands and Typical Distances

Range	Bands
Nearby	VHF/UHF
Regional	VHF/UHF (via repeaters) and HF (direct contact)
Country-wide	HF
World-wide	HF (direct contact), VHF/UHF (Internet links)

You may find it useful to start with one of our four common ham "profiles". Your operating needs could be similar to one of them. Once you choose a profile, fill in the details by reading the VHF/UHF or HF sections that follow for details on the various features and functions. Download the comparison form from the Buying Your First Radio web page to organize your shopping list.

Consult local hams to find out what bands are used most commonly. For example, in more sparsely populated areas, the 70 cm band may not be in widespread use. In other areas, another VHF or UHF band – such as 222 MHz – may be in regular use.

Emergency Communication Team Member or Personal Emergency Communications

For operation through easily accessible repeaters or over very short ranges, you'll need a VHF/UHF (dual-band) handheld radio with at least 50 memories and 3 to 5 watts of output power. A spare rechargeable *battery pack*, a pack that holds AA or AAA batteries, an automotive adapter, and a desktop *quick-charger* are must-have accessories. For operation without repeaters or through distant repeaters while mobile or in remote areas, add a VHF/UHF mobile radio with an output of 25 watts or more and a dual-band quarter-wave mobile whip antenna. Many hams have one of each type of radio.

Beginning Home HF Operation

HF transceivers with 100 watts of output and a built-in *antenna tuner* make an excellent entry-level radio. HF radios with VHF and even UHF coverage are available at higher cost. If the radio can't operate directly from ac power you'll need an external dc *power supply*. A multi-band dipole is an excellent and inexpensive antenna, connected to the radio through coaxial cable or open-wire feed line—the latter will require an external antenna tuner. With some compromise in performance, a multi-band, ground-independent vertical antenna is easy to set up and requires coaxial cable to connect to the radio. To try Morse code, add a straight key or *paddle* (most radios have a built-in *keyer*). To operate using digital modes, add a sound card *data interface* with cables made to connect to your radio.

Casual Local or Regional Operating

Start with a VHF/UHF mobile radio with an output of 25 watts or more. For mobile operation, add a dual-band quarter-wave mobile whip antenna. At home, a mobile antenna can be used indoors or a dual-band base antenna can be mounted outside, using coaxial cable to connect the radio and antenna. For operation from ac power, a power supply capable of supplying the radio's specified current consumption is required. If you need more portability, add a VHF/UHF handheld as a second radio—it is common (and practical) to have both types of radio.

Portable or Mobile Operating – All Bands

This type of operation is best supported with an "all-band, all-mode" mobile-sized transceiver that covers the HF and VHF bands, plus the 70 cm band in some models. You'll need two types of antenna; a dual-band quarter-wave mobile whip antenna for 2 meters and 70 cm (there are also tri-band antennas that include 6 meters), and a mobile HF antenna. A permanent or magnetic mount with one or more interchangeable single-band HF whip antenna is a good way to try operating from your vehicle.

Base or Fixed (a permanent location)
Mobile (in a vehicle)
Portable (temporary installation)
Handheld (carried by hand, pocket, or belt clip)

From Where Will You Be Operating?

Radios are designed with sets of features that favor the types of use listed at the side of this page, although they can be used in more than one way. Considering the circumstances in which you expect to do most of your operating and the ranges at which you expect to make contacts will help you decide what type of radio is right for you.

Radios designed for *base* or *fixed-station* are the most capable and powerful. They are also bigger and their standby power requirements are higher. Many of the manufacturers of base-type radios also offer smaller, lighter models for *portable* operating in tight quarters, from a vehicle, while camping or otherwise away from home and with limited power availability. These radios generally don't have the same level of performance or as many front-panel controls as base station models. A *mobile radio* is one intended for use in vehicles, usually on FM voice in the VHF and UHF bands. Mobile "rigs" with a dc power supply make a good VHF/UHF FM voice base station, too. *Handheld* radios are intended to be carried while in use, operate from rechargeable battery packs, and have FM voice transmitter outputs of a few watts on the VHF and UHF bands.

Ready, Set, Go!

Begin by reading this guide from start to finish, even if you think you already have a good idea of what radio you want. You might find some material that changes your thinking a little bit, possibly saving you some money or increasing your long-term satisfaction with the equipment you select. Once you've finished, focus on the frequency bands (VHF/UHF or HF) and type of operating (base, mobile, portable, or handheld). Then make your list of "must-haves" and "nice-to-haves" (useful features that aren't absolutely necessary). Armed with catalogs and a web browser, you're ready to start shopping or take in a *hamfest* (a ham radio flea market and convention)! A blank form to help you compare different radios is available for downloading from the Buying Your First Radio web page.

VHF/UHF EQUIPMENT

The VHF and UHF bands above 30 MHz are available to all classes of ham licensees and are widely used for local and regional contacts. The majority of activity uses FM voice on repeaters and simplex channels at the higher frequency segment of each band. The lowest frequency segments in each band are set aside for *weak-signal* operating where longer range contacts are made using SSB voice and Morse code. (SSB and Morse offer better performance than FM for contacts made over long distances without the aid of a repeater to relay weak signals.) In adjacent segments you'll also find Amateur Radio satellite signals and data communication using digital modes, among other activities.

For FM voice in an area with good repeater coverage, a handheld radio provides the maximum flexibility in operating—from home, a vehicle, or while on foot. It's more practical, however, to purchase a higher-powered mobile radio that can be used at home, too. Most hams have a mobile radio in their vehicle plus a handheld radio for portable operation.

If you are going to operate primarily from home, an *all-mode* radio designed for fixed- or base station use may be a better choice. While larger and more expensive than mobile and handheld radios, they also operate on SSB and CW (Morse code) and offer better receiver performance, larger displays, and easier access to many controls and functions.

VHF/UHF GLOSSARY

Airband: VHF channels for aviation air-to-air and air-to-ground communications

All-mode: radio that can operate on AM, SSB, CW, digital modes and FM

APRS: Automatic Packet Reporting System

ARES®: Amateur Radio Emergency Service, sponsored by the ARRL's Field Organization

Attenuate (attenuation): reduce in strength

Auto-patch: connection ("patch") between a radio and the telephone system

Automotive adapter: device that plugs into a vehicle lighter socket and supplies power to a radio or electronic device

Battery pack: several battery cells connected together to act as a single, larger battery

Beam: antenna with gain primarily in one direction

Charger: device for recharging batteries

Cloning: duplicating the memory contents of a radio in another radio

Cross-band: receiving on one band and transmitting on another

CW (Continuous-wave): Morse Code

dB (decibels): logarithmic method of comparing two signal strengths (power, voltage, current)

Digital mode: communication method that exchanges characters instead of voice or CW

DTMF: Dual-Tone, Multi-Frequency, signaling tones ("TouchTones®") used by telephone systems

Duplex: transmitting on one frequency and receiving on another in the same band

Energy density: amount of energy a battery stores per amount of weight or volume

Feed line: cable used to transfer radio-frequency energy

FRS: Family Radio Service, short-range handheld radios that can be used without a license

GMRS: General Mobile Radio Service, a no-test licensed service in the UHF range for family use

Gain (antenna): antenna's ability to concentrate received or transmitted energy in a preferred direction

Ground-plane: conductive surface that acts as an electrical mirror. A ground plane antenna is an antenna that requires a ground plane to operate

HF: High Frequency (3- 30 MHz)

Impedance: opposition to ac current flow by a circuit, feed line, or antenna

IRLP: Internet Repeater Linking Project, a system of accessing and linking repeaters through the Internet

MARS: Military Affiliate Radio System in which hams communicate with military stations

Menu: list of selectable control or configuration functions or options

Monitor: listen without transmitting or disable a radio's squelch to listen for weak signals

Mount (mag, trunk, lip, mirror): a method of attaching an antenna to a vehicle

Overload: signal so strong that circuits begin to operate improperly

Packet: amateur digital data system that communicates using VHF and UHF frequencies

Polarization: orientation of radio waves with respect to the surface of the Earth (vertical, horizontal polarization)

Power supply: device that changes ac power into dc power

Range: distance over which communication can take place

Rotate (batteries): to take in and out of service, preventing continuous use

Scanning: monitor a range of frequencies or a set of memory channels for activity

Simplex: transmitting and receiving on the same frequency

Shortwave: see *HF*

Sub-audible: audio frequencies below the usual communication range of 300 – 3000 Hz

Trunking (trunked) systems: VHF/UHF systems used by commercial and government agencies, sharing a few channels among many users by using computers to control the radio's frequencies

UHF: Ultra High Frequency (300 MHz – 3 GHz)
VFO: Variable Frequency Oscillator, refers to a radio's continuous tuning mechanism
VHF: Very High Frequency (30 MHz – 300 MHz)
Wall wart: light-duty power supply plugged directly into the an ac outlet
Whip: antenna made from a long, thin metal rod
Winlink (Winlink 2000): system for sending and receiving email via Amateur Radio

FREQUENCIES AND MODES

If your interest lies primarily in using the local FM repeaters and simplex channels, information about what frequencies are most used is available from local radio clubs and newsletters, the local frequency coordinator's website, the *ARRL Repeater Directory* (**www.arrl.org/shop**), and emergency communications teams. Ask about the coverage (*range*) of the repeaters, as well.

The most popular VHF/UHF radios for FM voice and data are *dual-band*, meaning that they can transmit on both the 2 meter (144-148 MHz) and 70 cm (420-450 MHz) bands. Higher-end units allow you to monitor several bands at once and listen on one band while transmitting on the other.

Whether you need the radio to cover the 50 MHz, 222 MHz or 1.2 GHz bands will depend on activity in your area. The first two are often used by emergency communications teams and the last is more likely to be used in densely populated areas. *Cross-band* repeater capability (receiving on one band and retransmitting the signal on the other band) is a "nice-to-have", but not usually needed.

Receiver coverage comes in three flavors. "*Ham band receive*" can tune all of the ham bands, plus narrow ranges above and below the band edges for amateur auxiliary services, such as MARS. "*Extended receive*" includes reception of selected frequency ranges such as those in **Table 2**. "*Wideband receive*" covers a wide range of continuous frequencies—good for using the radio as a scanner. Being able to receive commercial broadcasts and non-ham services is very helpful during emergencies although this capability may make the receiver more susceptible to overload and interference from strong nearby transmitters.

Caution: Even though some VHF/UHF amateur radios

Table 2

Useful Non-Amateur Bands

AM Broadcast	550 kHz – 1.8 MHz
Shortwave Broadcast	3 MHz – 25 MHz
Low-band VHF	30 – 50 MHz
FM Broadcast	88 – 108 MHz
Aviation (AM & FM)	118 - 144 MHz
High-band VHF	148 – 174 MHz
Marine	156 – 158 MHz
NOAA Weather	162.4 – 162.55 MHz
Military Aviation	225 – 389 MHz
Government	406 – 420 MHz
UHF	450 – 470 MHz

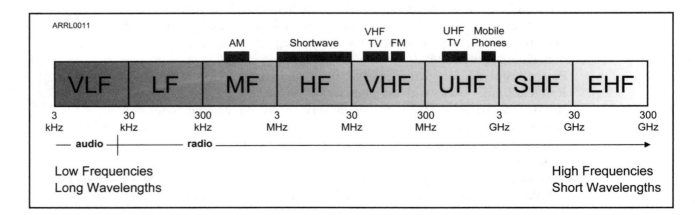

with wideband receive can tune in AM broadcast, *shortwave broadcast* (SW BC) and aviation AM signals, most cannot receive SSB or CW signals and thus are unsuitable for use on the HF ham bands. Receiving VHF AM aviation signals (sometimes called *airband*) is also handy for tracking down and identifying sources of interfering noise, such as from power lines or appliances.

Many municipal and government systems use *trunking* systems in the VHF and UHF bands. (Trunking systems share a few channels among many users by changing the frequency of individual radios under the control of a central station.) These can not be received by ham radios or scanners without trunking features.

POWER – INPUT AND OUTPUT

Transmitter output power is crucial to successful communication because, along with antenna choice, it affects communications range. Check with local hams about the power levels required to "hit" (access) popular repeaters and for effective simplex contacts. The local terrain should be considered—flat, open areas require less power. A handheld radio should be capable of 3 to 5 watts of output for consistent coverage. Mobile radios (10 – 50 watts) have far better coverage, but require more current, such as from a car battery or power supply. Most radios have variable power settings to conserve power and battery capacity.

Handheld radios use sealed and rechargeable multi-cell *battery packs* that fit a specific radio and are rarely interchangeable between models. It is wise to purchase a spare pack and *rotate* (swap) the packs regularly. Packs that hold regular alkaline AAA or AA cells are important to have for operation away from home or in emergencies when recharging power may not be available. (Remember to recycle dead and weak battery packs properly!) Several battery options are described in **Table 3**.

The higher a battery's *energy density or specific energy*, the longer a battery of a given size will last. Energy density drops with temperature, so a cold battery won't deliver as much energy as one at room temperature. Most batteries are rated to be stored at temperatures from -20 to +45° C.

Handheld radios are sold with a wall-mounted supply or "wall wart" that charges the batteries slowly. This may be inconvenient during emergency or heavy use. A *desk* or *drop-in charger* charges the packs quickly and holds the radio

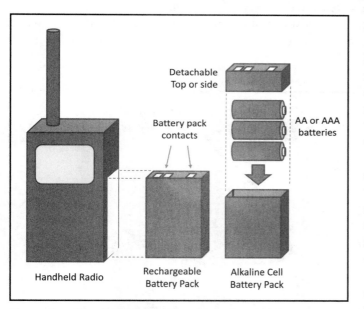

Figure 1 — Handheld radios use sealed and rechargeable multi-cell *battery packs* that fit a specific radio and are rarely interchangeable between models.

Table 3
Types of Batteries and Battery Packs

Technology	Energy Density	Limitations	Cost
NiCd	Low	Low energy ratings	Low to medium
NiMH	High	Tend to self-discharge	Medium
Li-Ion	Highest	Tend to self-discharge	Medium to High
Alkaline	High	Not rechargeable	Low

upright. If the radio can charge from 12 V, it can also be charged in a vehicle with an *automotive adapter*. An automotive adapter that allows your handheld radio to charge (and possibly operate) from a vehicle's cigarette lighter or other 12-volt source is useful if ac power is unavailable. Not all handheld radios can transmit while connected to chargers—be sure to check the manual!

Radios for base, portable, or mobile use require an external power source. This may be a vehicle's electrical system, a deep-cycle battery, or an ac-operated power supply. Mobile radios may require 20 amps or more at full power output, so be sure your power source can deliver enough current.

Power supplies and the cables used to connect the radio must be able to supply the maximum required current for your radio at the proper voltage. Check the radio's specifications for input current to find out how much current is needed. The radio manual should also specify what wire size for a given cable length is needed to insure adequate voltage at the radio under full load.

SQUELCH AND TONES

Squelch is the circuitry in FM radios that prevents the user from having to listen to noise or unwanted signals by muting the radio's audio output unless the proper type of signal is received. The radio's manual will explain how to use a squelch level control knob (like the typical knob in the illustration) or *menu* setting, common on pocket-sized handhelds. These are the different types of squelch and associated features:

• *Carrier squelch*, the simplest and most common type of squelch, mutes the radio when no signal stronger than a preset settable threshold is present.

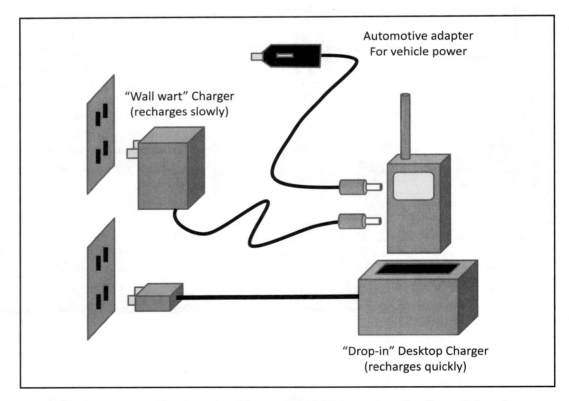

Figure 2 — Handheld radios are sold with a wall-mounted supply or "wall wart" that charges the batteries slowly. A *desk* or *drop-in charger* charges the packs quickly and holds the radio upright. If the radio can charge from 12 V, it can also be charged in a vehicle with an *automotive adapter.*

• *Tone receive squelch* requires a specific *sub-audible* tone for the received audio to be heard, as when accessing a repeater. This keeps you from hearing other users on the same frequency, just like "privacy codes" on the popular handheld radios that use the FRS and GMRS channels. (You still have to listen to the channel before transmitting.)

• *Digital Code Squelch (DCS)* – a continuous sequency of sub-audible tones must be received during a transmission to keep the output audio turned on. DCS is used by groups sharing a frequency so that they only hear audio from other group members. (Like tone-receive squelch, listen before transmitting.)

• *Call sign squelch* – digital systems, such as D-STAR, send the call sign of the receiving station along with the transmitted signal. The station called will then be alerted to the incoming call.

• *Attenuation* – some radios *attenuate* the received signal when the squelch threshold is set to high levels. This reduces interference from *overload* interference where strong paging and commercial signals are present.

• *Monitor* – an FM radio's *monitor* button or key temporarily defeats or "opens" the squelch so that you can hear any station using the channel. This is used to listen for weak signals or for other stations before transmitting.

All new radios can generate sub-audible repeater access tones. (These are also called PL™ or CTCSS tones.) Some have a feature called *tone scan* that enables the radio to determine what access tones a repeater requires by listening to the stations using it. This is very useful when traveling or accessing an unfamiliar repeater.

• *DTMF* or *Touch Tone*™ dual-frequency tones are used to dial phone numbers through a repeater's *auto-patch* or to enter IRLP and Echolink access codes. A radio's ability to store and play back sequences of DTMF tones saves a lot of time when using either service.

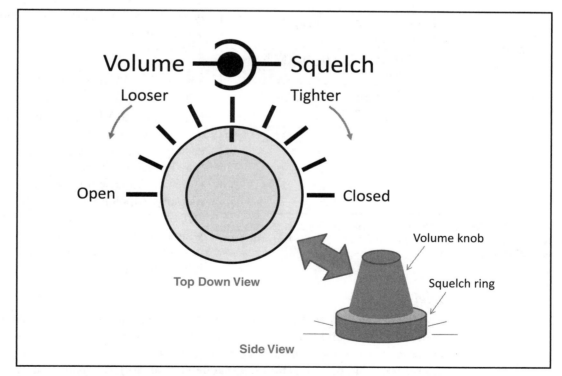

Figure 3 — Squelch is the circuitry in FM radios that prevents the user from having to listen to noise or unwanted signals by muting the radio's audio output unless the proper type of signal is received.

MEMORIES AND VFOS

Ads for radios tout the number of *memory channels* or *memories*—each can store the complete radio configuration to access a favorite repeater or channel. Having lots of memories means you can dedicate a group to your most-used repeaters and simplex channels while other groups can be used for channels you use for special events, training, and vacation or travel.

How many do you really need? Start by making a list of all local and regional repeaters and simplex channels on the bands covered by your radio. (A club or ARES team can be quite helpful in making the list.) If your radio has wideband- or extended-receive, add some AM and FM broadcast stations and the primary frequencies used by public safety and service agencies in your area. Don't forget the NOAA weather stations and if you are near water, the common VHF marine channels. Increase that total by about one third and you have a pretty good idea of how many memories you'll need.

Some memory channels have special functions. *Call channels* provide easy recall of your favorite channels. *Scan control* channels store frequency limits for scanning functions, if your radio can act as a scanner.

The ability to receive on two channels at once ("*dual receive*") is very useful. "*Priority channel*" and "*Channel watch*" monitor a channel for activity at all times and switch to that channel when a signal is present. This is useful during disaster response and public service activities or if you want to monitor a "home" repeater while operating elsewhere.

All radios have at least one VFO that tunes to any frequency the radio covers. VFOs on FM-only radios usually tune in discrete steps (e.g., every 2.5, 5 or 10 kHz) rather than continuously. On radios with more than one VFO, each can separately set the access tone, transmit offset, and other operating parameters. Once a VFO is configured, the information is transferred or *programmed* into (stored in) a memory. VFOs can act as temporary memories, too. A second VFO is a "nice-to-have" for flexibility.

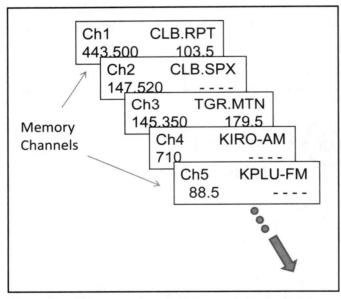

Memory Channels

Figure 4 — *Memory channels* or *memories* can store the complete radio configuration to access a favorite repeater or channel.

PROGRAMMING MEMORIES

Look in the radio's manual at the method used to program the VFO's information into a memory channel. You may find that some methods are easier to remember or perform. Some radios may have a *quick-program* function that quickly stores the VFO settings in an un-programmed memory. Alphanumeric channel labels (such as "CLB_RPT") make it much easier to remember which channel is which. (See the section on "Programming and configuration software" for more information.)

Digital modes

Exchanging text, email, graphics, and files is an important part of today's emergency communications and other applications of radio. The modes that transmit and receive data are referred to as *digital modes*. On VHF and UHF, the most common digital mode is *packet radio* or "packet". (**tapr.org/packetradio.html**) The name comes from data being transmitted in groups of characters that are called *packets*. Packet is also known as *AX.25*, the designator of the technical standard that describes it. To use packet, you'll need an FM

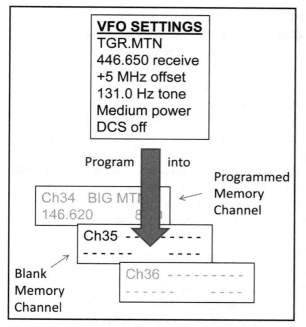

VFO SETTINGS
TGR.MTN
446.650 receive
+5 MHz offset
131.0 Hz tone
Medium power
DCS off

Program ──────► into

Ch34 BIG MTN
146.620

Ch35 - - - - - - - - - - - -
- - - - - - - -   - - - -

Ch36 - - - - - - - - - -
- - - - - - - - - - - -

Programmed
Memory
Channel

Blank
Memory
Channel

Figure 5 — Once a VFO is configured, the information is transferred or *programmed* into (stored in) a memory.

radio, a special interface called a *terminal node controller* (TNC), and a computer as shown in **Figure 6**. Some radios have TNCs built in.

Packet provides "keyboard-to-keyboard" communication a bit like instant messaging. It is also used to send email from your computer via *Winlink* system mailboxes. The *Automatic Packet Reporting System* (APRS - **www.aprs.org**) uses packet radio to report your position and other information over the Internet. A few radios have features designed for use with APRS, such as special text displays, a data interface to communicate with a GPS receiver, or built-in GPS receivers.

Packet commonly operates at two speeds; 1200 bits/second (bps) and 9600 bps (about 120 and 960 characters/second, respectively). At the slower speed, the TNC converts characters from the computer's serial or USB port into audio tones fed to the radio's microphone input. Any FM voice radio is suitable for packet radio use at 1200 bps.

To operate at the higher data rate, the radio must have a special connection specifically for use with digital modes. 9600 bps is more demanding of the radio and not all radios work well at that speed. Check the radio's specifications to see if it is rated for 9600 bps data. Unless you specifically need to use high-speed data communication, you don't need 9600 bps capability. Most packet operation takes place at 1200 bps.

D-STAR is a digital data system based on a standard from Japan. Equipment is currently available from Icom (look for the D-Star link at **icomamerica.com/en/amateur**) along with some third-party converters and accessory adapters. D-STAR radios can talk directly to each other or through networks of repeaters linked over the Internet. On 2 meters and 70 cm, D-STAR radios provide a low-speed data connection (about 80 bps) directly from the radio to your computer's USB or serial port—no TNC is required. On 1.2 GHz, D-STAR radios provide a network-style Ethernet connection to your computer, compatible with web browsing and other Internet applications. The speed is about the same as a 56 kbps dial-up connection.

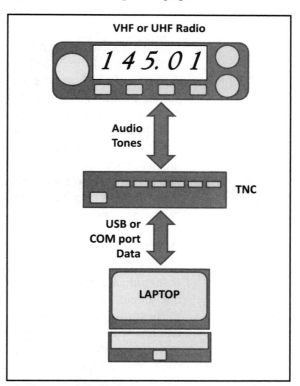

VHF or UHF Radio

145.01

Audio Tones

TNC

USB or COM port Data

LAPTOP

Figure 6 — To use packet, you'll need an FM radio, a special interface called a *terminal node controller* (TNC) and a computer.

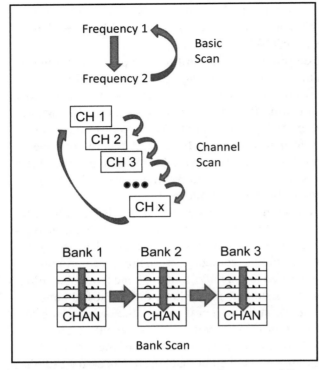

Figure 7 — Radios on the market today can rapidly switch from frequency to frequency to look for signals. This is called *scanning*. This allows the operator to monitor many different frequencies without having to manually tune to each one. There are three types of scanning (see text).

Scanning

Radios on the market today can rapidly switch from frequency to frequency to look for signals. This is called *scanning*. It allows the operator to monitor many different frequencies without having to manually tune to each one.

There are three types of scanning. Basic scanning consists of the receiver starting from the *scan start* frequency and tuning continuously to the *scan stop* frequency. If a signal is detected, scanning is stopped or paused for the operator to listen in. This is most useful in the weak-signal or satellite segments of the VHF and UHF bands and when listening for simplex FM signals.

Channel scanning is jumping from channel to channel in sequence. Again, scanning stops or pauses when a signal is detected. Channel scanning requires a *start channel* and a *stop channel*. Channels may be designated to be skipped, as well. This type of scanning is the most useful for watching many repeater or simplex channels.

Groups of channels may be organized in *bank*s. *Bank scanning* scans all the channels in a bank before moving to the next bank. Not all radios organize their memory channels as banks. Banks are useful for grouping channels together by function; police, fire, aircraft, ham, etc.

Programmed scan is the most flexible of all and allows the user to set up lists of channels to be scanned. This is most conveniently done with a PC instead of the radio keypad. Channels can generally be scanned in any order. The PC software often allows the operator to set up "profiles" of programmed scans (for example, parade communication or ARES drills in which different sets of channels are in use) so that the radio can be configured quickly and efficiently.

Scanning is much more convenient to use if you have control over *scan delay* and *scan resume*. Scan delay is the time the radio spends listening to each channel before moving to the next one. A longer delay catches more activity and lets you listen longer to each channel, but slows down the overall scanning process. Scan resume tells the radio to how long to pause or to stop if a signal is detected. Some radios offer *voice detection* to distinguish between a voice signal and a steady tone or noise that may result from interference.

Antennas

Antenna choice is the single biggest factor in determining whether you'll be able to communicate effectively with any type of radio. Handheld radios come with a stubby, flexible antenna ("rubber duck") that attaches directly to the radio for convenience. These are fairly sturdy, but are not very efficient. Consider purchasing a more efficient mobile *whip* antenna with the necessary connector or adaptor for your radio. At home you can attach a mobile antenna to a metal surface or structure. Longer, more efficient antennas to replace the rubber duck are also available. For repeater and FM voice communication, the antenna should be oriented vertically ("vertically polarized") to match the signals from repeaters and other hams.

Mobile antennas are mounted on the outside of a car. Temporary mounts using magnets (*mag-mounts*) or clamps (*lip-, trunk-, mirror-mount, etc*) and permanent through-the-body mounts provide a *ground-plane* for the antenna. You can purchase the complete antenna system, including the whip, mount, and cable or you can purchase a separate mount with cable. Antennas attach to the mount by one of several different methods: a threaded base, PL-259/SO-239, and NMO are the most common. Your antenna will need to match the mount. An *on-glass* antenna does not need the metal ground plane, using an adhesive pad to attach to the vehicle's window, but is usually less efficient and may not work with some types of auto glass.

Antennas are generally rated in terms of *gain* – how well they concentrate signals in a preferred direction. Gain is specified in *dB* and every 3 dB of gain doubles your signal's strength. An *omnidirectional* antenna or "omni" radiates equally well in all horizontal directions and can be used with base, mobile, and handheld radios. The gain of an omni antenna concentrates the signal towards the horizon.

At home, you may want to install a permanent antenna such as the common ground-plane antenna with three or four radials. Mounted in the clear, a ground-plane will give good performance. The *J-pole* antenna operates similarly to the ground-plane, but does not require radials. *Collinear* antennas with several sections working together look like ground-planes, but are longer and have higher gain. A *Yagi* antenna is a type of rotatable *beam* antenna that has gain in one direction and is used to communicate with stations that are out of range of simple omnidirectional antennas. A *rotator* is needed to point a beam in the desired direction.

If you purchase an antenna without the connecting *feed line*, use a high-quality cable that has low losses at VHF and UHF frequencies. (All cables have increasing loss with frequency.) For distances shorter than 50 feet, RG-8 or RG-213 are fine. At 2 meters, RG-8X can be used up to 50 feet, but no longer than 25 feet at 70 cm. You won't need ultra-low-loss "hard-line" or specialty cables to get started. For longer "runs" of cable, ask for help from your Elmer, from a radio store, or from the manufacturer's specifications to choose the right cable. While cables are available with connectors pre-installed, you should eventu-

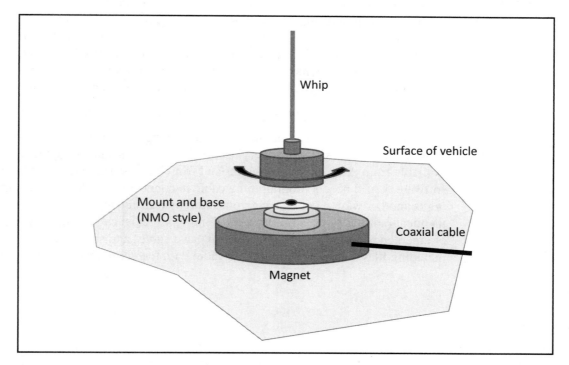

Whip

Surface of vehicle

Mount and base
(NMO style)

Coaxial cable

Magnet

Figure 8 – One popular type of temporary mobile antenna mount uses magnets and is often referred to as a *mag-mount*.

ally learn how to install your own connectors—both to have the skill and to save money! The ARRL's online Technical Information Service **www.arrl.org/technical-information-service** and the *ARRL Antenna Book* and *ARRL Handbook* contain methods of installing connectors on coaxial cable.

Accessories & Special Features

Along with the items included with your radio, other common accessories can be a great help in the convenient and effective use of your radio. The most important accessory for a mobile radio is the antenna—we covered that previously. For a handheld radio, antennas and batteries make the most difference.

• *Battery chargers*—Recharging a battery with a wall wart supply can take hours for one of the larger packs. This is often unacceptably long and is a good argument for having a spare battery pack. Charging time can be reduced dramatically by using a *quick charger* or *smart charger*. Your radio's manufacturer will probably offer one as an accessory. A desk or "drop-in" charger holds the radio conveniently upright while charging, too.

• *Detachable front panels*—Some radios can operate with their front panel detached from the body of the radio and mounted in a convenient location with the radio out of sight below a desk or seat. An accessory *control cable* is required to connect them. Some radios require the microphone to be connected to the radio and not the control panel, so check carefully before planning where to mount the radio!

• *Smart microphones*—Handheld microphones or *hand mikes* for mobile and base radios are available with enough keys and buttons to act as miniature front panels of their own. There may be several variations of microphones available for your radio.

• *Headsets*—Base station radios come with a hand mike, but third-party desk microphones and headphone-boom microphone combinations called *boomsets* may be more convenient and provide somewhat higher quality transmitted audio. Headphones can help you hear other stations more clearly, particularly in noisy environments such as a busy emergency-operations center or an outdoor event. (Driving with a headset or headphones on is illegal in many areas—check your local regulations.)

• To tune your antenna, an *SWR bridge* or *RF power meter* is a "nice-to-have", measuring the amount of power flowing to and from the antenna. By watching the meter when you transmit, you can tell when an antenna is not tuned properly, whether the wrong antenna has been attached, or whether some part of the antenna system is broken. Be sure the bridge or meter is designed for the frequency you'll be using—VHF/UHF or HF.

PROGRAMMING AND CONFIGURATION SOFTWARE

With so many memory channels and radio configuration settings to manage, having some software to assist you is very useful. Programming and configuration software is available from the radio manufacturer and from independent authors. Along with the software, you'll need a programming cable to connect the radio to the computer. *Cloning* is another way of configuring your radio by transferring the memory contents from an identical radio. If your radio supports cloning, a special cable is usually required. Your club or emergency communications team may have cables and software for common radios, including files that will program your radio with the common channels used by members.

The Icom IC-7100 transceiver.

HF EQUIPMENT

The HF or "short-wave bands" are important because of their long-range capability. When HF conditions are favorable, contacts around the world are possible without the need for repeaters or the use of Internet-based systems, such as IRLP or Echolink. The traditional amateur HF bands include 160, 80, 40, 20, 15 and 10 meters (1.8 MHz to 29.7 MHz). In the early 1980s, the 30, 17 and 12 meter bands (10, 18, and 24 MHz) were added and are sometimes referred to as the "WARC" bands (in reference to the World Administrative Radio Conference at which frequencies are allocated). The 60 meter band (5.5 MHz) was added more recently.

Many models of transceivers are available for the HF bands from portable, mobile and fixed stations. There is also a wider price range than for VHF and UHF radios. As you look through the catalogs, you'll see large radios and small radios with much the same specifications—this section will help you understand the real differences.

Portable radios are designed to be compact, lightweight and power-efficient. They are available with power outputs of 5 to 100 watts and cover all of the HF bands—some even operate on VHF and UHF bands. Low-power models may have an internal battery pack. They have fewer features than most fixed-station radios and receiver performance is generally not as good. Their smaller front panels mean they have fewer controls and often use menus for some functions.

The Yaesu FT-897 transceiver.

Mobile radios are intended to be operated in a vehicle, but they can make an excellent base-station radio, especially if you have limited space. These radios consume somewhat less current than fixed station radios when not transmitting. This may be important if you intend to use the radio for emergency communications and expect to be operating on battery power occasionally. If not used in a vehicle, a dc power supply will be needed. The same concerns about small front panels and ease of use apply.

Radios for base stations are available in many different price ranges because of the great differences in performance and features. Some will operate directly from ac power outlets. Most of the performance differences are associated with the receiver. High-performance receivers have better *selectivity* and *strong-signal performance* as discussed later. Some radios even have two receivers! It is normally best to start with a basic radio and develop a feel for what is important to you.

HF GLOSSARY

All-mode: radio that can operate on AM, SSB, CW, Digital, and FM

Attenuate (attenuation): reduce in strength

Balun: stands for "balanced-to-unbalanced", provides a transition from parallel wire feed lines or antennas to coaxial feed lines.

Beam: antenna with gain primarily in one direction

Crystal filter: filters that use quartz crystals to reject unwanted signals in receivers

CW (continuous wave): Morse Code

dB (decibels): logarithmic method of comparing two signal strengths (power, voltage, current)

Data interface: a device for connecting a computer to a radio

Digital mode: communication method that exchanges characters instead of voice or Morse Code

Dipole: a simple wire antenna ½-wavelength long with feed line attached in the middle

Directional wattmeter: a wattmeter that can measure power flowing in both directions

Emcomm: abbreviation for "emergency communications"

Feed line (transmission line): cable used to transfer radio-frequency energy

Gain (antenna): antenna's ability to receive or transmit energy in a preferred direction

Ground plane: (1) conductive surface that acts as an electrical mirror; (2) an antenna that requires a ground plane to create an electrical image

Half-wave: ½ wavelength

HF: High Frequency (3 MHz to 30 MHz)

Impedance: a measure of how easily power can be transferred into a load or through a feed line

Keyer: an electronic device that generates Morse code elements.

Linear: an amplifier that boosts the power output from a radio without distorting the signal

Menu: list of selectable control or configuration functions or options to select from

Overload: a signal so strong that circuits begin to operate improperly

Paddle: used with a keyer to send Morse code

Power supply: device that changes ac power into dc power

QRP: very low-power operating (less than 5 watts on CW and 10 watts (peak) on phone)

Quarter-wave: ¼ wavelength

Range: distance over which communication can take place

Scanning: monitor a range of frequencies or a set of memory channels for activity

Screwdriver: a tunable mobile whip, refers to electric screwdriver motors used on early models

Selectivity: a receiver's ability to receive only the desired signal and reject all others

Sensitivity: a receiver's ability to detect weak signals

Signal-to-noise ratio: a comparison of a signal's strength compared to background noise

Strong-signal performance: the ability to withstand overload and distortion from strong signal

SWR: Standing Wave Ratio, indicates how much power is transferred to a load or antenna

VFO: Variable Frequency Oscillator, refers to a radio's continuous tuning mechanism

WARC: World Administrative Radio Conference at which frequency allocations are determined

Weak-signal: making long-distance SSB and CW contacts with low signal-to-noise ratios

Whip: antenna made from a long, thin metal rod

Frequencies and Modes

Commercially available HF transceivers cover all of the amateur bands described in the introduction to this section. Some add the 6 meter band from 50 – 54 MHz and even VHF/UHF bands from 2 meters to 23 cm (1.2 GHz). These radios are "all-mode", using AM, SSB, CW, FM, and digital modes. You can use these radios for everything from contacts on your local repeater to long-distance contacts on the HF bands where SSB and CW are the most popular modes. The *ARRL Operating Manual* (**www.arrl.org/shop**) discusses the characteristics of the different HF and VHF/UHF bands.

The longer wavelength HF bands (160 – 30 meters) are generally used for local and regional contacts through the day, but support long-distance (*DX*) contacts at night. Shorter wavelength bands (20 – 10 meters) "open" and provide long-distance contacts through the day, but "close" at night when the signals are no longer reflected back to Earth by the ionosphere.

The VHF and UHF bands support line-of-sight regional contacts and long-distance contacts via several interesting means of propagation. This is called *weak-signal* operating and is conducted on 50, 144, and 432 MHz mostly using SSB and CW because those modes work better at low *signal-to-noise ratios*, while FM requires stronger signals to be effective.

While able to transmit only in the ham bands, HF radios typically include *general coverage* receivers that can tune from a few hundred kHz to 30 MHz. (*Medium wave (MW)* stations use frequencies from 300 kHz – 3 MHz and *long wave (LW)* below 300 kHz.) *Ham band only* receivers don't offer this coverage. *Wideband receive* coverage extends above 30 MHz into the VHF and UHF range. The Wikipedia entry on "shortwave" (**en.wikipedia.org/wiki/Shortwave**) will give you an idea of what you can listen to. The VHF/UHF section on frequency coverage will help you decide if wideband receive is useful for you.

Power – Input and Output

Transmitter output power of HF transceivers ranges from 100 to 200 watts with most between 100 and 150 watts. The extra power will not make a dramatic difference on the air, however. Power amplifiers (a.k.a. – *linears* or *linear amplifiers*) are available to increase the output power to 600 to 1500 watts (maximum legal power for hams) for more demanding conditions and activities.

Beginning HF operators will have the most success using power levels around 100 watts.

The Kenwood TS-590S transceiver.

Running "barefoot" (without an amplifier) with a modest antenna is often sufficient for lots of contacts, including world-wide DX during favorable conditions. More power, provided by an amplifier, will extend your transmit range and enable you to maintain contact longer as conditions change, but adds significant expense and raises the demands placed on station equipment. Lower power, or *QRP* operating, is something you can try as you gain experience, turning the radio's output power down whenever you'd like to give it a try!

A radio with a built-in ac power supply does not need an external dc supply. Most radios do require an external dc supply, though. Be sure to select a power supply that is rated to continuously provide the maximum current specified in the radio's manual. Because most radio accessories operate from the same voltage as the radio (13.8 V for most "12 V" radios) it is wise to add a few more amperes of current output to power them. For example, if a radio requires 24 amps at full power, choose a supply that can deliver 30 amps or more.

Filters & Receiver Performance

The HF bands are a challenging environment for receivers; signal strengths vary from barely detectable to extremely strong. The bands are often crowded with strong signals. The most important receiver characteristics are *selectivity* (the ability to reject unwanted signals) and *strong-signal performance.*

Selectivity is created by filters with different bandwidths suited to the signal's mode. You'll encounter radios that use discrete electronic *crystal filters* and software-based *DSP (Digital Signal Processing) filters.* HF radios that use discrete filters will come with a medium-bandwidth filter for SSB (about 2 kHz), an AM filter (6 kHz), and an FM filter (15 kHz). A CW filter (500 Hz) or RTTY filter (250-300 Hz) are needed if you expect to use those modes a lot. Filters can be added after you purchase the radio.

DSP filters are created by software in the radio's controlling microprocessors, reducing or eliminating the need for separate electronic filters. The radio will come pre-programmed with several common filter types. You can create new filter types and even adjust them while you are using the radio.

The radio's ability to operate properly in the presence of strong signals is measured primarily in two ways; *blocking dynamic range (BDR)* and *3rd-order intercept point (TOI* or *IP3)*. In both cases, higher figures are better. BDR in decibels (dB) describes the receiver's ability to ignore unwanted signals. TOI in dBm (a power level) describes the receiver's

To operate a 100-watt transceiver, you need a power supply that can provide the necessary current, typically 30 amps or more.

reaction to multiple strong signals. A difference of 6 dB or more represents a noticeable change in performance. While commercial transceivers have adequate strong-signal performance, the more you expect to operate on HF, the more important these figures will become. The ARRL's *Product Reviews* are a valuable source of information about receiver performance and are free to ARRL members on the ARRL website and in *QST* magazine.

Four other features help a receiver reject interference and other unwanted signals. Having these controls on the front panel is very useful when operating on a crowded band.

Receive Incremental Tuning (RIT), sometimes called "Clarifier", changes the receive frequency without affecting the transmit signal. *Passband tuning* or *passband shift*, controls where the receiver's filters are tuned relative to the main receive frequency. This allows you to reduce off-frequency interference without changing the receiver's main tuning frequency.

Notch filters can reduce interference appearing as continuous tones, not uncommon on the HF bands, created by unwanted signals or commercial broadcast stations. A notch filter removes a very narrow slice of the audio range that can be adjusted to match the tone of the interference. This either eliminates the tone or reduces it to a tolerable level.

A *noise blanker* suppresses *impulse noise* created by motors and vehicle ignition systems.

Digital Signal Processing

Many new radios employ *Digital Signal Processing (DSP)* to perform many functions in software that were previously performed by electronic circuits. Special microprocessors do the job inside the radio. The latest generation of radios using DSP has very good performance. DSP filtering was discussed in the preceding section.

DSP is also employed to get rid of unwanted noise. *Noise reduction (NR)* is used to reduce the hissing and crackling of static and other random noise present in the audio of received SSB and CW signals. This function is usually available with several levels of

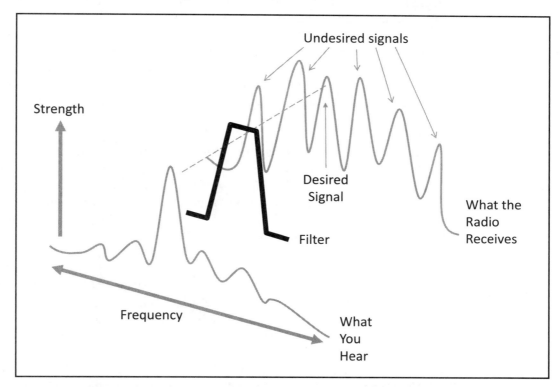

Figure 9 — Selectivity is created by filters with different bandwidths suited to the signal's mode.

processing, with more aggressive levels also giving the signals a "digital" sound. *Noise blanking (NB)* is used to remove repetitive noise pulses, such as those from an automobile's ignition system or power-line noise from nearby ac utility lines. Noise blanking is very useful in mobile HF radios and noise reduction can be used for all SSB and CW HF operating.

DSP can also be used to create a notch filter. Not only can the filter's notch frequency be manually adjusted, just like an electronic notch filter, but many DSP notch filters can automatically detect the frequency of the interfering tone and tune the notch filter (*Automatic Notch Filter* or *ANF*). Some radios even have the ability to "notch out" more than one tone at a time!

Radios with more DSP functions also allow you to control the *filter response*—how aggressively nearby signals are rejected. A filter that allows some of those signals to be heard has a "soft" response, while one that rejects the signals more completely is "sharp" or "steep". Soft filters result in more natural sounding audio, but do not reject interfering signals as well. If the DSP functions are also applied to the output signal, you may also be able to tailor your transmitted audio to suit your voice, adding more average power and increasing the intelligibility of your signal.

Memories

HF transceivers use memories and VFOs in much the same way that VHF/UHF models do. Start by reading the VHF/UHF section on memories. HF radios often offer *band memories* as well, sometimes referred to as *band-stacking registers*. These memories store one to several settings of the VFO on each of the HF bands. Pressing a band-select key on the radio's front panel cycles through these memories. This is useful if you use the same frequency on a regular basis.

Scratchpad memory is a set of temporary memories for use when you are just tuning around the bands and want to save a frequency and mode setting without dedicating a memory. Like VHF/UHF radio memories, the memories store not just the frequency, but also the mode, filter and other settings.

To find out how many memories you'll want on your HF radio, count all the nets and emergency frequencies you might use, some of the frequencies for WWV and WWVH (**www.nist.gov**—the US time and frequency standard stations), calling frequencies for your most-used modes, propagation beacons, and any other frequencies of interest. Add some extra memories for expansion.

Scanning

Scanning is also provided in mid- to top-scale HF radios, just as in the VHF/UHF radios (read the VHF/UHF scanning section for a discussion of the terms). The most popular scanning mode on HF is *range scanning* in which the radio tunes continuously across a portion of the band, looking for any signal. Scanning is less useful on HF, however, because of the higher noise levels and the wider range of usable signals, making it difficult to set a single level to use as the scanning threshold.

Antennas

If antennas are the most important part of an amateur station, that is doubly true on HF where long-distance contacts place a premium on the antenna's ability to transmit and receive efficiently. Antennas are generally rated in terms of *gain*—how well they concentrate signals in a preferred direction. Gain is specified in dB and every 3 dB of gain doubles your signal's strength.

The simplest antenna (and a very effective one) is a *half-wave dipole* made of wire, one-half wavelength long, and installed horizontally. You can build it yourself as de-

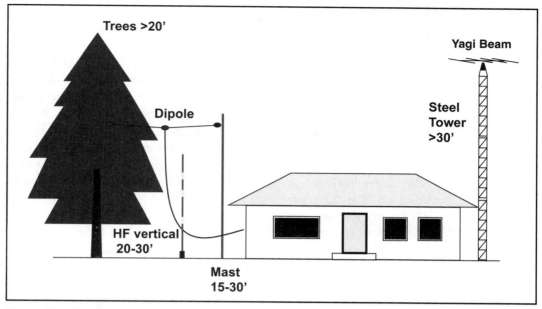

Figure 10 – You have many antenna options to choose from, depending on your budget and how much space you have available.

scribed on the ARRL's Technical Information Service (TIS) web page or you can buy one or any of several common variations; off-center-fed, multi-wire, end-fed, and G5RV antennas are popular. While a dipole's gain is low, it's efficient and hard to beat for the price. You'll need one or two supports (trees work well) at least 20 feet tall. Large-scale versions of the Yagi beam antennas mentioned in the VHF/UHF section can be placed atop steel towers, although this is not required to get started.

Vertical antennas are also popular, particularly where a horizontal antenna may be difficult to put up, for portable use, and where a "low profile" antenna will be more suitable. The simplest is a *quarter-wave vertical* made of metal tubing and *radial wires* fanning out from the base to act as a ground plane. To use it on several of the HF bands will require an *antenna tuner* described in the next section. *Multi-band verticals* are constructed to operate on several bands without the antenna tuner. *Ground-independent verticals* are available that operate without the radial wires.

Mobile antennas for HF use come in two common styles; fixed-tuned and tunable whips. A fixed-tuned whip is adjusted to present the proper load to the transmitter on one band or over a portion of a band. You will need one for each band you intend to use, but they are inexpensive. The whips have a ⅜"-24 threaded base that screws in to the antenna mount on the vehicle, similarly to what is shown in the VHF/UHF section. A tunable whip with an internal coil (called a "screwdriver" antenna) can vary its length continuously to tune up on nearly any HF frequency. A controller is mounted in the vehicle. Only one tunable whip is required, although they are much more expensive than the fixed-tune whips.

Mobile antennas can be mounted on the vehicle temporarily or permanently. A permanent mount generally results in a better electrical ground connection to the vehicle, which is important for the HF antennas to work well. Temporary mounts, such as larger versions of the VHF/UHF *magnet* or *mag-mounts*, are usable for most purposes

The High Sierra HS-1800 is a so-called "screwdriver" style HF mobile antenna that provides continuous coverage for all HF bands.

but have lower efficiency than mounts attached directly to the vehicle and can be knocked loose. If you purchase the mount separately from the antenna, make sure the mount and antenna have the same type of mechanical connection!

The most popular antenna feed line is *coaxial cable* or *coax*. There are many types, but the most common are (from smallest to largest) are RG-58, RG-8X, and RG-8 or RG-213. Use RG-58 only for short (50 feet or less) distances due to its higher losses and never at the output of an amplifier. RG-8X will carry the full legal power, but is not a good choice for feed lines longer than 100 feet or mistuned antennas. RG-213 is suitable for all amateur HF uses, except for extremely long feed lines. At HF, the standard connectors are the UHF-family of connectors which include the PL-259 (cable plug) and SO-239 (equipment receptacle) illustrated in the VHF section. Cable is available with connectors pre-installed or with a little "Elmering" you can learn how to install them yourself as described on the ARRL TIS website.

The other type of feed line is *open-wire, ladder,* or *window line* consisting of two parallel wires coated with plastic insulation. Open-wire line has very low losses, but is not as convenient to use as coaxial cable and requires an antenna tuner or some other kind of *impedance transformer* to work with most HF radios along with a *balun* to convert the open-wire line to coax that connects to the radio.

Accessories & Special Features

Antenna tuners are known by several names; *impedance matching unit, tuning unit, matchbox, transmatch,* etc. They do not actually tune the antenna, but convert whatever impedance is presented by the antenna system at the input to the feed line into a 50-ohm load so that your transmitter will deliver maximum power output. Some radios have an automatic antenna tuner or ATU built-in, but if yours doesn't and your antenna's *SWR* is much higher than 2:1 on a frequency you wish to use, then you'll need an external antenna tuner, either automatic or manually-adjusted. A model rated at 300 watts will accommodate the output of a 100-watt transceiver with room to spare. Manual antenna tuners often include an *SWR meter* or *directional wattmeter*, but these can also be purchased as individual items and are very handy shack accessories. An SWR meter can be used as an antenna system test instrument or to monitor the state of an antenna. Directional wattmeters measure the power flowing back and forth in your feed line and may also be calibrated to show SWR. Both power and SWR meters are designed to be

Two examples of PL-259 coaxial cable connectors.

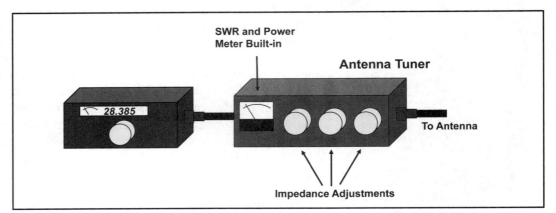

Figure 11 — *Antenna tuners* are known by several names; *impedance matching unit, tuning unit, matchbox, transmatch,* etc. They convert whatever impedance is presented by the antenna system at the input to the feed line into a 50-ohm load so that your transmitter will deliver maximum power output.

Some transceivers include the ability to separate the front panel (the "control head") from the rest of the unit for easier mobile installations.

used at either HF or VHF and will provide uncalibrated readings at other frequencies.

Antenna switches allow you to select different antennas quickly. The "common" port of the switch connects to the output of the radio or antenna tuner with a short *jumper* of coax and the antenna system coaxial cables then connects to the selectable ports of the switch. Some radios have an antenna switch built-in. As long as we're on the subject, it's useful for an HF transceiver to have a receive antenna (RX ANT) input for a special receiving antennas or external filters.

All manufacturers offer separate speakers for their radios that sound somewhat better than the radio's small built-in speakers. For the highest quality "copy" of signals, though, use a pair of headphones. A set intended for music will do or you can buy headphones designed specifically for radio communication. A *boom-set* is headphones combined with a *boom microphone*. The boom mike frees you from having to hold a hand mike or speak into a desk microphone. The voice-operated transmit (VOX) option of most radios frees your hands from having to press a Push-to-Talk (PTT) switch.

Computer Interfaces

A headset/microphone combination, or *boomset*, combines the micro-phone and earphones in one unit.

What if you want to make digital mode contacts using a PC or just control the rig from the PC keyboard? There are several software packages available and some are even free! There are three types of computer interface connections to ham radios.

The first is a *control port* by which a PC can read, change, and operate many, if not all, of the radio's control functions. This allows a PC to monitor and record your radio's configuration for logging contacts and other useful functions. Radios "speak" a control proto-col that is unique to each manufacturer. For many years, most radio control ports were a serial or COM port for communicating with the PC, but newer models often have USB ports. (Older Icom radios use a proprietary interface called CI-V.) PC manufacturers have dropped the or serial port, but USB-to-serial port converters are available to solve that problem.

The second type are called *keying* interfaces or *voice keyers* that allow the PC to key the radio (to send CW) from the keyboard or send recorded speech through the microphone jack. These are often used during competitive events called *contests* or *radiosport*.

The third type of interface supports digital modes, such as radio-

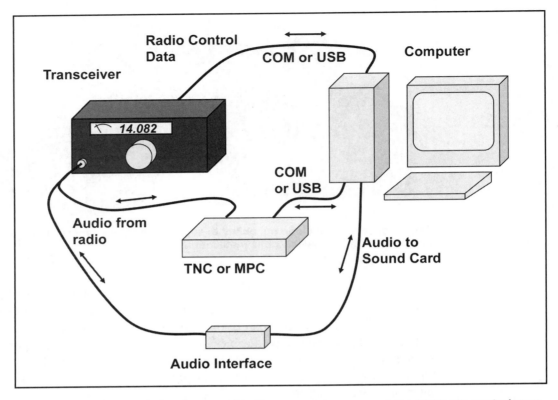

Figure 12 – With the right hardware and software, you can use your computer to control your transceiver and even make contacts using a variety of digital operating modes.

teletype (RTTY) or PACTOR, the mode used by the ham radio email network known as Winlink. The computer's sound card is connected to the microphone and headphone jacks through a *data interface*, available from several manufacturers. These devices *isolate* the radio's input from the sound card's output, preventing hum, RF feedback, and other problems. If your radio has a *data port*, the computer can send and receive data directly from the radio without using audio to or from your sound card. Interface cables are usually available for specific radios from the interface manufacturer.

MANUFACTURER'S DIRECTORY

These are the websites for the primary manufacturers of Amateur Radio transceivers:

Alinco	**www.alinco.com**
Elecraft	**www.elecraft.com**
FlexRadio Systems	**www.flex-radio.com**
Icom	**www.icomamerica.com**
Kenwood	**www.kenwoodusa.com**
MFJ Enterprises	**www.mfjenterprises.com**
TEN-TEC	**www.tentec.com**
Yaesu	**www.yaesu.com**

Many other manufacturers and distributors are listed in the pages of *QST* magazine. You can also find detailed information about manufacturers and products in the Product Reviews available to ARRL members at **www.arrl.org/product-review**.

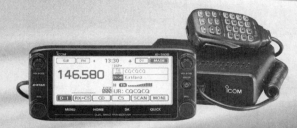

YAESU
The radio

FTDX1200 100W HF + 6M Transceiver

• Triple Conversion Receiver With 32-bit Floating Point DSP • 40 MHz 1st IF with selectable 3 kHz, 6kHz & 15 kHz Roofing Filters • Optional FFT-1 Supports AF-FFT Scope, RTTY/PSK31 Encode/Decode, CW Decode/Auto Zero-In • Full Color 4.3" TFT Display

FTDX5000MP 200W HF + 6M Transceiver

• Station Monitor SM-5000 (Included) • 0.05ppm OCXO (Included) • 300Hz, 600Hz & 3KHz Roofing filters (Included)

FTDX3000 100W HF + 6M Transceiver

• 100 Watt HF/6 Meters • Large and wide color LCD display • High Speed Spectrum Scope built-in • 32 bit high speed DSP /Down Conversion 1st IF

Call For Low Pricing!

FT-450D 100W HF + 6M Transceiver

• 100W HF/6M • Auto tuner built-in • DSP built-in • 500 memories • DNR, IF Notch, IF Shift

Call Now For Pricing!

FT-897D VHF/UHF/HF Transceiver

• HF/6M/2M/70CM • DSP Built-in • HF 100W (20W battery) • Optional P.S. + Tuner • TCXO Built-in

Call Now For Our Low Pricing!

FT-857D Ultra Compact HF/VHF/UHF

• 100w HF/6M, 50W 2M, 20W UHF • DSP included • 32 color display • 200 mems • Detachable front panel (YSK-857 required)

Call For Low Price!

FT-7900R 2M/440 Mobile

• 50W 2M, 45W on 440MHz • Weather Alert • 1000+ Memories • WIRES capability • Wideband receiver (cell blocked)

Call Now For Your Low Price!

FT-8800R 2M/440 Mobile

• V+U/V+V/U+U operation • V+U full duplex • Cross Band repeater function • 50W 2M 35W UHF • 1000+ memory channels • WIRES ready

Call Now For Low Pricing!

FTM-400DR 2M/440 Mobile

• Color display-green, blue, orange, purple, gray • GPS/APRS • Packet 1200/9600 bd ready • Spectrum scope • Bluetooth • MicroSD slot • 500 mem per band

FT1DR C4FM FDMA 144/430 5W Digital Xcvr

• 1200/9600bps AX.25 APRS & GPS Recvr Built-in • Dual Band Operation w/Dual Recvrs (V+V/U+U/V+U) • Wideband Receive/AM Bar Antenna/ Aircraft Receive • 1266 Memory Channels w/16 Char Alpha Tagging

Also Available in Silver!

VX-6R 2M/220/440 HT

• Wideband RX – 900 memories • 5W 2/440, 1.5W 220 MHz TX • Li-ION Battery - EAI system • Fully submersible to 3 ft. • CW trainer built-in

New Low Price!

VX-8DR

50/144/220/440 • 5W (1W 222 MHz) • Bluetooth optional • Waterproof/ submersible (3' for 30 min) • GPS APRS operation optional • Li-ion Hi-capacity battery • Wide band Rx

FT-60R 2M/440 5W HT

• Wide receiver coverage • AM air band receive • 1000 memory channels w/alpha labels • Huge LCD display • Rugged die-cast, water resistant case • NOAA severe weather alert with alert scan

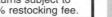

The perfect complement to this book:

HamTestOnline™
The software that knows you™

Online courses for the ham radio exams

The fastest and easiest way to prepare for the exams

- *Top-rated* on eHam.net, with more *5-star* user reviews than all of our competitors *combined*!
- Fastest way to learn — most students pass easily after:
  - 10 hours of study for the Technician exam.
  - 20 hours of study for the General exam.
  - 30 hours of study for the Extra exam.
- More effective than books:
  - Study materials — so you learn, not memorize.
  - Integrated question drill with *intelligent repetition* to focus on *your* weak areas.
  - Practice exams with memory, and more.
- More flexible than classes — online 24x7, study when you want, at your own pace, all platforms.
- 100% guaranteed — pass the exam or get a full refund!
- Try our free trial!

Use a *more effective* study method and *achieve your dream!*

www.hamtestonline.com

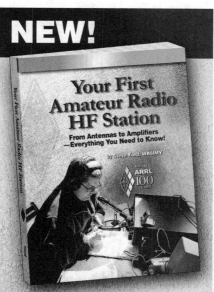

Join or Renew your ARRL Membership
www.arrl.org/join

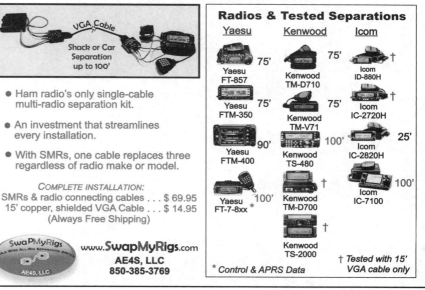